Birkhäuser

Advances in Mechanics and Mathematics

Advances in Continuum Mechanics

Volume 52

This series includes collective thematic books and monographs on advanced research topics in continuum mechanics. It is a unique forum for reference reviews discussing critically foundational issues in classical field theories under the light of up-to-date problems raised by the offspring of condensed matter physics and industrial needs, including those concerning the mechanics of microstructures and the relevant multi-field (or phase-field) representations. The series aims also to show new mathematical methods in continuum mechanics and/or recent developments of existing powerful practice. Also, the series pays attention to applications with concrete conceptual content. The series editors select editors of books on specific themes, which may split into sequential volumes.

All manuscripts are peer-reviewed to meet the highest standards of scientific literature:

- Series editors provide their final decision for inclusion in the series after manuscripts have been fully peer-reviewed;
- For contributed volumes, volume editors conduct the anonymous peer-review process before final submission;
- The peer-review process is anonymous and confidential in order to ensure reviewers provide impartial opinions

Pierluigi Benevieri • Jaqueline Godoy Mesquita
Editors

Functional Differential Equations and Dynamic Equations on Time Scales

With Applications to Continuum Mechanics

Birkhäuser

Editors
Pierluigi Benevieri
Instituto de Matemática e Estatística
Universidade de São Paulo
São Paulo, Brazil

Jaqueline Godoy Mesquita
Department of Mathematics
University of Brasília
Brasilia, Brazil

ISSN 1571-8689 ISSN 1876-9896 (electronic)
Advances in Mechanics and Mathematics
ISSN 2524-4639 ISSN 2524-4647 (electronic)
Advances in Continuum Mechanics
ISBN 978-3-031-83326-7 ISBN 978-3-031-83327-4 (eBook)
https://doi.org/10.1007/978-3-031-83327-4

Mathematics Subject Classification: 34-XX, 34Kxx

This book is published under the imprint Birkhäuser, www.birkhauser-science.com by the registered company Springer Nature Switzerland AG
The registered company address is: Gewerbestrasse 11, 6330 Cham, Switzerland

Preface

Continuum Mechanics is a branch of mechanics that deals with the behavior of materials which are assumed to be continuous, meaning they can be modeled without considering individual particles or atoms. Instead, the material is viewed as a continuous substance, and its behavior is described using mathematical models that consider properties such as strain, deformation, and motion. It encompasses a wide range of physical phenomena, including the motion of solids and fluids, the propagation of mechanical waves, the deformation of materials, and the interaction between different bodies. It is essential for understanding and predicting the behavior of structures, machines, devices, and systems in various areas of engineering, materials science, geophysics, and biomechanics, among others.

This book intends to investigate the advances in the field of dynamic equations on time scales and functional differential equations, as well as to understand how these types of equations can be used to describe phenomena in continuum mechanics, in order to understand them better.

The theory of dynamic equations on time scales is very recent, it was introduced by Stefan Hilger in 1988 in his PhD thesis, and it has been developed by many authors since then, due to many applications that can be addressed using this theory. It happens since these equations provide a unified framework for studying systems that exhibit both continuous and discrete behavior, as well as solely one of these behaviors. On the other hand, it is a known fact that Continuum Mechanics can be viewed, from the mathematical point of view, as representations of natural phenomena involving bodies extended in space and considered at a scale in which the quantum structure of the matter can be reasonably left momentarily apart. This fact emphasizes the importance of dynamic equations on time scales to describe the phenomena in Continuum Mechanics, since the quantum scale is also crucial for the understanding of such phenomena.

Further, functional differential equations are essential for capturing delayed feedback or propagation in mechanical systems and to describe the time in between cause and effect. In Continuum Mechanics, delays often arise due to finite signal propagation speeds, material response times, or the time required for information to travel through a medium. These delays can significantly influence the behavior of the mechanical systems such as stability, oscillatory, and periodicity, among others, changing completely the behavior of the solutions of the system. For example, in structural dynamics, delay differential equations are used to model the vibrations of beams or plates with time-dependent boundary conditions or material properties. In fluid dynamics, delay effects may arise due to the finite time it takes for pressure disturbances to propagate through a fluid medium, leading to phenomena such as wave propagation or flow instabilities. Also, the memory of the object can be described by delay differential equations.

This book is divided in 11 chapters in the following way: Chaps. 1–3, 5–9 are related to recent results for the theory of dynamic equations on time scales. Some results appear in an indirect way, using equations that generalize dynamic equations on time scales such as measure differential equations and generalized ODEs. On the other hand, Chap. 4 presents an overview of important results concerning functional analysis, which serves as basis to investigate many types of equations; Chaps. 10 and 11 are focused in working with equations involving delay and memory terms. Also, they present new and recent results in the literature. Below we describe the contents of all the chapters in details.

Chapter 1 aims to present some recent results for φ-laplacian equations on time scales, showing differences depending on the chosen time scale. Using topological methods, the authors investigate the existence and multiplicity of solutions, considering different boundary conditions.

Chapter 2 is devoted to study Ulam stability for a class of dynamic equations on isolated time scales. It is important to mention that quantum scale is also included in this setting. The authors apply a new concept of periodicity on isolated time scales that was recently introduced by Bohner, Mesquita, and Streipert in order to study the Ulam stability. Their results show the following properties: if the graininess function (step size) of an isolated time scale is bounded by a finite constant, then the linear $\omega-$periodic dynamic equations are Ulam stable if and only if the exponential function has modulus different from unity; on the other hand, if the graininess function increases at least linearly to infinity, then the $\omega-$periodic dynamic equations are not Ulam stable. The authors present several examples containing specific isolated time scales to illustrate their results.

Chapter 3 provides two new Leibniz formulas for conformable differentiation of classical and conformable delta integrals, and a Variation of Parameters Formula and Green's function for delta conformable Sturm–Liouville initial value and boundary value problems. In the chapter, it is also presented two dynamic BVPs relating to heat transfer of a fin and axial bending of a rod.

Chapter 4 presents some new inequalities of Hermite-Hadamard-type for r-convex functions on an arbitrary time scale using the diamond-α dynamic integrals, which is defined as a linear combination of the δ and ∇ integrals. Dealing with diamond-α is very complex, since one needs to work with δ and ∇ operators. The results presented in this chapter improve some known dynamic inequalities on time scales, unify, and extend some continuous inequalities and their corresponding discrete analogues.

Chapter 5 is devoted to the study of the qualitative properties for a special class of measure differential equations. It is important to mention that measure differential equations encompass dynamic equations on time scales and impulsive differential equations. Therefore, all the results of this chapter can be easily transferred to dynamic equations on time scales and equations with impulses as an immediate consequence. The authors prove many results on the existence of global solutions, including the existence of regulated, continuous, differentiable, and S–asymptotically ω–periodic solutions. To obtain their results, they employ the extensions of Krasnoselskii's fixed-point theorem. Furthermore, in this chapter, the authors present a result on continuous dependence of solutions in terms of parameters, generalizing previous results in the literature.

Chapter 6 deals with a general class of equations called generalized ordinary differential equations (GODEs, for short). This class of equations encompasses many types of equations such as ordinary differential equations, impulsive equations, measure differential equations, dynamic equations on time scales, and delay equations, among others. This fact turns this type of equations very attractive to work. In this chapter, the author investigates steady-state oscillation (in the sense of LaSalle) in the framework of generalized ordinary differential equations (generalized ODEs in short), and its applications to measure differential equations. Also, the author discusses about some applications of the presented results.

Chapter 7 works with singular control systems on time scales and justifies the importance of such systems for applications in different field of knowledge. The authors provide results to ensure the existence of solutions for homogeneous and nonhomogeneous singular systems on time scales. They apply the Variation of Constants Formula to get a characterization for the controllability of singular control systems on time scales in terms of its parameters.

Chapter 8 aims to prove a new result concerning the existence of solutions of a dynamic equation on time scales and also, a property on uniqueness of solutions for this type of equations. These results provide a solid basis to prove Massera's type theorems for dynamic equations on time scales. More precisely, they guarantee the existence of periodic solutions for a dynamic equation on time scales defined by a periodic forcing function and taking values in an infinite dimensional Banach space. Also, the authors investigate the existence of almost periodic solutions of dynamic equations on time scales defined by a periodic forcing function. Here, the concept of almost periodic functions is the classical one for time scales, and therefore, the authors work with invariant under translations time scales.

Chapter 9 is devoted to investigate Bresse systems, a robust model for circular beams, given by a set of three coupled wave equations. The goal here is to establish the existence of global attractors for the dynamics of semilinear problems with localized damping. In order to deal with localized damping, a unique continuation property (UCP) is needed. Due to this, the authors provide a suitable UCP for Bresse systems, and they consider the problem in a Riemannian geometry framework and analyze the system as a single equation with different Riemann metrics. Finally, using Carleman-type estimates, they prove the main results.

Chapter 10 presents a spectral decomposition of the Sobolev Space $H^1(\mathbb{R}^n)$. To obtain their results, the authors give a thorough construction stating the foundations of functional calculus for operators and present a proof of the Spectral Theorem for unbounded self-adjoint operators. Despite this last mentioned result is a classical result in functional analysis, there are few references that build all arguments from scratch, and it is exactly what the authors explore in the chapter. It serves as an important overview for classical results in functional analysis.

Lastly, Chap. 11 investigates the harmonic solutions of periodically perturbed implicit integro-differential equations of a particular form obtaining, under appropriate topological conditions, a "branching-type" result for the nontrivial periodic solutions. To obtain the results, the author performs the theory of topological degree, functional differential equations, and differential-algebraic equations.

We gratefully acknowledge Professor Paolo Maria Mariano of the University of Florence, Italy, for the invitation to edit this volume and the support he gave us throughout our work. We also thank the staff of Springer-Birkäuser for the help we received and the solution to all the technical problems we encountered. In particular, we would like to mention Vignesh Viswanathan, Jeffrey Taub, Britta Rao, and Chris Eder. Also, we are very grateful to the authors of each chapter that sent their valuable contribution to the book.

São Paulo, Brazil — Pierluigi Benevieri
Brasilia, Brazil — Jaqueline Godoy Mesquita
April 2024

Contents

Contributors

Editors

Pierluigi Benevieri Instituto de Matemática e Estatística, Universidade de São Paulo, São Paulo, Brazil

Jaqueline Godoy Mesquita Department of Mathematics, University of Brasília, Brasília, Brazil

Authors

Pablo Amster Departamento de Matemática, Universidad de Buenos Aires & IMAS-CONICET, Buenos Aires, Argentina

Douglas R. Anderson Department of Mathematics, Concordia College, Moorhead, MN, USA

Fatma Ayça Çetinkaya Department of Mathematics, Faculty of Science, Mersin University, Mersin, Turkey
Department of Mathematics, University of Tennessee at Chattanooga, Mersin, Türkiye

Tom Cuchta Department of Mathematics and Physics, Marshall University, Huntington, WV, USA

Ahmed A. El-Deeb Department of Mathematics, Faculty of Science, Al-Azhar University, Nasr City, Cairo, Egypt

Claudio A. Gallegos Departamento de Matemáticas, Universidad de Chile, UCH, Santiago, Chile

Rogelio Grau Acuña Departamento de Matemáticas y Estadística, Universidad del Norte, Barranquilla, Atlántico, Colombia

Hernán Henríquez ***(In Memoriam)*** Departamento de Matemática, Universidad de Santiago de Chile, Santiago, Chile

Mariel P. Kuna Departamento de Matemática, Universidad de Buenos Aires & IMAS-CONICET, Buenos Aires, Argentina

To Fu Ma Departamento de Matemática, Universidade de Brasília, Brasília, Brazil

Jaqueline Godoy Mesquita Department of Mathematics, University of Brasília, Brasília, Brazil

Rodrigo N. Monteiro Department of Mathematics, State University of Londrina, Londrina, PR, Brazil

Masakazu Onitsuka Department of Applied Mathematics, Okayama University of Science, Okayama, Japan

Sandra Pinelas Departamento de Ciências Exatas e Engenharia, Amadora, Portugal
Departamento de Matemática, Universidade de Aveiro, Aveiro, Portugal

César Augusto Rubim Departamento de Matemática, Universidade Federal de Juiz de Fora, Minas Gerais, Brazil

Dionicio P. Santos Departamento de Matemática, Facultad de Ciencias Exactas, Universidad Nacional del Centro de la Provincia de Buenos Aires, Buenos Aires, Argentina

Paulo N. Seminario-Huertas Departamento de Matemática Aplicada a las Tecnologías de la Información y las Comunicaciones, Universidad Politécnica de Madrid, Madrid, Spain

Marco Spadini Dipartimento di Matematica e Informatica "Ulisse Dini", Università degli Studi di Firenze, Firenze, Italy

Eduard Toon Departamento de Matemática, Universidade Federal de Juiz de Fora, Minas Gerais, Brazil

Nick Wintz Department of Mathematics, Computer Sciences, and Informational Technology, Lindenwood University, Saint Charles, MO, USA

Acronyms

BVP	Boundary value problem
IVP	Initial value problem
ODE	Ordinary differential equations
GODE	Generalized ordinary differential equation
MDE	Measure differential equation
UCP	Unique continuation property

Chapter 1
Recent Results for Liénard Like Equations with φ-Laplacian on Time Scales

Pablo Amster, Mariel P. Kuna, and Dionicio P. Santos

Abstract Some recent results for φ-Laplacian equations on time scales are described. Using topological methods, the existence and multiplicity of solutions under different boundary conditions is analyzed.

Keywords Time scales · Functional dynamic equations · Leray-Schauder degree · Boundary value problems · Periodic solutions · Upper and lower solutions

MSC 2020 34N05, 34K10, 34C25, 47H11

1.1 Introduction

The study of dynamic equations on time scales (measure chains) has been an active area of research since the celebrated works of Hilger [13, 14]. The literature on time scales is very vast; for example, we may refer the reader to the already classical monographs [7, 8].

In this survey chapter, we shall briefly describe the main results obtained in [2–4] for second order dynamic equations on time scales of φ-Laplacian type. We remark that the problems studied in each case are of different nature, and consequently, the techniques employed for their resolution vary from one problem to another.

P. Amster (✉) · M. P. Kuna
Departamento de Matemática, Universidad de Buenos Aires & IMAS-CONICET, Buenos Aires, Argentina
e-mail: pamster@dm.uba.ar; mpkuna@dm.uba.ar

D. P. Santos
Departamento de Matemática, Facultad de Ciencias Exactas, Universidad Nacional del Centro de la Provincia de Buenos Aires, Buenos Aires, Argentina
e-mail: dpdallos@exa.unicen.edu.ar

P. Benevieri, J. G. Mesquita (eds.), *Functional Differential Equations and Dynamic Equations on Time Scales*, Advances in Continuum Mechanics 52,
https://doi.org/10.1007/978-3-031-83327-4_1

In the first place, we shall be concerned with mixed Hilger derivatives, specifically the problem

$$(\varphi(u^{\Delta}(t)))^{\nabla} = f(t, u(t)), \ t \in [0, T]_{\mathbb{T}}, \tag{1.1}$$

under Dirichlet, Neumann, or periodic boundary conditions, that is,

$$u(\rho(0)) = u(\sigma(T)) = 0; \tag{1.2}$$

$$u^{\Delta}(\rho(0)) = u^{\Delta}(T) = 0; \tag{1.3}$$

$$u(\rho(0)) = u(\sigma(T)), \ u^{\Delta}(\rho(0)) = u^{\Delta}(T), \tag{1.4}$$

respectively. Here, $\varphi : \mathbb{R} \to \mathbb{R}$ is an increasing homeomorphism with $\varphi(0) = 0$. A well-known example of such mapping is the so-called p-Laplacian, namely $\varphi(s) := |s|^{p-2}s$, which reduces to the standard linear case when $p = 2$. In this work, we shall present an existence result based on the method of upper and lower solutions, adapted to this particular situation. Furthermore, we shall define monotone iterative schemes that converge to extremal solutions. Moreover, the Leray-Schauder continuation method will allow us to formulate a version of the 3-solutions theorem, directly deduced from the excision property of the degree. It is worth mentioning that, due to the mixed derivatives, the boundary conditions are different from those considered in problems with two delta (forward) derivatives. Moreover, unlike the continuous case $\mathbb{T} = \mathbb{R}$, the periodic condition (1.4) is not equivalent to the search of periodic solutions, in the sense that $u(t + T) \equiv u(t)$ for some $T > 0$, which is the subject of the other two aforementioned papers. Specifically, in [3], the periodic problem for a pendulum equation with relativistic effects was considered. This means that the involved homeomorphism is the relativistic operator, given by

$$\varphi(x) := \frac{x}{\sqrt{1 - \frac{x^2}{c^2}}}, \tag{1.5}$$

where the constant c represents the light speed. A crucial difference with the previous case is the fact that, here, φ is defined only over the bounded interval $(-c, c)$. This situation is known in the literature as *singular* and usually requires a different treatment with respect to the standard case. In this work, we shall consider the forced pendulum equation with friction

$$\varphi(u^{\nabla}(t))^{\Delta} + au^{\Delta}(t) + b \sin u(t) = p_0(t) + s, \tag{1.6}$$

where $a \geq 0$ and $b > 0$ are constants, p_0 is a continuous and T-periodic function with zero average defined on the time scale, and s is a parameter. An obvious assumption is that the time scale is T-periodic, that is, $T + \mathbb{T} = \mathbb{T}$. We recall that, for the continuous classical case (i.e., $\mathbb{T} = \mathbb{R}$ and $\varphi(x) = x$), the preceding statement of the problem in terms of s was firstly studied by Castro [10] for $a = 0$ using variational methods and extended to the non-variational case $a > 0$ by Fournier and Mawhin [12]. In a previous work [1], the variational setting was successfully reproduced for time scales for the closely related problem

$$u^{\Delta\Delta}(t) + b \sin u(\sigma(t)) = p_0(t) + s,$$

and Castro's result was extended to the time scales framework; however, the relativistic case with a friction term on a time scale was only treated in [3]. Mimicking the results for the classical case, we prove the existence of a nonempty compact solvability interval $I(p_0)$, which depends continuously on p_0, such that the problem has a periodic solution if and only if $s \in I(p_0)$. Since the problem is not necessarily variational if $a > 0$ or $\mathbb{T} \neq \mathbb{R}$, it is not guaranteed that $0 \in I(p_0)$; here, we shall give a sufficient condition which, for the continuous case $\mathbb{T} = \mathbb{R}$, improves the previously known bounds obtained successively in [6, 18], and [11] for $a > 0$.

A more general situation was explored in [4], where a Liénard equation for an arbitrary singular operator is considered in the presence of a time delay, that is,

$$(\varphi(u^{\Delta}(t)))^{\Delta} + h(u(t))u^{\Delta}(t) + g(u(t-r)) = p(t) \quad t \in \mathbb{T}. \tag{1.7}$$

Interestingly, when combined with periodicity, the delay imposes, for $\mathbb{T} \neq \mathbb{R}$, the extra assumption that r is commensurable with T. The main goal in [4] is to provide sufficient conditions for the multiplicity of T-periodic solutions of (1.7) by finding disjoint bounded open subsets of the space of T-periodic continuous functions defined over $\mathbb{T}$, in which the associated solution operator has nonzero degree. Unlike the standard case $\mathbb{T} = \mathbb{R}$, here the Liénard term $h(u)u^{\Delta}$ does not necessarily vanish upon integration, because the usual version of the chain rule is no longer true.

The chapter is organized as follows. In the next section, we introduce the basic definitions and notations that shall be used throughout the manuscript. In Sect. 1.3, the method of upper and lower solutions and a variant of the 3-solutions theorem is presented for a problem with mixed delta-nabla derivatives. Next, in Sect. 1.4 the solvability problem for the relativistic pendulum equation shall be discussed. Finally, a multiplicity result for a general singular φ-Laplacian equation will be presented in the last section.

1.2 Preliminaries

For the reader's convenience, let us recall some basic definitions concerning time-scales that shall be used in this work. A more detailed exposition can be found in [7, 8].

A time scale $\mathbb{T}$ is a nonempty closed subset of $\mathbb{R}$, with the induced metric. For the sake of simplicity, we shall always assume that $0 \in \mathbb{T}$. For $a, b \in \mathbb{T}$ with

$a \le b$, we shall denote $[a,b]_{\mathbb{T}} := [a,b] \cap \mathbb{T}$. The forward/backward jump operators $\sigma, \rho : \mathbb{T} \to \mathbb{T}$ are defined by

$$\sigma(t) := \inf\{s \in \mathbb{T} : s > t\}, \qquad \rho(t) := \sup\{s \in \mathbb{T} : s < t\}.$$

A function u is delta/nabla differentiable at $t \in \mathbb{T}$ if there exists a number (denoted, respectively, by $u^{\Delta}(t)$ and $u^{\nabla}(t)$) with the property that given any $\epsilon > 0$ there is a neighborhood U of t such that, for all $s \in U$,

$$\left|u(\sigma(t)) - u(s) - u^{\Delta}(t)(\sigma(t) - s)\right| \le \epsilon \left|\sigma(t) - s\right|$$

or

$$\left|u(s) - u(\rho(t)) - u^{\nabla}(t)(s - \rho(t))\right| \le \epsilon \left|s - \rho(t)\right|,$$

respectively. As usual, we shall say that u is delta/nabla differentiable if its delta/nabla derivative exists for all t.

Regarding integration, it is not difficult to prove that a standard Lebesgue approach is possible, although the usual definition is given through antiderivatives, namely by the Barrow formula

$$\int_{t_0}^{t} u(s)\Delta s = U(t) - U(t_0) \quad \text{for all } t \in \mathbb{T},$$

where $U^{\Delta} = u$. It is readily verified that such an (unique up to a constant) antiderivative exists for a wide range of functions, which includes the continuous ones.

The sets of continuous and continuously differentiable functions defined over an interval $J \subset \mathbb{T}$ shall be denoted, respectively, by C and C^1. For the T-periodic problem, we shall always assume that $T + \mathbb{T} = \mathbb{T}$ and define $C_T = C_T(\mathbb{T}, \mathbb{R})$ as the Banach space of all continuous T-periodic real functions on $\mathbb{T}$ endowed with the uniform norm. We shall denote by $\tilde{C}_T$ the subspace of those elements of C_T having zero average and by C_T^1 the subspace of those functions that are continuously Δ-differentiable, which is a Banach space when equipped with the norm

$$\|u\|_1 := \max\{\|u\|_{\infty}, \|u^{\Delta}\|_{\infty}\}.$$

For $u \in C_T$, the average, the maximum value, and the minimum value of u shall be denoted, respectively, by $\overline{u}$, $u_{\max}$, and $u_{\min}$, namely

$$\overline{u} := \frac{1}{T}\int_0^T u(t)\Delta t, \qquad u_{\max} := \max_{t \in [0,T]_{\mathbb{T}}} u(t), \qquad u_{\min} := \min_{t \in [0,T]_{\mathbb{T}}} u(t).$$

1.3 A Problem with Mixed Delta-Nabla Derivatives

This section is devoted to the study of problem (1.1). We shall restrict ourselves to the condition (1.4); analogous techniques may be implemented also for (1.2) and (1.3), among many other possible boundary conditions.

The key result for several applications is based on the adaptation to the time scales context of the well-known method of upper and lower solutions, which, roughly speaking, ensures the existence of a solution between a lower solution α and an upper solution β such that $\alpha \leq \beta$. But, in fact, a stronger statement may be proven, namely, that $\deg(I - K, \Omega_{\alpha,\beta}, 0) = -1$, where deg denotes de Leray-Schauder degree, $\Omega_{\alpha,\beta}$ is the open set

$$\Omega_{\alpha,\beta} := \{u \in C^1 : \alpha(t) < u(t) < \beta(t), t \in [\rho(0), \sigma(T)]_{\mathbb{T}}\}$$

and K is an appropriate fixed point operator associated with (1.1).

The proof follows the steps of the usual procedures within continuation methods, although the context of time scales introduces some technicalities that have been treated in [2]. Combined with the excision property of the degree, a version of the 3-solutions theorem is readily deduced:

Theorem 1.1 *For $i = 1, 2$, assume there exist strict lower solutions α_i and upper solutions β_i of problem (1.1)–(1.4) such that $\alpha_i(t) < \beta_i(t)$, $\alpha_1(t) < \alpha_2(t)$, and $\beta_1(t) < \beta_2(t)$ for all $t \in [\rho(0), \sigma(T)]_{\mathbb{T}}$. Furthermore, assume that $\alpha_2(t) > \beta_1(t)$ for some $t \in [\rho(0), \sigma(T)]_{\mathbb{T}}$. Then (1.1)–(1.4) has at least three different solutions u_1, u_2, u_3 with*

$$\alpha_1(t) < u_1(t) < \beta_1(t), \quad \alpha_2(t) < u_2(t) < \beta_2(t), \quad \alpha_1(t) < u_3(t) < \beta_2(t)$$

for all $t \in [\rho(0), \sigma(T)]_{\mathbb{T}}$.

For a proof, it suffices to consider the open sets

$$\Omega_{\alpha_1,\beta_1}, \Omega_{\alpha_2,\beta_2} \subset \Omega_{\alpha_1,\beta_2}.$$

Because the two subsets on the left-hand side are disjoint, we already deduce the existence of $u_1 \neq u_2$ with $u_j \in \Omega_{\alpha_j,\beta_j}$, and moreover, the excision property implies

$$\deg(I - K, \Omega_{\alpha_1,\beta_2} \backslash (\overline{\Omega}_{\alpha_1,\beta_1} \cup \overline{\Omega}_{\alpha_2,\beta_2}), 0) =$$

$$\deg(I - K, \Omega_{\alpha_1,\beta_2}, 0) - \deg(I - K, \Omega_{\alpha_1,\beta_1}, 0) - \deg(I - K, \Omega_{\alpha_2,\beta_2}, 0) = 1.$$

This proves the existence of a solution $u_3 \notin \overline{\Omega}_{\alpha_1,\beta_1} \cup \overline{\Omega}_{\alpha_2,\beta_2}$, which clearly differs from u_1 and u_2.

Furthermore, under a one-side growth condition on f, a monotone iterative method yields the existence of extremal solutions of (1.1)–(1.4).

Theorem 1.2 *Let $f : [\rho(0), \sigma(T)]_{\mathbb{T}} \times \mathbb{R} \to \mathbb{R}$ be continuous, and let $\alpha(t)$ and $\beta(t)$ be lower and upper solutions of (1.1)–(1.4), respectively, with $\alpha \le \beta$ over $[\rho(0), \sigma(T)]_{\mathbb{T}}$. Assume that there exists $M > 0$ such that for all $t \in [0, T]_{\mathbb{T}}$ and any $u, w \in \mathbb{R}$ with $\alpha(t) \le w \le u \le \beta(t)$ it is verified that*

$$f(t, u) - f(t, w) \le M(u - w). \tag{1.8}$$

Then problem (1.1)–(1.4) has a maximal solution u^* and a minimal solution u_* with $u_* \le u^*$ in the sector

$$[\alpha, \beta]_1 := \{w \in C^1 : \alpha \le w \le \beta\}.$$

In particular, assumption (1.8) is satisfied when f is locally Lipschitz in its second variable; for example, this is obviously true if f is of class C^1.

The iterative scheme reads as follows. Starting at $u_0 := \alpha$ or $u_0 := \beta$, define u_{n+1} as the unique solution of the linear problem

$$\begin{cases} (\varphi(u_{n+1}^{\Delta}(t)))^{\nabla} = f(t, u_n(t)) + M(u_{n+1}(t) - u_n(t)), \ t \in [0, T]_{\mathbb{T}} \\ u_{n+1}(\rho(0)) = u_{n+1}(\sigma(T)), \ u_{n+1}^{\Delta}(\rho(0)) = u_{n+1}^{\Delta}(T). \end{cases} \tag{1.9}$$

It is not difficult to prove that $\{u_n\}$ is monotone increasing when $u_0 = \alpha$ and decreasing when $u_0 = \beta$. Moreover, it is also verified that the sequence is bounded which, in turn, implies pointwise convergence. Using the integral equivalent form of the equation, it is finally proven that $\{u_n\}$ converges uniformly to a solution of the problem, which is minimal when $u_0 = \alpha$ and maximal when $u_0 = \beta$.

1.4 The Solvability Interval for the Relativistic Pendulum

The periodic problem for a forced pendulum equation with relativistic effects has attracted the interest of researchers in the last decades. However, to our knowledge, it was only in [3] when the problem for a general time scale was considered for the first time. For the continuous case $\mathbb{T} = \mathbb{R}$, many relevant works have been published, e.g., [5, 6, 9, 16, 18, 19]. In particular, the works [6, 19] are devoted to the so-called *solvability set*, namely, the set $\mathcal{I}(p_0)$ of values of s for which (1.6) admits at least one T-periodic solution. Clearly, if u is a T-periodic solution, so is $u + 2k\pi$ for all $k \in \mathbb{Z}$; for this reason, the literature refers to *geometrically distinct* to those solutions not differing in a multiple of 2π.

The solvability set for the classical pendulum equation with no friction was firstly studied in the pioneering work [10], where it was shown by variational methods that $\mathcal{I}(p_0) \subset [-b, b]$ is a nonempty compact interval containing 0, which depends continuously on p_0. These results were partially extended to the relativistic case in [9]. However, when $a > 0$ the problem is no longer variational, and topological methods are applied instead in [12] for the classical equation and in [19] for the relativistic case.

Here, following the outline of [3], we shall present a generalization of the main results of [6] and [19] to the context of time scales. With this in mind, it is worth noticing, in the first place, that the set of all possible T-periodic solutions of (1.6) for arbitrary $s \in \mathbb{R}$ can be identified with the set of solutions of the integrodifferential equation

$$\varphi(u^\nabla(t))^\Delta + au^\Delta(t) + b\sin u(t) = p_0(t) + s(u), \qquad t \in (0,T)_\mathbb{T} \tag{1.10}$$

satisfying $u(0) = u(T)$, where $s(u) := \frac{b}{T}\int_0^T \sin u(t)\Delta t$ plays the role of the Δ-average of the function $b\sin u$.

Next, a straightforward application of the Schauder fixed point theorem guarantees that the set

$$S_\eta := \{u \in C[0,T]_\mathbb{T} : u \text{ solves (1.10) and } u(0) = u(T) = \eta\}$$

is nonempty and, consequently,

$$\mathcal{I}(p_0) = \bigcup_{\eta\in\mathbb{R}} \{s(u) : u \in S_\eta\} \neq \emptyset.$$

Furthermore, it is clear that $\mathcal{I}(p_0)$ is contained in the interval $[-b, b]$, and its compactness follows from the periodicity of the problem, since it suffices to consider, in the latter formula, $\eta \in [0, 2\pi]$.

The fact that $\mathcal{I}(p_0)$ is connected is directly deduced from the method of upper and lower solutions, adapted to the present context. Thus, the conclusion is immediate because if $s_1 < s_2$ belong to $\mathcal{I}(p_0)$, then the corresponding solutions u_1 and u_2 are, respectively, an upper and a lower solution for any $s \in (s_1, s_2)$. Adding a multiple of 2π if necessary, it may also be assumed that $u_2 < u_1$, so the existence of a solution u for the value s follows. Furthermore, the argument in the previous section allows us to deduce the existence of an extra solution that is geometrically different from u. Indeed, without loss of generality we may suppose $u_2 < u_1 \not\leq u_2 + 2\pi$, so the 3-solutions theorem applies for the couples (α_i, β_i), with

$$\alpha_i := u_2 + 2\pi(i-1), \qquad \beta_i := u_1 + 2\pi(i-1), \qquad i = 1, 2.$$

Summarizing, we are able to formulate a general result:

Theorem 1.3 *Assume that $p_0 \in C_T$ has zero average. Then, there exist numbers $d(p_0)$ and $D(p_0)$, with $-b \leq d(p_0) \leq D(p_0) \leq b$, such that (1.6) has at least one T-periodic solution if and only if $s \in [d(p_0), D(p_0)]$. Furthermore, for $s \in (d(p_0), D(p_0))$ the problem has at least two geometrically distinct solutions.*

A standard continuous dependence argument allows showing that the functions $d, D\colon \tilde{C_T} \to \mathbb{R}$ are continuous. However, it was already mentioned that, in contrast to the continuous case $\mathbb{T} = \mathbb{R}$, the problem has not necessarily variational structure even when $a = 0$. Consequently, it is not guaranteed that $d(p_0) \le 0 \le D(p_0)$.

In the continuous case, the problem of proving whether or not the latter inequalities hold when $a > 0$ has been extensively discussed. For the classical pendulum, there exist well-known results proving that for arbitrary $a, b > 0$ there exists p_0 such that $0 \notin \mathcal{I}(p_0)$; however, for the relativistic operator such examples cannot exist when the period is sufficiently small, independently of the friction term. Indeed, it was proven in [6] that $0 \in \mathcal{I}(p_0)^\circ$ when $cT < \sqrt{3}\pi$. This bound was improved in [11], where a condition in terms of a, b, and $\|p_0\|_{L^1}$ was established, from which the uniform condition $cT \le 2\pi$ is deduced. When applied to the continuous case, the result from [3] described below constitutes a slight improvement of the preceding bound. Specifically, there exists T^* with $cT^* > \pi$ such that $0 \in \mathcal{I}(p_0)$ for $T \le T^*$, and if the inequality is strict, then 0 is an interior point. A lower bound for T^* can be obtained as a zero of a real function; for the continuous case $\mathbb{T} = \mathbb{R}$, it was numerically shown that $cT^* > 6.318$. We remark that the computation is independent of p_0; in consequence, the range of the relativistic pendulum operator $\mathcal{P} := \varphi(u^\nabla(t))^\Delta + au^\Delta + b\sin u$ for $T < T^*$ contains a strip of the form $\tilde{C_T} + (-\varepsilon, \varepsilon)$ for some $\varepsilon > 0$.

In order to state the results, it proves convenient to introduce the quantity $k(\mathbb{T})$, defined as the best possible constant of the inequality

$$\|u - \overline{u}\|_\infty \le k\|u^\Delta\|_\infty, \qquad u \in C^1_T.$$

We remark that, for arbitrary $\mathbb{T}$, the rough estimate $k(\mathbb{T}) \le \frac{T}{2}$ is readily deduced from the periodicity and the fact that u^Δ has zero average:

$$u_{\max} - u_{\min} \le \int_{t_{\min}}^{t_{\max}} [u^\Delta(t)]^+ \Delta t \le \int_0^T [u^\Delta(t)]^+ \Delta t = \frac{1}{2}\int_0^T |u^\Delta(t)|\Delta t \le \frac{T}{2}\|u^\Delta\|_\infty.$$

We remark that, in the continuous case, the (optimal) Sobolev inequality $\|x - \overline{x}\|_\infty \le \sqrt{\frac{T}{12}}\|x'\|_2$ implies $k(\mathbb{R}) \le \frac{T}{2\sqrt{3}}$, while in the discrete case $\mathbb{T} = \mathbb{Z}$ the Sobolev inequality obtained in [17] leads to the estimate

$$k(\mathbb{Z}) \le \sqrt{\frac{T^2 - 1}{6}}.$$

Our result in [3] reads as follows.

Theorem 1.4 *Assume that $ck(\mathbb{T}) < \pi$, and define the function*

$$\psi(\delta) := 2\delta\cos(\delta) + (cT - 2\delta)\cos\left(ck(\mathbb{T})\right).$$

If $\psi(\delta) \geq 0$ for some $\delta \in (0, \frac{\pi}{2})$, then $0 \in \mathcal{I}(p_0)$. Furthermore, if the previous inequality is strict, then $0 \in \mathcal{I}(p_0)^\circ$.

The proof is based on the Leray-Schauder continuation method which, adapted to this situation, only requires to verify that the equation

$$\left(\frac{u^\nabla(t)}{\sqrt{1-\frac{u^\nabla(t)^2}{c^2}}}\right)^\Delta = \lambda[p_0(t) - au^\Delta(t) - b\sin u(t)] \tag{1.11}$$

has no T-periodic solutions with average $\pm\frac{\pi}{2}$. This is clear when $ck(\mathbb{T}) \leq \frac{\pi}{2}$, because in this case the function $\sin u(t)$ has a constant sign, and upon integration, a contradiction arises, namely

$$0 = b\int_0^T \sin u(t)\Delta t \neq 0.$$

In the continuous case, this already retrieves the condition $cT \leq \sqrt{3}\pi$, but the estimate can be sharpened as follows.

From the previous considerations, we may already assume that $\frac{\pi}{2} < ck(\mathbb{T}) < \pi$. If u is a solution of (1.11) such that $\overline{u} = \frac{\pi}{2}$, then

$$u(t) \in \left[\frac{\pi}{2} - ck(\mathbb{T}), \frac{\pi}{2} + ck(\mathbb{T})\right] \subset \left(-\frac{\pi}{2}, \frac{3\pi}{2}\right)$$

for all $t \in \mathbb{T}$, whence

$$\sin u(t) \geq -\sin(A) > -1, \qquad \text{where } A = ck(\mathbb{T}) - \frac{\pi}{2}.$$

For fixed $\delta \in (0, \frac{\pi}{2})$, consider the set

$$C_\delta = \left\{t \in [0, T]_\mathbb{T} : \left|u(t) - \frac{\pi}{2}\right| \leq \delta\right\};$$

then

$$\begin{aligned} 0 = \textstyle\int_0^T \sin u(t)\Delta t &\geq \textstyle\int_{C_\delta}(\sin u(t) + \sin A)\Delta t - T\sin A \\ &\geq \left[\sin(\tfrac{\pi}{2} - \delta) + \sin A\right]\mathfrak{m}(C_\delta) - T\sin A \\ &= \cos(\delta)\mathfrak{m}(C_\delta) - [T - \mathfrak{m}(C_\delta)]\sin A, \end{aligned} \tag{1.12}$$

where $\mathfrak{m}(C_\delta)$ denotes the measure of the set C_δ associated with the Δ-integral, that is, $\mathfrak{m}(C_\delta) = \int_{C_\delta}\Delta t$.

Next, observe that if $u(t_0) \leq \frac{\pi}{2}$ and $t_1 > t_0$ is such that $u(t_1) \geq \frac{\pi}{2} + \delta$, then

$$\delta \leq u(t_1) - u(t_0) = \int_{t_0}^{t_1} u^{\Delta}(s)\Delta s < c(t_1 - t_0).$$

In the same way, if $t_0 < t_1$ are such that $u(t_0) \geq \frac{\pi}{2}$ and $u(t_1) \leq \frac{\pi}{2} - \delta$, then $c(t_1 - t_0) > \delta$. Thus, by periodicity, it is seen that $\mathfrak{m}(C_\delta) > \frac{2\delta}{c}$. The same conclusions are obtained if $\overline{u} = -\frac{\pi}{2}$; hence, the assumption $\psi(\delta) \geq 0$ produces a contradiction in (1.12). In other words, (1.11) cannot have solutions with average $\pm\frac{\pi}{2}$, and the conclusion follows. Furthermore, if $\psi(\delta) > 0$, then we may add a small parameter s to the function p_0 in (1.11) and still get a contradiction as in (1.12), that is, 0 is an interior point.

Corollary 1.1 *There exists a constant T^* with $cT^* > \pi$ such that $0 \in \mathcal{I}(p_0)$ for all $p_0 \in \tilde{C}_T$ if $T \leq T^*$, and it is an interior point if $T < T^*$.*

Proof For arbitrary $\mathbb{T}$, we know already that $k(\mathbb{T}) \leq \frac{T}{2}$; then a sufficient condition when $cT \in (\pi, 2\pi)$ is the existence of $\delta \in (0, \frac{\pi}{2})$ such that $\Psi(\delta, T) \geq 0$, where

$$\Psi(\delta, T) := 2\delta\cos(\delta) + (cT - 2\delta)\cos\left(\frac{cT}{2}\right).$$

The result now follows trivially from the fact that $\Psi(\delta, \frac{\pi}{c}) = 2\delta\cos(\delta)$.

□

It may be noticed that a similar reasoning for $\mathbb{T} = \mathbb{R}$ is possible, now taking

$$\Psi(\delta, T) := 2\delta\cos(\delta) + (cT - 2\delta)\cos\left(\frac{cT}{2\sqrt{3}}\right).$$

This leads to a direct proof of the fact that $cT^* > \sqrt{3}\pi$ although, as mentioned, numerical experiments in [3] reveal that a sufficient condition is $cT \leq 6.318$, which is slightly better than the bound $cT \leq 2\pi$ given in [11]. For arbitrary $\mathbb{T}$, numerical experiments show that $0 \in \mathcal{I}(p_0)^\circ$ for $cT \leq 4.19$.

To conclude this section, let us mention a well-known question which still remains unanswered after some decades since it was proposed: Is it possible to find $p_0 \in \tilde{C}_T$ such that $\mathcal{I}(p_0)$ reduces to a single point? Such a situation is called *degenerate*, and up to present times, for the classical pendulum it is known that $\mathcal{I}(p_0)$ is nontrivial for a large (open and dense) subset of $\tilde{C}_T$, see, e.g., [15]. To our knowledge, the degeneracy problem has not been considered in the literature for the relativistic framework, not even in the continuous case $\mathbb{T} = \mathbb{R}$. The following toy-model may shed some light on the subject.

Example 1.1 Let $\mathbb{T} = \mathbb{Z}$ and $T = 2$; then p_0 is characterized by

$$p_0(0) = \alpha = -p_0(1)$$

for some $\alpha \in \mathbb{R}$. Setting $x := u$ and $y := \varphi(u^\Delta)$, the equation is transformed into the system

$$\begin{cases} x^\nabla = \psi(y), \\ y^\Delta = p_0 + s - b\sin x, \end{cases}$$

where $\psi = \varphi^{-1}$, i.e.,

$$\psi(w) = \frac{w}{\sqrt{1 + \frac{w^2}{c^2}}}.$$

Due to 2-periodicity, the preceding problem reduces to a four-dimensional system for $x_i := x(i)$, $y_i := y(i)$ with $i = 0, 1$:

$$\begin{cases} x_0 = x_1 + \psi(y_0) \\ y_1 = y_0 + \alpha + s - b\sin x_0 \\ x_1 = x_0 + \psi(y_1) \\ y_0 = y_1 - \alpha + s - b\sin x_1. \end{cases}$$

By adding the first and the third equation, it is deduced that $\psi(y_0) + \psi(y_1) = 0$ and, by symmetry, $y_1 = -y_0$. This is obviously not a surprise, because $\psi(y)$ is the (discrete) derivative of a 2-periodic function. Replacing into the previous system, the problem is reduced to find x_0 such that

$$y_0 := \frac{b\sin x_0 - \alpha - s}{2} = \frac{-b\sin(x_0 - \psi(y_0)) - \alpha + s}{2}.$$

Thus, we may define the function

$$F(x_0, s) = b[\sin x_0 + \sin(x_0 + \psi(y_0))] - 2s$$

with y_0 as before. A straightforward computation shows that, even if the problem is non-variational, there exists always a solution for $s = 0$. Indeed, the equation $F(x_0, 0) = 0$ yields

$$x_0 - \pi = x_0 + \psi\left(\frac{b\sin x_0 - \alpha}{2}\right) + 2k\pi$$

or

$$-x_0 = x_0 - \psi\left(\frac{b\sin x_0 - \alpha}{2}\right) + 2k\pi,$$

and although the first equality might not be solvable, the second one admits a solution for arbitrary k. Clearly, when searching for geometrically distinct solutions, it suffices to consider $k = 0, 1$. Moreover, observe that if x_0 is a solution of the latter equation, then

$$\frac{1}{b}\frac{\partial F}{\partial x_0}(x_0, 0) = \cos x_0 + \cos(-x_0)\left(1 - \psi'\left(\frac{b\sin x_0 - \alpha}{2}\right)\frac{b}{2}\cos x_0\right)$$

$$= \cos x_0 \left(2 + \frac{b\cos x_0}{2\left(1 - \frac{(b\sin x_0 - \alpha - s)^2}{4c^2}\right)^{3/2}}\right).$$

It is verified, for almost all choices of α, that $\frac{\partial F}{\partial x_0}(x_0, 0) \neq 0$; thus, the implicit function theorem guarantees the existence of a smooth curve defined in a neighborhood of $s = 0$ such that $F(x_0(s), s) = 0$. In particular, it is deduced that 0 is an interior point of $\mathcal{I}(p_0)$.

1.5 Multiplicity Results for a Delayed Liénard Equation

In this section we shall present some multiplicity results for Eq. (1.7) where $h, g\colon \mathbb{R} \to \mathbb{R}$ are continuous and $\varphi : (-a, a) \to \mathbb{R}$ is an arbitrary singular operator, namely an increasing homeomorphism with $0 < a < +\infty$ and $\varphi(0) = 0$. It shall be assumed that $\mathbb{T}$ is T-periodic and $p(t + T) = p(t)$ is continuous in $\mathbb{T}$ with zero average. The presence of the delay $r > 0$ requires the extra assumption that $\mathbb{T} - r \subset \mathbb{T}$.

As mentioned in the introduction, when $\mathbb{T} \neq \mathbb{R}$, its periodicity implies that $r = qT$ for some positive $q \in \mathbb{Q}$. Indeed, this is simply due to the fact that, since $\mathbb{T} - r\mathbb{Z} \subset \mathbb{T}$, it is deduced that $\mathbb{T}/T\mathbb{T}$ is a dense subset of $\mathbb{R}/T\mathbb{Z}$, and the conclusion follows from the fact that $\mathbb{T}$ is closed.

The main result of [4] reads as follows:

Theorem 1.5 *Assume there exists a strictly increasing sequence* $\{\alpha_j\}_{j=0}^{n}$ *such that*

$$(-1)^j \int_0^T h(u(t))u^\Delta(t) + g(u(t))\Delta t < 0 \text{ if } u(0) = \alpha_j, \ \left\|u^\Delta\right\|_\infty < a,$$

for every j *and each* C^1 *and* T*-periodic function* $x(t)$*. Then, for any continuous* T*-periodic function* p *with zero average, problem (1.7) has at least* n *different* T*-periodic solutions.*

The proof relies on the Leray-Schauder degree theory which, adapted to the present context, yields the following continuation theorem:

Theorem 1.6 *Assume that* Ω *is an open bounded set in* C^1_T *such that the following conditions hold:*

1. *For each* $\lambda \in (0, 1)$ *the problem*

$$(\varphi(u^{\Delta}(t)))^{\Delta} = \lambda N(u) \tag{1.13}$$

has no solution on $\partial\Omega$, *where*

$$N(u)(t) := p(t) - h(u(t))u^{\Delta}(t) - g(u(t-r)).$$

2. *The equation*

$$g(y) = 0$$

has no solution on $\partial\Omega \cap \mathbb{R}$, *where we consider the natural identification of* $\mathbb{R}$ *with the subspace of constant functions of* C_T^1.

3. *The Brouwer degree of* g *satisfies*

$$\deg_B(g, \Omega \cap \mathbb{R}, 0) \neq 0.$$

Then problem (1.7) has at least one T*-periodic solution.*

It is seen, in Theorem 1.5, that the condition involves the integral of the Liénard term $h(u)u^{\Delta}$, which may differ from 0 due to the failure of the chain rule when $\mathbb{T} \neq \mathbb{R}$. However, the following auxiliary result may be of help when monotonicity conditions on h are verified.

Lemma 1.1 *Assume that* h *is nondecreasing (respectively, nonincreasing) over the range of* $x \in C_T^1$. *Then*

$$\int_0^T h(x(t))x^{\Delta}(t)\Delta t \leq 0 \quad (\text{respectively}, \geq 0).$$

With the help of the preceding lemma, a particular instance of the multiplicity theorem occurs when g is oscillatory over $\mathbb{R}$ and h is locally monotone. Indeed, it is seen that (1.7) has infinitely many different T-periodic solutions, provided that the oscillations are sufficiently slow. Analogous conclusions may be obtained when h is locally close to a constant; more generally, the following corollaries are established:

Corollary 1.2 *Assume there exists a strictly increasing sequence* $\{\alpha_j\}_{j=0}^{n}$ *such that*

$$(-1)^j g > 0 \text{ and } (-1)^j h \text{ is nonincreasing over } \left(\alpha_j - \frac{aT}{2}, \alpha_j + \frac{aT}{2}\right).$$

Then, for any continuous T*-periodic function* p *with zero average, problem (1.7) has at least* n *different* T*-periodic solutions.*

Corollary 1.3 *Assume there exist a strictly increasing sequence* $\{\alpha_j\}_{j=0}^n$ *and constants* γ_j *such that*

$$a|h(x) - \gamma_j| < (-1)^j g(x) \text{ for all } x \in \left(\alpha_j - \frac{aT}{2}, \alpha_j + \frac{aT}{2}\right).$$

Then, for any continuous T-periodic function p with zero average, problem (1.7) has at least n different T-periodic solutions.

Example 1.2 Consider the generalized sunflower equation with relativistic effects,

$$\left(\frac{u^\Delta(t)}{\sqrt{1 - \frac{u^\Delta(t)^2}{c^2}}}\right)^\Delta + h(u(t))u^\Delta(t) + \sin u(t-r) = p(t) \quad t \in \mathbb{T}, \tag{1.14}$$

where $p \in C_T$ has zero average and $h : \mathbb{R} \to \mathbb{R}$ is continuous and may depend on r. According to the computations of the previous section, it is verified that if $T \leq T^*$, then problem (1.14) has infinitely many T-periodic solutions under one of the following assumptions:

1. $(-1)^j h$ is nonincreasing in $\left(\alpha_j - \frac{cT}{2}, \alpha_j + \frac{cT}{2}\right)$, where $\alpha_j = (2j+1)\frac{\pi}{2}$ for $j \in \mathbb{Z}$.
2. $c\left|h(x) - \gamma_j\right| < |\sin x|$ for $x \in \left(\alpha_j - \frac{cT}{2}, \alpha_j + \frac{cT}{2}\right)$ for $j \in \mathbb{Z}$ and some constants γ_j.

Both conditions are fulfilled when h is constant; however, in this case we cannot ensure that the multiple solutions are geometrically distinct.

As a final remark, let us mention that the idea of "slow oscillations" may be illustrated by a slightly different equation

$$\left(\frac{u^\Delta(t)}{\sqrt{1 - \frac{u^\Delta(t)^2}{c^2}}}\right)^\Delta + hu^\Delta(t) + \sin \sqrt[3]{u(t-r)} = p(t),$$

where, as before, p has zero average and, for simplicity, we assume that h is constant. Because the length of the gaps between zeros of the nonlinear term $\sin \sqrt[3]{u}$ tends to infinity as $u \to \infty$, the assumptions of the multiplicity theorem are satisfied, and the existence of infinitely many T-periodic solutions is deduced.

Acknowledgments This work was partially supported by projects UBACyT 20020190100039BA and PIP 11220200100175CO, CONICET.

References

1. P. Amster, P. De Nápoli, C.C. Tisdell, Variational methods for two resonant problems on time scales. Int. J. Differ. Equ. **2**(1), 1–12 (2007)
2. P. Amster, M.P. Kuna, D.P. Santos, Multiple solutions of boundary value problems on time scales for a φ-Laplacian operator. Opuscula Math. **40**(4), 405–425 (2020)
3. P. Amster, M.P. Kuna, D.P. Santos, On the solvability of the periodically forced relativistic pendulum equation on time scales. Electron. J. Qual. Theory Differ. Equations **2020**(62), 1–11 (2020)
4. P. Amster, M.P. Kuna, D.P. Santos, Multiplicity of periodic solutions for dynamic Liénard equations with delay and singular φt-Laplacian of relativistic type. Fixed Point Theory **25**(1), 31–42 (2024)
5. C. Bereanu, J. Mawhin, Existence and multiplicity results for some nonlinear problems with singular ϕ-laplacian. J. Differ. Equ. **243**, 536–557 (2007)
6. C. Bereanu, P. Jebelean, J. Mawhin, Periodic solutions of pendulum-like perturbations of singular and bounded ϕ-Laplacians. J. Dyn. Differ. Equ. **22**, 463–471 (2010)
7. M. Bohner, A. Peterson, *Dynamic Equations on Time Scales* (Birkhauser, Boston, 2001)
8. M. Bohner, A. Peterson (eds.), *Advances in Dynamic Equations on Time Scales* (Birkhauser, Boston, 2003)
9. H. Brezis, J. Mawhin, Periodic solutions of the forced relativistic pendulum. Differ. Integr. Equ. **23**(9), 801–810 (2010)
10. A. Castro, Periodic solutions of the forced pendulum equation. Differential Equations, Academic Press, 149–160 (1980)
11. J.A. Cid, On the existence of periodic oscillations for pendulum-type equations. Adv. Nonlinear Anal. **10**, 121–130 (2021)
12. G. Fournier, J. Mawhin, On periodic solutions of forced pendulum-like equations. J. Differ. Equ. **60**(3), 381–395 (1985)
13. S. Hilger, Ein Maβkettenkalkül mit Anwendung auf Zentrumsmanningfaltingkeiten. PhD thesis, Universität Würzburg (1988)
14. S. Hilger, Analysis on measure chains – A unified approach to continuous and discrete calculus. Results Math. **18**(1–2),18–56 (1990)
15. J. Mawhin, *Global Results for the Forced Pendulum Equation*. Handbook of Differential Equations: Ordinary Differential Equations, vol. 1 (Elsevier/North-Holland, Amsterdam, 2004), pp. 533–589
16. J. Mawhin, Periodic solutions of the forced pendulum: classical vs relativistic. Le Matematiche (Catania) **65**(2), 97–107 (2010)
17. X. Tang, X. Zhang, Periodic solutions for second-order discrete Hamiltonian systems. J. Differ. Equ. Appl. **17**, 1413–1430 (2011)
18. P. Torres, Periodic oscillations of the relativistic pendulum with friction. Phys. Lett. A. **372**(42), 6386–6387 (2008)
19. P. Torres, Nondegeneracy of the periodically forced Liénard differential equation with ϕ-Laplacian. Commun. Contemp. Math. **13**(2), 283–292 (2011)

Chapter 2
Ulam Stability and Instability of First-Order Linear ω-Periodic Dynamic Equations on Isolated Time Scales

Douglas R. Anderson and Masakazu Onitsuka

Abstract We apply a new definition of periodicity on isolated time scales introduced by Bohner, Mesquita, and Streipert to the study of Ulam stability. If the graininess (step size) of an isolated time scale is bounded by a finite constant, then the linear ω-periodic dynamic equations are Ulam stable if and only if the exponential function has modulus different from unity. If the graininess increases at least linearly to infinity, the ω-periodic dynamic equations are not Ulam stable. Several examples of specific isolated time scales are also supplied.

Keywords Ulam stability · Periodic · Best constant · Time scales · Perturbation

MSC 2020 34N05, 39A30, 39A23, 39A06, 39A13

2.1 Introduction

Bohner, Mesquita, and Streipert [12] recently introduced the idea of periodicity of functions on isolated time scales by focusing on repeated area under the curve rather than repeated function values; see also the earlier paper by Bohner and Chieochan [9], which establishes this concept first for q-difference equations. The nature of this new notion of periodicity is distinct once the graininess (step size) of the isolated time scale is nonconstant, making even 1-periodic functions new and interesting. For example, see Bohner, Cuchta, and Streipert [11].

D. R. Anderson (✉)
Department of Mathematics, Concordia College, Moorhead, MN, USA
e-mail: andersod@cord.edu

M. Onitsuka
Department of Applied Mathematics, Okayama University of Science, Okayama, Japan
e-mail: onitsuka@ous.ac.jp

P. Benevieri, J. G. Mesquita (eds.), *Functional Differential Equations and Dynamic Equations on Time Scales*, Advances in Continuum Mechanics 52,
https://doi.org/10.1007/978-3-031-83327-4_2

In this chapter, we explore the Ulam stability and instability of first-order linear dynamic equations over isolated time scales, where the coefficient function is ω-periodic in the sense of [12]. Stability analysis that variously goes by the names Ulam stability [31], Hyers–Ulam stability [18], Hyers–Ulam–Rassias stability [28] is an area of interest that differs from Lyapunov stability analysis. Brillouët–Belluot, Brzdęk, and Ciepliński [13] give an overview on some recent developments in Ulam-type stability, while Brzdęk, Popa, Raşa and Xu [15] have a book on the Ulam stability of operators. Brzdęk and Wójcik [14] write on approximate solutions of some difference equations. Popa [25, 26] studies Hyers–Ulam and Hyers–Ulam–Rassias stability of the linear recurrence, while Rasouli, Abbaszadeh, and Eshaghi [27] study approximately linear recurrences.

Jung and Nam [19] are interested in the Hyers–Ulam stability of the Pielou logistic difference equation. Nam [20–22] has explored the Hyers–Ulam stability of hyperbolic, elliptic, and loxodromic Möbius difference equations, respectively. Fukutaka and Onitsuka [16] and Onitsuka [23, 24] have studied Hyers–Ulam stability for first-order homogeneous linear differential equations with a periodic coefficient, first-order homogeneous linear difference equations, and second-order nonhomogeneous linear difference equations, respectively. Anderson and Onitsuka [2, 4, 5] investigate Hyers–Ulam stability for Cayley quantum equations and its application to h-difference equations, while Anderson, Onitsuka, and Rassias [7] find the best constant for Ulam stability of first-order h-difference equations with periodic coefficient. András and Mészáros [8] study Ulam–Hyers stability of dynamic equations on time scales via Picard operators. Hua, Li, and Feng [17] investigate Hyers–Ulam stability of dynamic integral equation on time scales. Shen [29] has done work on the Ulam stability of first-order linear dynamic equations on time scales, while Shen and Li [30] examine Hyers–Ulam stability of first-order nonhomogeneous linear dynamic equations on time scales.

The outline of the paper is the following. In Sect. 2.2, a brief review of time scales and the corresponding notation is provided. In Sect. 2.3, we investigate the stability and instability of first-order dynamic equations on isolated time scales with a 3-periodic coefficient. In Sect. 2.4, we investigate the stability and instability of first-order dynamic equations on isolated time scales in the general case, with a ω-periodic coefficient, for any $\omega \in \mathbb{N}$. In Sect. 2.5, we provide a few examples of isolated time scales and apply our theorems to them to remark on their stability/instability.

2.2 Brief Review of Time Scales

A time scale $\mathbb{T}$ is any closed subset of the real line [10]. For $t \in \mathbb{T}$, define the forward and backward jump operators $\sigma, \rho : \mathbb{T} \to \mathbb{T}$ by

$$\sigma(t) := \inf\{s \in \mathbb{T} : s > t\} \quad \text{and} \quad \rho(t) := \sup\{s \in \mathbb{T} : s < t\},$$

respectively, and the graininess function $\mu : \mathbb{T} \to [0, \infty)$ by $\mu(t) := \sigma(t) - t$. A time scale $\mathbb{T}$ is isolated if $\sigma(t) > t$ and $\rho(t) < t$ both hold for all $t \in \mathbb{T}$. For $f : \mathbb{T} \to \mathbb{C}$, the derivative of f at $t \in \mathbb{T}$ is $f^{\Delta}(t) = \frac{f(\sigma(t))-f(t)}{\mu(t)}$. For an isolated time scale $\mathbb{T} = \{t_n\}_{n=0}^{\infty}$ that is unbounded above, the derivative can also be written as

$$f^{\Delta}(t_n) = \frac{f(t_{n+1}) - f(t_n)}{t_{n+1} - t_n} = \frac{f(t_{n+1}) - f(t_n)}{\mu(t_n)},$$

where $\mu(t_n) = t_{n+1} - t_n$ in this case. Throughout this chapter, we will assume $\mathbb{T}$ is an isolated time scale that is unbounded above.

2.3 Investigating Stability of First-Order Dynamic Equations with Period-3 Coefficient

Consider an arbitrary isolated time scale $\mathbb{T} = \{t_n\}_{n\in\mathbb{N}_0}$, where $\mathbb{N}_0 := \{0, 1, 2, 3, \cdots\}$. Recently, the authors discussed the Ulam stability and instability of first-order linear 1-periodic and 2-periodic dynamic equations on arbitrary isolated time scales [6]. In this section we consider the Ulam stability on $\mathbb{T}$ of the first-order linear homogeneous periodic dynamic equation on isolated time scales, with 3-periodic coefficient, represented by

$$z^{\Delta}(t) - p(t)z(t) = 0, \tag{2.1}$$

where $p : \mathbb{T} \to \mathbb{C}$ is 3-periodic [9, Definition 3.1] if

$$p(t_{n+3}) = \frac{\mu(t_n)p(t_n)}{\mu(t_{n+3})}, \qquad \forall t = t_n \in \mathbb{T} = \{t_n\}_{n=0}^{\infty}. \tag{2.2}$$

Let $p_k = p(t_k)$, $\mu_k = \mu(t_k)$, and $p_k \in \mathbb{C}\backslash\left\{\frac{-1}{\mu_k}\right\}$ for $k \in \{0, 1, 2\}$. Through iteration on (2.2), one sees that p is given by

$$p(t_n) := \frac{1}{\mu(t_n)} \begin{cases} \mu_0 p_0 & \text{if} \quad n \equiv 0 \pmod 3, \\ \mu_1 p_1 & \text{if} \quad n \equiv 1 \pmod 3, \\ \mu_2 p_2 & \text{if} \quad n \equiv 2 \pmod 3. \end{cases} \tag{2.3}$$

Remark 2.1 (Exponential Function) Let $p_k = p(t_k)$, $\mu_k = \mu(t_k)$, and $p_k \in \mathbb{C}\backslash\left\{\frac{-1}{\mu_k}\right\}$ for $k = 0, 1, 2$, and set

$$e_p(t_n,t_0) := \prod_{j=0}^{n-1}\left[1+\mu(t_j)p(t_j)\right], \quad \text{where} \quad \prod_{j=0}^{-1}\left[1+\mu(t_j)p(t_j)\right] \equiv 1. \tag{2.4}$$

By the 3-periodic nature of p given in (2.2) and (2.3), the exponential function (2.4) takes the form

$$e_p(t_n,t_0) = \begin{cases} [1+\mu_0p_0]^{\frac{n}{3}}\,[1+\mu_1p_1]^{\frac{n}{3}}\,[1+\mu_2p_2]^{\frac{n}{3}} & \text{if} \quad n \equiv 0 \pmod 3, \\ [1+\mu_0p_0]^{\frac{n+2}{3}}\,[1+\mu_1p_1]^{\frac{n-1}{3}}\,[1+\mu_2p_2]^{\frac{n-1}{3}} & \text{if} \quad n \equiv 1 \pmod 3, \\ [1+\mu_0p_0]^{\frac{n+1}{3}}\,[1+\mu_1p_1]^{\frac{n+1}{3}}\,[1+\mu_2p_2]^{\frac{n-2}{3}} & \text{if} \quad n \equiv 2 \pmod 3. \end{cases} \tag{2.5}$$

Note that the conditions $p_k \in \mathbb{C}\backslash\left\{\frac{-1}{\mu_k}\right\}$ for $k \in \{0,1,2\}$ are really regressive conditions, since $p(t_k) = p_k \neq \frac{-1}{\mu_k}$ is equivalent to $1+\mu_kp_k \neq 0$, for $k = 0, 1, 2$.

For notational convenience throughout the remainder of this section, define the key constants as

$$E_k := 1+\mu_kp_k, \qquad k = 0, 1, 2. \tag{2.6}$$

It is left to the reader to check that e_p indeed satisfies (2.1), with $e_p(t_0,t_0) = 1$.

Definition 2.1 (Ulam Stability) Let $\xi : \mathbb{T} \to \mathbb{C}$ be a perturbation. Equation (2.1) has Ulam stability if and only if there exists a constant $\mathcal{U} > 0$ such that, for arbitrary $\varepsilon > 0$, if a function $\phi : \mathbb{T} \to \mathbb{C}$ satisfies

$$\phi^{\Delta}(t) - p(t)\phi(t) = \xi(t), \quad |\xi(t)| \leq \varepsilon, \qquad t \in \mathbb{T}, \tag{2.7}$$

then there exists a solution $z : \mathbb{T} \to \mathbb{C}$ of (2.1) such that $|\phi(t) - z(t)| \leq \mathcal{U}\varepsilon$ for all $t \in \mathbb{T}$. If (2.1) is Ulam stable, the constant $\mathcal{U}$ is called an Ulam stability constant for (2.1) on $\mathbb{T}$.

Theorem 2.1 *Assume* $\mathbb{T} = \{t_n\}_{n=0}^{\infty}$ *is an isolated time scale with graininess function* $\mu > 0$ *such that* $\lim_{n\to\infty} t_n = \infty$*. If* p *is a* 3*-periodic function, that is,* p *satisfies* (2.3)*, and*

$$|E_0E_1E_2| = |(1+\mu_0p_0)(1+\mu_1p_1)(1+\mu_2p_2)| = 1$$

for $p_k = p(t_k)$, $\mu_k = \mu(t_k)$, $p_k \in \mathbb{C}\backslash\left\{\frac{-1}{\mu_k}\right\}$ *with* E_k *given by* (2.6) *for* $k \in \{0,1,2\}$, *then* (2.1) *is not Ulam stable.*

Proof Suppose $|E_0E_1E_2| = 1$, and let e_p be given by (2.5) with (2.6). It follows in this case that

$$\left|e_p(t_n,t_0)\right| = \begin{cases} 1 & \text{if } \; n \equiv 0 \pmod 3, \\ |E_0| & \text{if } \; n \equiv 1 \pmod 3, \\ |E_0E_1| & \text{if } \; n \equiv 2 \pmod 3. \end{cases}$$

Given arbitrary $\varepsilon > 0$, set

$$m := \min\{1, |E_0|, |E_0E_1|\}, \qquad M := \max\{1, |E_0|, |E_0E_1|\},$$

and let $\phi(t_n) := \frac{\varepsilon t_n}{M} e_p(t_n, t_0)$. Then, ϕ satisfies

$$\left|\phi^\Delta(t_n) - p(t_n)\phi(t_n)\right| = \frac{\varepsilon}{M}\left|e_p(\sigma(t_n), t_0)\right| \leq \varepsilon$$

for all $t_n \in \mathbb{T}$, so that (2.7) holds, but

$$\begin{aligned}|\phi(t_n) - z(t_n)| &= \left|\frac{\varepsilon t_n}{M} e_p(t_n, t_0) - z_0 e_p(t_n, t_0)\right| = |e_p(t_n, t_0)|\left|\frac{\varepsilon t_n}{M} - z_0\right| \\ &\geq m\left|\frac{\varepsilon t_n}{M} - z_0\right| \to \infty\end{aligned}$$

as $n \to \infty$, making (2.1) unstable in the Ulam sense, where $z(t_n) = z_0 e_p(t_n, t_0)$ is the general solution of (2.1). □

Theorem 2.2 *Let $\mathbb{T} = \{t_n\}_{n=0}^\infty$ be an isolated time scale with graininess function $\mu > 0$ such that $\lim_{n\to\infty} t_n = \infty$. Let $p_k = p(t_k)$, $\mu_k = \mu(t_k)$, and $p_k \in \mathbb{C}\backslash\left\{\frac{-1}{\mu_k}\right\}$ for $k = 0, 1, 2$, and let E_k be given by (2.6) for $k \in \{0, 1, 2\}$. If μ satisfies*

$$\mu(t_n) \leq \mu_{\sup} := \sup_{t\in\mathbb{T}} \mu(t) < \infty, \tag{2.8}$$

and p is a 3-periodic function, then (2.1) is Ulam stable if and only if

$$|E_0E_1E_2| = |(1+\mu_0 p_0)(1+\mu_1 p_1)(1+\mu_2 p_2)| \neq 1,$$

with Ulam stability constant $\mathcal{U}$ given by

$$\mathcal{U} = \max\left\{\widetilde{E}_{12}, \widetilde{E}_{20}, \widetilde{E}_{01}\right\}\frac{\mu_{\sup}}{||E_0E_1E_2| - 1|}, \tag{2.9}$$

where $\widetilde{E}_{ij} := |E_iE_j| + |E_j| + 1$.

Proof Recall the criteria for Ulam stability given in Definition 2.1. Since p is 3-periodic in this case, p satisfies (2.2) and has the form (2.3). Let E_k take the form (2.6) for $k = 0, 1, 2$. If $|E_0E_1E_2| = 1$, then (2.1) is not Ulam stable by Theorem 2.1. Otherwise, we consider the following two cases.

(i) First, suppose $|E_0E_1E_2| > 1$, where $E_0E_1E_2 = (1+\mu_0p_0)(1+\mu_1p_1)(1+\mu_2p_2)$. For arbitrary $\varepsilon > 0$, suppose a function ϕ satisfies (2.7) for all $t \in \mathbb{T}$. Additionally, by (2.5), the exponential function has the form

$$e_p(t_{3m}, t_0) = (E_0E_1E_2)^m, \quad e_p(t_{3m+1}, t_0) = E_0(E_0E_1E_2)^m,$$
$$e_p(t_{3m+2}, t_0) = E_0E_1(E_0E_1E_2)^m, \quad m \in \mathbb{N}_0,$$

and from the variation of constants formula, we have that ϕ has the form

$$\phi(t_n) = \phi_0 e_p(t_n, t_0) + e_p(t_n, t_0)\sum_{j=0}^{n-1}\frac{\mu(t_j)\xi(t_j)}{e_p(t_{j+1}, t_0)}, \quad |\xi(t_n)| \le \varepsilon \tag{2.10}$$

for all $t = t_n \in \mathbb{T}$, where $\phi(t_0) = \phi_0 \in \mathbb{C}$. From this expression for $\phi(t_n)$, we have

$$\left|\frac{\phi(t_n)}{e_p(t_n, t_0)}\right| \le |\phi_0| + \varepsilon\mu_{\sup}\sum_{j=0}^{n-1}\frac{1}{|e_p(t_{j+1}, t_0)|},$$

as $|\xi(t_j)| \le \varepsilon$, and μ satisfies (2.8) for all $t = t_j \in \mathbb{T}$ in this case. These facts allow us to rewrite ϕ as

$$\phi(t_n) = \left[\phi_0 + \sum_{j=0}^{\infty}\frac{\mu(t_j)\xi(t_j)}{e_p(t_{j+1}, t_0)}\right]e_p(t_n, t_0) - e_p(t_n, t_0)\sum_{j=n}^{\infty}\frac{\mu(t_j)\xi(t_j)}{e_p(t_{j+1}, t_0)}, \tag{2.11}$$

where

$$z_0 := \phi_0 + \sum_{j=0}^{\infty}\frac{\mu(t_j)\xi(t_j)}{e_p(t_{j+1}, t_0)} \in \mathbb{C}$$

exists and is a finite number due to $e_p(t_n, t_0)$ given in (2.5), the assumption $|E_0E_1E_2| > 1$, and the absolute convergence of the infinite series. It is straightforward to see that $z(t_n) := z_0 e_p(t_n, t_0)$ for $t_n \in \mathbb{T}$ is a solution of (2.1), and

$$\lim_{n\to\infty}\frac{\phi(t_n)}{e_p(t_n, t_0)} = \phi_0 + \sum_{j=0}^{\infty}\frac{\mu(t_j)\xi(t_j)}{e_p(t_{j+1}, t_0)} = z_0$$

exists. Consequently,

$$z(t_n) = \left(\lim_{n\to\infty}\frac{\phi(t_n)}{e_p(t_n, t_0)}\right)e_p(t_n, t_0),$$

and

$$
\begin{aligned}
|\phi(t_n) - z(t_n)| &= \left| -e_p(t_n, t_0) \sum_{j=n}^{\infty} \frac{\mu(t_j)\xi(t_j)}{e_p(t_{j+1}, t_0)} \right| \\
&\leq \varepsilon\mu_{\sup} |e_p(t_n, t_0)| \sum_{j=n}^{\infty} \frac{1}{|e_p(t_{j+1}, t_0)|} \\
&= \frac{\varepsilon\mu_{\sup}}{|E_0E_1E_2| - 1} \begin{cases} (|E_1E_2| + |E_2| + 1) & \text{if} \quad n \equiv 0 \pmod 3 \\ (|E_0E_2| + |E_0| + 1) & \text{if} \quad n \equiv 1 \pmod 3 \\ (|E_0E_1| + |E_1| + 1) & \text{if} \quad n \equiv 2 \pmod 3 \end{cases}
\end{aligned}
$$

holds for all $t = t_n \in \mathbb{T} = \{t_n\}_{n=0}^{\infty}$. Therefore, (2.1) is Ulam stable with Ulam stability constant $\mathcal{U}$ given by (2.9) for $E_0E_1E_2 = (1+\mu_0 p_0)(1+\mu_1 p_1)(1+\mu_2 p_2)$ and $|E_0E_1E_2| > 1$.

(ii) Second, suppose $0 < |E_0E_1E_2| < 1$, where E_k satisfy (2.6) for $k = 0, 1, 2$. As before, we know that a function ϕ that satisfies (2.7) takes the form (2.10). Let $z = z(t_n) = \phi_0 e_p(t_n, t_0)$ for $t = t_n \in \mathbb{T}$ be the solution of (2.1) such that $z(t_0) = \phi_0$. Then we have

$$
\phi(t_n) - z(t_n) = e_p(t_n, t_0) \sum_{j=0}^{n-1} \frac{\mu(t_j)\xi\left(t_j\right)}{e_p\left(t_{j+1}\right)},
$$

and thus

$$
\begin{aligned}
|\phi(t_n) - z(t_n)| &\leq \varepsilon\mu_{\sup} |e_p(t_n, t_0)| \sum_{j=0}^{n-1} \frac{1}{|e_p(t_{j+1}, t_0)|} \\
&\leq \frac{\varepsilon\mu_{\sup}}{1 - |E_0E_1E_2|} \begin{cases} (|E_1E_2| + |E_2| + 1) & \text{if} \quad n \equiv 0 \pmod 3 \\ (|E_0E_2| + |E_0| + 1) & \text{if} \quad n \equiv 1 \pmod 3 \\ (|E_0E_1| + |E_1| + 1) & \text{if} \quad n \equiv 2 \pmod 3 \end{cases}
\end{aligned}
$$

for all $t = t_n \in \mathbb{T}$. Therefore, (2.1) is Ulam stable with Ulam stability constant

$$
\mathcal{U} = \max\left\{\widetilde{E}_{12}, \widetilde{E}_{20}, \widetilde{E}_{01}\right\} \frac{\mu_{\sup}}{||E_0E_1E_2| - 1|}
$$

for $0 < |E_0E_1E_2| < 1$, where E_k is given in (2.6) and $\widetilde{E}_{ij} := |E_iE_j| + |E_j| + 1$. This ends case (ii). Overall, we have shown that (2.1) is Ulam stable if and only if $|E_0E_1E_2| = |(1 + \mu_0 p_0)(1 + \mu_1 p_1)(1 + \mu_2 p_2)| \neq 1$, with Ulam constant given by (2.9), completing the proof. □

Example 2.1 (h-Difference Equation) Let $h > 0$, and set $\mathbb{T} = \{0, h, 2h, 3h, \ldots\}$. If p is 3-periodic, then $p(t + 3h) = p(t)$ for all $t \in \mathbb{T}$, so

$$p(t_n) = p(nh) := p_k \ \text{ if } \ n \equiv k \pmod 3$$

for $p_k \in \mathbb{C}\backslash\{-\frac{1}{h}\}$ with $k \in \{0, 1, 2\}$. Since $\mu(t) \equiv h$ for all $t \in \mathbb{T}$, by Theorem 2.2, (2.1) is Ulam stable if and only if $|(1 + hp_0)(1 + hp_1)(1 + hp_2)| \neq 1$. Letting $E_k = (1 + hp_k)$ for $k \in \{0, 1, 2\}$, the Ulam constant is given by

$$\mathcal{U} = \max\left\{(|E_1E_2| + |E_2| + 1)\,, (|E_0E_2| + |E_0| + 1)\,,\right.$$
$$\left.(|E_0E_1| + |E_1| + 1)\right\}\frac{h}{||E_0E_1E_2| - 1|}.$$

This agrees with [7, Theorem 4.3]. Moreover, if $|(1+hp_0)(1+hp_1)(1+hp_2)| > 1$, then $\mathcal{U}$ is the best (minimal) Ulam constant by [7, Theorem 4.3 (i)].

Example 2.2 (Two Step Sizes) Let $\eta, \tau > 0$ be two step sizes, and let the isolated time scale $\mathbb{T}$ built upon them be denoted by

$$\mathbb{T}_{\eta,\tau} := \{0, \eta, \eta + \tau, (\eta + \tau) + \eta, 2(\eta + \tau), 2(\eta + \tau) + \eta, \cdots\} = \{t_n\}_{n=0}^{\infty}.$$

This time scale has been studied in [1, 3], but only for the constant coefficient case. Let $p(t_0) = p_0 \in \mathbb{C}\backslash\left\{\frac{-1}{\eta}\right\}$, $p(t_1) = p_1 \in \mathbb{C}\backslash\left\{\frac{-1}{\tau}\right\}$, and $p(t_2) = p_2 \in \mathbb{C}\backslash\left\{\frac{-1}{\eta}\right\}$. Through iteration on (2.2), one sees that p is 3-periodic if it is given by

$$p(t_n) := \begin{cases} p_0 & \text{if } \ n \equiv 0 \pmod 6, \\ p_1 & \text{if } \ n \equiv 1 \pmod 6, \\ p_2 & \text{if } \ n \equiv 2 \pmod 6, \\ \frac{\eta}{\tau}p_0 & \text{if } \ n \equiv 3 \pmod 6, \\ \frac{\tau}{\eta}p_1 & \text{if } \ n \equiv 4 \pmod 6, \\ \frac{\eta}{\tau}p_2 & \text{if } \ n \equiv 5 \pmod 6. \end{cases} \tag{2.12}$$

Let

$$E_0 = (1 + \eta p_0), \quad E_1 = (1 + \tau p_1), \quad E_2 = (1 + \eta p_2).$$

By Theorem 2.2, we have the new result that (2.1) is Ulam stable if and only if $|E_0E_1E_2| \neq 1$, with Ulam constant given by

$$\mathcal{U} = \max\left\{(|E_1E_2| + |E_2| + 1)\,, (|E_0E_2| + |E_0| + 1)\,,\right.$$
$$\left.(|E_0E_1| + |E_1| + 1)\right\}\frac{\max\{\eta, \tau\}}{||E_0E_1E_2| - 1|}.$$

Theorem 2.3 *Let $\mathbb{T} = \{t_n\}_{n=0}^{\infty}$ be an isolated time scale. Let $p = p(t)$ be given by* (2.3) *for all $t = t_n \in \mathbb{T}$; videlicet, p is 3-periodic. If there exist real constants $a > 0$ and $b \geq 0$ such that*

$$\mu(t_n) \geq an + b, \tag{2.13}$$

then (2.1) *is not Ulam stable.*

Proof Let $\varepsilon > 0$ be a fixed arbitrary constant throughout the proof. Let $p_k = p(t_k)$, $\mu_k = \mu(t_k)$, and $p_k \in \mathbb{C}\backslash \left\{\frac{-1}{\mu_k}\right\}$, and let E_k be given by (2.6), all for $k = 0, 1, 2$. If $|E_0 E_1 E_2| = 1$, then (2.1) is not Ulam stable by Theorem 2.1. Otherwise, by the definition of Ulam stability given in Definition 2.1, we need to find a function $\phi : \mathbb{T} \to \mathbb{C}$ fulfilling (2.7), such that there is no solution z of (2.1) satisfying $|\phi(t) - z(t)| \leq \mathcal{U}\varepsilon$ for all $t \in \mathbb{T}$, for any constant $\mathcal{U} > 0$. Suppose a function ϕ fulfills (2.7) for all $t \in \mathbb{T}$ and has the form (2.10). Take $\phi(t_0) = 0$, and

$$\xi(t_j) = \frac{\varepsilon(aj + b)e_p(t_{j+1}, t_0)}{\mu(t_j)\left|e_p(t_{j+1}, t_0)\right|}, \qquad \left|\xi(t_j)\right| \leq \varepsilon,$$

while recalling that $(aj + b) \leq \mu(t_j)$ for $t_j \in \mathbb{T}$ from (2.13). For the purpose of simplifying the following notation in this proof, define

$$E_{ij} := E_i E_j \quad \text{and} \quad E_{012} := E_0 E_1 E_2.$$

We consider the following cases, where z is a solution of (2.1):

$$|\phi(t_{3m}) - z(t_{3m})| = \varepsilon|E_{012}|^m \left|\left(\frac{b}{|E_0|} + \frac{a+b}{|E_{01}|} + \cdots + \frac{(3m-1)a+b}{|E_{012}|^m}\right) - \frac{z_0}{\varepsilon}\right|,$$

$$\begin{aligned}
&|\phi(t_{3m+1}) - z(t_{3m+1})| \\
&= \varepsilon|E_0||E_{012}|^m \left|\left(\frac{b}{|E_0|} + \frac{a+b}{|E_{01}|} + \cdots + \frac{(3m)a+b}{|E_0||E_{012}|^m}\right) - \frac{z_0}{\varepsilon}\right|, \\
&|\phi(t_{3m+2}) - z(t_{3m+2})| \\
&= \varepsilon|E_{01}||E_{012}|^m \left|\left(\frac{b}{|E_0|} + \frac{a+b}{|E_{01}|} + \cdots + \frac{(3m+1)a+b}{|E_{01}||E_{012}|^m}\right) - \frac{z_0}{\varepsilon}\right|.
\end{aligned}$$

For any $z_0 \in \mathbb{C}$, pick $n \in \mathbb{N}$ such that $(n-1)a + b > \frac{|z_0|}{\varepsilon}\max\{1, |E_0|, |E_{01}|\}$, which is possible since $a > 0$ and $b \geq 0$. Then, for any of the three cases just highlighted, we have

$$|\phi(t_n) - z(t_n)| > \varepsilon\left|(n-1)a + b - \frac{|z_0|}{\varepsilon}\max\{1, |E_0|, |E_{01}|\}\right| \to \infty \quad \text{as} \quad n \to \infty,$$

when $|E_{012}| < 1$.

Suppose $|E_{012}| > 1$. If $t_n = t_{3m}$, then

$$|E_{012}|^m \left(\frac{b}{|E_0|} + \frac{a+b}{|E_{01}|} + \cdots + \frac{(3m-1)a+b}{|E_{012}|^m} \right)$$
$$= \sum_{j=1}^{m} \frac{(3j-3)a+b}{|E_{012}|^{j-1-m}|E_0|} + \sum_{j=1}^{m} \frac{(3j-2)a+b}{|E_{012}|^{j-1-m}|E_{01}|} + \sum_{j=1}^{m} \frac{(3j-1)a+b}{|E_{012}|^{j-m}},$$

so we can write

$$\frac{1}{\varepsilon}|\phi(t_{3m}) - z(t_{3m})|$$
$$= |E_{012}|^m \left| \left(\frac{b}{|E_0|} + \frac{a+b}{|E_{01}|} + \cdots + \frac{(3m-1)a+b}{|E_{012}|^m} \right) - \frac{z_0}{\varepsilon} \right|$$
$$= \left| \frac{|E_{12}|\,(b + 3a(m-1) - (3am+b)|E_{012}| + |E_{012}|^m(3a - b + b|E_{012}|))}{(|E_{012}|-1)^2} \right.$$
$$+ \frac{|E_2|\,(b + a(3m-2) - (3am+a+b)|E_{012}| + |E_{012}|^m(2a - b + (a+b)|E_{012}|))}{(|E_{012}|-1)^2}$$
$$+ \frac{b + a(3m-1) - ((3m+2)a+b)|E_{012}| + |E_{012}|^m(a - b + (2a+b)|E_{012}|)}{(|E_{012}|-1)^2}$$
$$\left. - \frac{z_0|E_{012}|^m}{\varepsilon} \right|.$$

Notice the coefficient of $|E_{012}|^m$ is

$$\frac{|E_{12}|(3a - b + b|E_{012}|) + |E_2|(2a - b + (a+b)|E_{012}|) + a - b + (2a+b)|E_{012}|}{(|E_{012}|-1)^2}$$
$$- \frac{z_0}{\varepsilon},$$

so z_0 could be chosen to zero this out, but we would still have

$$\frac{1}{\varepsilon}|\phi(t_{3m}) - z(t_{3m})| = \left| \frac{|E_{12}|\,(b + 3a(m-1) - (3am+b)|E_{012}|)}{(|E_{012}|-1)^2} \right.$$
$$+ \frac{|E_2|\,(b + a(3m-2) - (3am+a+b)|E_{012}|)}{(|E_{012}|-1)^2}$$
$$\left. + \frac{b + a(3m-1) - ((3m+2)a+b)|E_{012}|}{(|E_{012}|-1)^2} \right|,$$

where the numerator coefficient of m is $3a(1+|E_2|+|E_{12}|)(1-|E_{012}|) \neq 0$. Either way,

$$\frac{1}{\varepsilon}|\phi(t_{3m}) - z(t_{3m})| \to \infty \quad \text{as} \quad m \to \infty,$$

and (2.1) is not Ulam stable.

If $t_n = t_{3m+1}$, then

$$|E_0||E_{012}|^m \left(\frac{b}{|E_0|} + \frac{a+b}{|E_{01}|} + \cdots + \frac{(3m)a+b}{|E_0||E_{012}|^m} \right)$$
$$= \sum_{j=0}^{m} \frac{3ja+b}{|E_{012}|^{j-m}} + \sum_{j=0}^{m-1} \frac{(3j+1)a+b}{|E_{012}|^{j-m}|E_1|} + \sum_{j=0}^{m-1} \frac{(3j+2)a+b}{|E_{012}|^{j-m}|E_{12}|},$$

so we can write

$$\frac{1}{\varepsilon}|\phi(t_{3m+1}) - z(t_{3m+1})|$$
$$= |E_0||E_{012}|^m \left| \left(\frac{b}{|E_0|} + \frac{a+b}{|E_{01}|} + \cdots + \frac{(3m)a+b}{|E_0||E_{012}|^m} \right) - \frac{z_0}{\varepsilon} \right|$$
$$= \left| \frac{b + 3am + |E_{012}|(-b - 3a(m+1) + |E_{012}|^m(3a - b + b|E_{012}|))}{(|E_{012}| - 1)^2} \right.$$
$$- \frac{z_0|E_{012}|^m|E_0|}{\varepsilon}$$
$$+ \frac{|E_{02}|\,(b+a(3m-2)-(a+b+3am)|E_{012}| + |E_{012}|^m(2a-b+(a+b)|E_{012}|))}{(|E_{012}| - 1)^2}$$
$$\left. + \frac{|E_0|\,(b+a(3m-1)-(b+a(3m+2))|E_{012}| + |E_{012}|^m(a-b+(2a+b)|E_{012}|))}{(|E_{012}| - 1)^2} \right|.$$

Notice the coefficient of $|E_{012}|^m|E_0|$ is

$$\frac{|E_{12}|(3a - b + b|E_{012}|) + |E_2|(2a - b + (a+b)|E_{012}|) + a - b + (2a+b)|E_{012}|}{(|E_{012}| - 1)^2}$$
$$- \frac{z_0}{\varepsilon},$$

so z_0 could be chosen to zero this out, but we would still have

$$\frac{1}{\varepsilon}|\phi(t_{3m+1}) - z(t_{3m+1})| = \left| \frac{|E_{02}|\,(b + a(3m-2) - (a+b+3am)|E_{012}|)}{(|E_{012}| - 1)^2} \right.$$
$$+ \frac{|E_0|\,(b + a(3m-1) - (b + a(3m+2))|E_{012}|)}{(|E_{012}| - 1)^2}$$
$$\left. + \frac{b + 3am - (3a(m+1) + b)|E_{012}|}{(|E_{012}| - 1)^2} \right|,$$

where the numerator coefficient of m is $3a(1+|E_0|+|E_{02}|)(1-|E_{012}|) \neq 0$. Either way,

$$\frac{1}{\varepsilon}|\phi(t_{3m+1}) - z(t_{3m+1})| \to \infty \quad \text{as} \quad m \to \infty,$$

and (2.1) is not Ulam stable.

If $t_n = t_{3m+2}$, then

$$|E_{01}||E_{012}|^m \left(\frac{b}{|E_0|} + \frac{a+b}{|E_{01}|} + \cdots + \frac{(3m+1)a+b}{|E_{01}||E_{012}|^m} \right)$$
$$= |E_1| \sum_{j=0}^{m} \frac{3ja+b}{|E_{012}|^{j-m}} + \sum_{j=0}^{m} \frac{(3j+1)a+b}{|E_{012}|^{j-m}} + \sum_{j=0}^{m-1} \frac{(3j+2)a+b}{|E_{012}|^{j-m}|E_2|},$$

so we can write

$$\begin{aligned}
&\frac{1}{\varepsilon}|\phi(t_{3m+2}) - z(t_{3m+2})| \\
&= |E_{01}||E_{012}|^m \left| \left(\frac{b}{|E_0|} + \frac{a+b}{|E_{01}|} + \cdots + \frac{(3m+1)a+b}{|E_{01}||E_{012}|^m} \right) - \frac{z_0}{\varepsilon} \right| \\
&= \left| \frac{|E_1|\,(b + 3am + |E_{012}|(-b - 3a(m+1) + |E_{012}|^m(3a - b + b|E_{012}|)))}{(|E_{012}| - 1)^2} \right. \\
&\quad - \frac{z_0|E_{012}|^m|E_{01}|}{\varepsilon} \\
&\quad + \frac{a + b + 3am - (b + a(4+3m))|E_{012}| + |E_{012}|^{m+1}(2a - b + (a+b)|E_{012}|)}{(|E_{012}| - 1)^2} \\
&\quad \left. + \frac{|E_{01}|\,(b+a(3m-1)-(b+a(3m+2))|E_{012}|+|E_{012}|^m(a-b+(2a+b)|E_{012}|))}{(|E_{012}| - 1)^2} \right|.
\end{aligned}$$

Notice the coefficient of $|E_{012}|^m|E_{01}|$ is

$$\frac{|E_{12}|(3a-b+b|E_{012}|)+|E_2|(2a-b+(a+b)|E_{012}|)+a-b+(2a+b)|E_{012}|}{(|E_{012}|-1)^2}$$
$$-\frac{z_0}{\varepsilon},$$

so z_0 could be chosen to zero this out, but we would still have

$$\frac{1}{\varepsilon}|\phi(t_{3m+2}) - z(t_{3m+2})| = \left| \frac{|E_1|\,(b + 3am + |E_{012}|(-b - 3a(m+1)))}{(|E_{012}| - 1)^2} + \frac{a + b + 3am - (b + a(4+3m))|E_{012}|}{(|E_{012}| - 1)^2} + \frac{|E_{01}|\,(b + a(3m-1) - (b + a(3m+2))|E_{012}|)}{(|E_{012}| - 1)^2} \right|,$$

where the numerator coefficient of m is $3a(1+|E_1|+|E_{01}|)(1-|E_{012}|) \neq 0$. Either way,

$$\frac{1}{\varepsilon}|\phi(t_{3m+2}) - z(t_{3m+2})| \to \infty \quad \text{as} \quad m \to \infty,$$

and (2.1) is not Ulam stable. Thus, in all cases, (2.1) is not stable if time scale condition (2.13) holds. □

Example 2.3 (Triangular Equation) Let $\mathbb{N}_0 = \mathbb{N} \cup \{0\} = \{0, 1, 2, 3, \ldots\}$, and let

$$\mathbb{T} = \left\{ \frac{n(n+1)}{2} \right\}_{n=0}^{\infty} = \{0, 1, 3, 6, 10, \ldots\}$$

be the set of triangular numbers. This is an isolated time scale. It follows that

$$\mu(t_n) = t_{n+1} - t_n = n + 1, \quad n \in \mathbb{N}_0.$$

If p is a 3-periodic function on $\mathbb{T}$, then $\left(\frac{n+4}{n+1}\right) p(t_{n+3}) = p(t_n)$ for all $t = t_n \in \mathbb{T}$, and p takes the form

$$p(t_n) = \frac{1}{n+1} \begin{cases} p_0 & \text{if} \quad n \equiv 0 \pmod 3, \\ 2p_1 & \text{if} \quad n \equiv 1 \pmod 3, \\ 3p_2 & \text{if} \quad n \equiv 2 \pmod 3, \end{cases} \qquad n \in \mathbb{N}_0,$$

for $p_0 \in \mathbb{C}\backslash\{-1\}$, $p_1 \in \mathbb{C}\backslash\left\{-\frac{1}{2}\right\}$, and $p_2 \in \mathbb{C}\backslash\left\{-\frac{1}{3}\right\}$. Then, (2.1) is not Ulam stable over the triangular numbers by Theorem 2.3 and (2.13), since $\mu(t_n) = an+b$ for $a = 1 = b$ for all $n \in \mathbb{N}_0$.

Example 2.4 (*q*-Difference Equation) Let $q > 1$ and set $\mathbb{T} = \{1, q, q^2, q^3, \ldots\}$. Consider (2.1) in this case, namely the first-order homogeneous linear q-difference equation with 3-periodic coefficient

$$\mathcal{D}_q z(t) - p(t)z(t) = 0, \qquad \mathcal{D}_q z(t) := \frac{z(qt) - z(t)}{(q-1)t}, \tag{2.14}$$

where $n \in \mathbb{N}$ and $p : \mathbb{T} \to \mathbb{C}$ is 3-periodic if

$$p(t) = q^3 p\left(q^3 t\right), \qquad \forall t \in \mathbb{T}. \tag{2.15}$$

Then, $p : \mathbb{T} \to \mathbb{C}$ is given by

$$p(t) := \frac{q^j p_{j+1}}{t} \quad \text{if} \quad \log_q t \equiv j \pmod 3, \tag{2.16}$$

where $p_{j+1} = p(q^j)$ with $p_{j+1} \in \mathbb{C}\backslash\{\frac{-1}{q^j(q-1)}\}$ for $j \in \{0, 1, 2\}$, such that the coefficient function p is 3-periodic, and p is not periodic for any $k < 3$. The instability result follows directly from Theorem 2.3 and (2.13), where $\mu(q^n) \geq an$ for $a = (q-1)\ln q$ for all $n \in \mathbb{N}_0$.

2.4 Investigating Stability of First-Order Dynamic Equations with Period-ω Coefficient

In this section we consider the Ulam stability of Eq. (2.1) on the isolated time scale $\mathbb{T} = \{t_n\}_{n\in\mathbb{N}_0}$, with ω-periodic coefficient, where $\omega \in \mathbb{N}$, and $p : \mathbb{T} \to \mathbb{C}$ is ω-periodic [9, Definition 3.1]; that is, p satisfies

$$p(t_{n+\omega}) = \frac{\mu(t_n)p(t_n)}{\mu(t_{n+\omega})}, \qquad \forall t = t_n \in \mathbb{T} = \{t_n\}_{n=0}^{\infty}. \tag{2.17}$$

Let $p_k = p(t_k)$, $\mu_k = \mu(t_k)$, and $p_k \in \mathbb{C}\backslash\left\{\frac{-1}{\mu_k}\right\}$ for $k \in \{0, 1, 2, \cdots, \omega-1\}$. Through iteration on (2.17), one sees that p is given by

$$p(t_n) := \frac{1}{\mu(t_n)} \begin{cases} \mu_0 p_0 & \text{if} \quad n \equiv 0 \pmod{\omega}, \\ \mu_1 p_1 & \text{if} \quad n \equiv 1 \pmod{\omega}, \\ \mu_2 p_2 & \text{if} \quad n \equiv 2 \pmod{\omega}, \\ & \quad \vdots \\ \mu_{\omega-1} p_{\omega-1} & \text{if} \quad n \equiv \omega - 1 \pmod{\omega}. \end{cases} \tag{2.18}$$

Remark 2.2 (Exponential Function) Let $p_k = p(t_k)$, $\mu_k = \mu(t_k)$, and $p_k \in \mathbb{C}\backslash\left\{\frac{-1}{\mu_k}\right\}$ for $k \in \{0, 1, 2, \cdots, \omega-1\}$. Then we can find the exponential function as the form (2.4). Note that the conditions $p_k \in \mathbb{C}\backslash\left\{\frac{-1}{\mu_k}\right\}$ for $k \in \{0, 1, 2, \cdots, \omega-1\}$ are really regressive conditions, since $p(t_k) = p_k \neq \frac{-1}{\mu_k}$ is equivalent to $1 + \mu_k p_k \neq 0$, for $k \in \{0, 1, 2, \cdots, \omega-1\}$.

Define the key constants as

$$E_k := 1 + \mu_k p_k, \qquad k \in \{0, 1, 2, \cdots, \omega - 1\}. \tag{2.19}$$

By the ω-periodic nature of p given in (2.17) and (2.18), the exponential function (2.4) has the form

$$e_p(t_{\omega m+j}, t_0) = \left(\prod_{l=0}^{j-1} E_l\right)\left(\prod_{k=0}^{\omega-1} E_k\right)^m, \quad m \in \mathbb{N}_0,\ j \in \{0, 1, 2, \cdots, \omega - 1\}. \tag{2.20}$$

Theorem 2.4 *Assume* $\mathbb{T} = \{t_n\}_{n=0}^{\infty}$ *is an isolated time scale with graininess function* $\mu > 0$ *such that* $\lim_{n\to\infty} t_n = \infty$*. If* p *is a* ω*-periodic function, that is,* p *satisfies* (2.18), *and*

$$\left|e_p(t_\omega, t_0)\right| = \left|\prod_{k=0}^{\omega-1} E_k\right| = \left|\prod_{k=0}^{\omega-1}(1 + \mu_k p_k)\right| = 1$$

for $p(t_k) = p_k$, $\mu(t_k) = \mu_k$, $p_k \in \mathbb{C}\backslash\left\{\frac{-1}{\mu_k}\right\}$ *with* E_k *given by* (2.19) *for* $k \in \{0, 1, 2, \cdots, \omega - 1\}$*, then* (2.1) *is not Ulam stable.*

Proof Suppose $\left|\prod_{k=0}^{\omega-1} E_k\right| = 1$, and let e_p be given by (2.4) with (2.19). It follows from $\left|\prod_{k=0}^{\omega-1} E_k\right| = 1$ and (2.20) that

$$\left|e_p(t_n, t_0)\right| = \begin{cases} 1 & \text{if} \quad n \equiv 0 \pmod{\omega}, \\ |E_0| & \text{if} \quad n \equiv 1 \pmod{\omega}, \\ |E_0 E_1| & \text{if} \quad n \equiv 2 \pmod{\omega}, \\ & \quad \vdots \\ \left|\prod_{k=0}^{\omega-2} E_k\right| & \text{if} \quad n \equiv \omega - 1 \pmod{\omega}. \end{cases}$$

Given arbitrary $\varepsilon > 0$, set

$$m := \min_{j\in\{0,1,2,\cdots,\omega-1\}} \left|\prod_{k=0}^{j-1} E_k\right|, \qquad M := \max_{j\in\{0,1,2,\cdots,\omega-1\}} \left|\prod_{k=0}^{j-1} E_k\right|,$$

and let $\phi(t_n) = \frac{\varepsilon t_n}{M} e_p(t_n, t_0)$. Then, ϕ satisfies

$$\left|\phi^\Delta(t_n) - p(t_n)\phi(t_n)\right| = \frac{\varepsilon}{M}\left|e_p(\sigma(t_n), t_0)\right| \le \varepsilon$$

for all $t_n \in \mathbb{T}$, so that (2.7) holds. Let $z(t_n) = z_0 e_p(t_n, t_0)$ with $z_0 \in \mathbb{C}$. Since

$$|\phi(t_n) - z(t_n)| = \left| \frac{\varepsilon t_n}{M} e_p(t_n, t_0) - z_0 e_p(t_n, t_0) \right| = |e_p(t_n, t_0)| \left| \frac{\varepsilon t_n}{M} - z_0 \right|$$
$$\geq m \left| \frac{\varepsilon t_n}{M} - z_0 \right| \to \infty$$

as $n \to \infty$, we see that (2.1) unstable in the Ulam sense. □

Theorem 2.5 *Let $\omega \in \mathbb{N}$ and let $\mathbb{T} = \{t_n\}_{n=0}^{\infty}$ be an isolated time scale with graininess function $\mu > 0$ such that $\lim_{n\to\infty} t_n = \infty$. Let $p_k = p(t_k)$, $\mu_k = \mu(t_k)$, and $p_k \in \mathbb{C}\backslash \left\{ \frac{-1}{\mu_k} \right\}$ for $k \in \{0, 1, 2, \cdots, \omega - 1\}$, and let E_k be given by* (2.19) *for $k \in \{0, 1, 2, \cdots, \omega - 1\}$. If μ satisfies* (2.8) *for all $t = t_j \in \mathbb{T}$, and p is a ω-periodic function, then* (2.1) *is Ulam stable if and only if*

$$\left| e_p(t_\omega, t_0) \right| = \left| \prod_{k=0}^{\omega-1} E_k \right| = \left| \prod_{k=0}^{\omega-1} (1 + \mu_k p_k) \right| \neq 1,$$

with Ulam stability constant $\mathcal{U}$ given by

$$\mathcal{U} = \left(\max_{j\in\{0,1,2,\cdots,\omega-1\}} \sum_{l=0}^{\omega-1} \frac{1}{\left| \prod_{k=0}^{l} E_{c_{k+j}} \right|} \right) \frac{\mu_{\sup} \left| \prod_{k=0}^{\omega-1} E_k \right|}{\left| \left| \prod_{k=0}^{\omega-1} E_k \right| - 1 \right|}, \tag{2.21}$$

where c_m is defined by

$$c_m := m - \left\lfloor \frac{m}{\omega} \right\rfloor \omega, \quad m \in \mathbb{N}_0.$$

Remark 2.3 Note here that $\lfloor \cdot \rfloor$ means the floor function, that is, $\lfloor x \rfloor$ is defined by

$$\lfloor x \rfloor := \max\{n \in \mathbb{Z} : n \leq x\}, \quad x \in \mathbb{R}.$$

Hence we see that c_m for $m = 0, 1, 2, \cdots, \omega - 1, \omega, \omega + 1, \cdots$ are represented by

$$0, 1, 2, \cdots, \omega - 1, 0, 1, 2, \cdots, \omega - 1, 0, 1, 2, \cdots, \omega - 1, \cdots.$$

Namely, c_m is periodic with period ω.

Proof Recall the criteria for Ulam stability given in Definition 2.1. Since p is ω-periodic in this case, p satisfies (2.17) and has the form (2.18). Let E_k take the form (2.19) for $k \in \{0, 1, 2, \cdots, \omega - 1\}$. If $\left| \prod_{k=0}^{\omega-1} E_k \right| = 1$, then (2.1) is not Ulam stable by Theorem 2.4. Otherwise, we consider the following two cases.

(i) First, suppose $\left|\prod_{k=0}^{\omega-1} E_k\right| > 1$, where $\prod_{k=0}^{\omega-1} E_k = \prod_{k=0}^{\omega-1}(1+\mu_k p_k)$. For arbitrary $\varepsilon > 0$, suppose a function ϕ satisfies (2.7) for all $t \in \mathbb{T}$. By the variation of constants formula, we have (2.10) for all $t = t_n \in \mathbb{T}$, where $\phi(t_0) = \phi_0 \in \mathbb{C}$. Hence, by $\left|\xi(t_j)\right| \leq \varepsilon$ and (2.8), we obtain

$$\left|\frac{\phi(t_n)}{e_p(t_n, t_0)}\right| \leq |\phi_0| + \varepsilon\mu_{\sup}\sum_{j=0}^{n-1}\frac{1}{|e_p(t_{j+1}, t_0)|}.$$

Since

$$z_0 := \phi_0 + \sum_{j=0}^{\infty}\frac{\mu(t_j)\xi(t_j)}{e_p(t_{j+1}, t_0)} \in \mathbb{C}$$

is a well-defined finite number by (2.7), (2.8), and $\left|\prod_{k=0}^{\omega-1} E_k\right| > 1$, we see that ϕ is rewritten as (2.11). Moreover, we see that $z(t_n) := z_0 e_p(t_n, t_0)$ for $t_n \in \mathbb{T}$ is a solution of (2.1), and

$$\lim_{n\to\infty}\frac{\phi(t_n)}{e_p(t_n, t_0)} = \phi_0 + \sum_{j=0}^{\infty}\frac{\mu(t_j)\xi(t_j)}{e_p(t_{j+1}, t_0)} = z_0$$

exists. This implies that

$$z(t_n) = \left(\lim_{n\to\infty}\frac{\phi(t_n)}{e_p(t_n, t_0)}\right)e_p(t_n, t_0),$$

and

$$\begin{aligned}|\phi(t_n) - z(t_n)| &= \left|-e_p(t_n, t_0)\sum_{j=n}^{\infty}\frac{\mu(t_j)\xi(t_j)}{e_p(t_{j+1}, t_0)}\right| \\ &\leq \varepsilon\mu_{\sup}|e_p(t_n, t_0)|\sum_{j=n}^{\infty}\frac{1}{|e_p(t_{j+1}, t_0)|}\end{aligned}$$

holds for all $t = t_n \in \mathbb{T} = \{t_n\}_{n=0}^{\infty}$. Now we will calculate the right-hand side of this inequality from above. First, we consider the case $n \equiv 0 \pmod{\omega}$. Using (2.20) and $\left|\prod_{k=0}^{\omega-1} E_k\right| > 1$, we have

$$
\begin{aligned}
&|e_p(t_n,t_0)|\sum_{j=n}^{\infty}\frac{1}{|e_p(t_{j+1},t_0)|}\\
&=|e_p(t_n,t_0)|\left(\frac{1}{|e_p(t_{n+1},t_0)|}+\frac{1}{|e_p(t_{n+2},t_0)|}+\frac{1}{|e_p(t_{n+3},t_0)|}+\cdots\right)\\
&=\frac{1}{|E_0|}+\frac{1}{|E_0E_1|}+\cdots+\frac{1}{\left|\prod_{k=0}^{\omega-1}E_k\right|}+\frac{1}{\left|E_0\prod_{k=0}^{\omega-1}E_k\right|}+\frac{1}{\left|E_0E_1\prod_{k=0}^{\omega-1}E_k\right|}+\cdots\\
&=\left[1+\left(\frac{1}{\left|\prod_{k=0}^{\omega-1}E_k\right|}\right)+\left(\frac{1}{\left|\prod_{k=0}^{\omega-1}E_k\right|}\right)^2+\left(\frac{1}{\left|\prod_{k=0}^{\omega-1}E_k\right|}\right)^3+\cdots\right]\\
&\quad\sum_{l=0}^{\omega-1}\frac{1}{\left|\prod_{k=0}^{l}E_k\right|}\\
&=\frac{\left|\prod_{k=0}^{\omega-1}E_k\right|}{\left|\prod_{k=0}^{\omega-1}E_k\right|-1}\sum_{l=0}^{\omega-1}\frac{1}{\left|\prod_{k=0}^{l}E_k\right|}=\frac{|e_p(t_\omega,t_0)|}{|e_p(t_\omega,t_0)|-1}\sum_{l=0}^{\omega-1}\frac{1}{|e_p(t_{l+1},t_0)|}.
\end{aligned}
$$

Next, we consider the case $n\equiv 1 \pmod{\omega}$. Using (2.20) and $\left|\prod_{k=0}^{\omega-1}E_{c_{k+1}}\right|=\left|\prod_{k=0}^{\omega-1}E_k\right|>1$, we have

$$
\begin{aligned}
&|e_p(t_n,t_0)|\sum_{j=n}^{\infty}\frac{1}{|e_p(t_{j+1},t_0)|}\\
&=|e_p(t_n,t_0)|\left(\frac{1}{|e_p(t_{n+1},t_0)|}+\frac{1}{|e_p(t_{n+2},t_0)|}+\frac{1}{|e_p(t_{n+3},t_0)|}+\cdots\right)\\
&=\frac{1}{|E_1|}+\frac{1}{|E_1E_2|}+\cdots+\frac{1}{\left|\prod_{k=0}^{\omega-1}E_{c_{k+1}}\right|}+\frac{1}{\left|E_1\prod_{k=0}^{\omega-1}E_{c_{k+1}}\right|}\\
&+\frac{1}{\left|E_1E_2\prod_{k=0}^{\omega-1}E_{c_{k+1}}\right|}+\cdots\\
&=\left[1+\left(\frac{1}{\left|\prod_{k=0}^{\omega-1}E_{c_{k+1}}\right|}\right)+\left(\frac{1}{\left|\prod_{k=0}^{\omega-1}E_{c_{k+1}}\right|}\right)^2+\left(\frac{1}{\left|\prod_{k=0}^{\omega-1}E_{c_{k+1}}\right|}\right)^3+\cdots\right]\\
&\quad\sum_{l=0}^{\omega-1}\frac{1}{\left|\prod_{k=0}^{l}E_{c_{k+1}}\right|}
\end{aligned}
$$

$$= \left[\sum_{l=0}^{\infty} \left(\frac{1}{\left| \prod_{k=0}^{\omega-1} E_k \right|} \right)^l \right] \sum_{l=0}^{\omega-1} \frac{1}{\left| \prod_{k=0}^{l} E_{c_{k+1}} \right|} = \frac{\left| \prod_{k=0}^{\omega-1} E_k \right|}{\left| \prod_{k=0}^{\omega-1} E_k \right| - 1} \sum_{l=0}^{\omega-1} \frac{1}{\left| \prod_{k=0}^{l} E_{c_{k+1}} \right|},$$

where $c_k = k - \left\lfloor \frac{k}{\omega} \right\rfloor \omega$ for $k \in \mathbb{N}_0$. Similarly, we get

$$|e_p(t_n, t_0)| \sum_{j=n}^{\infty} \frac{1}{|e_p(t_{j+1}, t_0)|} = \frac{\left| \prod_{k=0}^{\omega-1} E_k \right|}{\left| \prod_{k=0}^{\omega-1} E_k \right| - 1} \sum_{l=0}^{\omega-1} \frac{1}{\left| \prod_{k=0}^{l} E_{c_{k+l}} \right|}$$

for the case $n \equiv l \pmod{\omega}$ with $l \in \{0, 1, 2, \cdots, \omega - 1\}$. Therefore, (2.1) is Ulam stable with Ulam stability constant $\mathcal{U}$ given by (2.21) for $\prod_{k=0}^{\omega-1} E_k = \prod_{k=0}^{\omega-1}(1 + \mu_k p_k)$ and $\left| \prod_{k=0}^{\omega-1} E_k \right| > 1$.

(ii) Second, suppose $0 < \left| \prod_{k=0}^{\omega-1} E_k \right| < 1$, where E_k satisfy (2.19) for $k \in \{0, 1, 2, \cdots, \omega - 1\}$. As before, we know that a function ϕ that satisfies (2.7) and takes the form (2.10). Let $z = z(t_n) = \phi_0 e_p(t_n, t_0)$ for $t = t_n \in \mathbb{T}$ be the solution of (2.1) such that $z(t_0) = \phi_0$. Then we have

$$\phi(t_n) - z(t_n) = e_p(t_n, t_0) \sum_{j=0}^{n-1} \frac{\mu(t_j)\xi(t_j)}{e_p(t_{j+1})}.$$

By (2.8), we get

$$|\phi(t_n) - z(t_n)| \leq \varepsilon \mu_{\sup} |e_p(t_n, t_0)| \sum_{j=0}^{n-1} \frac{1}{|e_p(t_{j+1}, t_0)|}$$

for all $t = t_n \in \mathbb{T}$. Now we will calculate the right-hand side of this inequality from above. Note that by (2.4) and (2.20), we see that

$$|e_p(t_{\omega m+j}, t_0)| \sum_{i=0}^{\omega m+j-1} \frac{1}{|e_p(t_{i+1}, t_0)|}$$

$$= \left| \prod_{l=0}^{j-1} E_l \right| \left| \prod_{k=0}^{\omega-1} E_k \right|^m \left(\frac{1}{|E_0|} + \frac{1}{|E_0 E_1|} + \cdots + \frac{1}{\left| \prod_{k=0}^{\omega-1} E_k \right|} \right.$$

$$+ \frac{1}{|E_0| \left| \prod_{k=0}^{\omega-1} E_k \right|} + \frac{1}{|E_0 E_1| \left| \prod_{k=0}^{\omega-1} E_k \right|} + \cdots + \frac{1}{\left| \prod_{k=0}^{\omega-1} E_k \right|^2} + \frac{1}{|E_0| \left| \prod_{k=0}^{\omega-1} E_k \right|^2}$$

$$+\frac{1}{|E_0E_1|\left|\prod_{k=0}^{\omega-1}E_k\right|^2}+\cdots+\frac{1}{\left|\prod_{l=0}^{j-1}E_l\right|\left|\prod_{k=0}^{\omega-1}E_k\right|^m}\Bigg) \tag{2.22}$$

for $j \in \{0, 1, 2, \cdots, \omega - 1\}$. First, we consider the case $n \equiv 0 \pmod{\omega}$. Using $0 < \left|\prod_{k=0}^{\omega-1} E_k\right| < 1$ and (2.22) with $j = 0$, we have

$$|e_p(t_{\omega m}, t_0)| \sum_{i=0}^{\omega m-1} \frac{1}{|e_p(t_{i+1}, t_0)|}$$

$$= \left|\prod_{k=0}^{\omega-1} E_k\right|^m \left(\sum_{l=0}^{\omega-1} \frac{1}{\left|\prod_{k=0}^{l} E_k\right|}\right)$$

$$\left(1+\frac{1}{\left|\prod_{k=0}^{\omega-1}E_k\right|}+\frac{1}{\left|\prod_{k=0}^{\omega-1}E_k\right|^2}+\cdots+\frac{1}{\left|\prod_{k=0}^{\omega-1}E_k\right|^{m-1}}\right)$$

$$= \left(\left|\prod_{k=0}^{\omega-1} E_k\right| \sum_{l=0}^{\omega-1} \frac{1}{\left|\prod_{k=0}^{l} E_k\right|}\right)\left(\sum_{l=0}^{m-1}\left|\prod_{k=0}^{\omega-1} E_k\right|^l\right)$$

$$< \left(\left|\prod_{k=0}^{\omega-1} E_k\right| \sum_{l=0}^{\omega-1} \frac{1}{\left|\prod_{k=0}^{l} E_k\right|}\right)\left(\sum_{l=0}^{\infty}\left|\prod_{k=0}^{\omega-1} E_k\right|^l\right) = \frac{\left|\prod_{k=0}^{\omega-1}E_k\right|}{1-\left|\prod_{k=0}^{\omega-1}E_k\right|}\sum_{l=0}^{\omega-1}\frac{1}{\left|\prod_{k=0}^{l}E_k\right|}.$$

Next, we consider the case $n \equiv 1 \pmod{\omega}$. Using $0 < \left|\prod_{k=0}^{\omega-1} E_{c_k+1}\right| = \left|\prod_{k=0}^{\omega-1} E_k\right| < 1$ and (2.22) with $j = 1$, we have

$$|e_p(t_{\omega m+1}, t_0)| \sum_{i=0}^{\omega m} \frac{1}{|e_p(t_{i+1}, t_0)|}$$

$$= \left|\prod_{k=0}^{\omega-1} E_k\right|^m \left(1+\frac{1}{|E_1|}+\frac{1}{|E_1E_2|}+\cdots+\frac{1}{\left|\prod_{k=0}^{\omega-1}E_{c_{k+1}}\right|}\right.$$

$$+\frac{1}{|E_1|\left|\prod_{k=0}^{\omega-1}E_{c_{k+1}}\right|}+\frac{1}{|E_1E_2|\left|\prod_{k=0}^{\omega-1}E_{c_{k+1}}\right|}+\cdots+\frac{1}{\left|\prod_{k=0}^{\omega-1}E_{c_{k+1}}\right|^2}$$

$$\left.+\frac{1}{|E_1|\left|\prod_{k=0}^{\omega-1}E_{c_{k+1}}\right|^2}+\frac{1}{|E_1E_2|\left|\prod_{k=0}^{\omega-1}E_{c_{k+1}}\right|^2}+\cdots+\frac{1}{\left|\prod_{k=0}^{\omega-1}E_{c_{k+1}}\right|^m}\right)$$

$$
\begin{aligned}
&= \left|\prod_{k=0}^{\omega-1} E_k\right|^m \left[1 + \left(\sum_{l=0}^{\omega-1} \frac{1}{\left|\prod_{k=0}^{l} E_{c_{k+1}}\right|}\right)\left(1 + \frac{1}{\left|\prod_{k=0}^{\omega-1} E_{c_{k+1}}\right|} + \frac{1}{\left|\prod_{k=0}^{\omega-1} E_{c_{k+1}}\right|^2}\right.\right.\\
&\qquad \left.\left. + \cdots + \frac{1}{\left|\prod_{k=0}^{\omega-1} E_{c_{k+1}}\right|^{m-1}}\right)\right]\\
&= \left|\prod_{k=0}^{\omega-1} E_k\right| \left[\left|\prod_{k=0}^{\omega-1} E_k\right|^{m-1} + \left(\sum_{l=0}^{\omega-1} \frac{1}{\left|\prod_{k=0}^{l} E_{c_{k+1}}\right|}\right)\left(\sum_{l=0}^{m-1}\left|\prod_{k=0}^{\omega-1} E_k\right|^l\right)\right]\\
&< \left|\prod_{k=0}^{\omega-1} E_k\right| \left(\sum_{l=0}^{\omega-1} \frac{1}{\left|\prod_{k=0}^{l} E_{c_{k+1}}\right|}\right)\left(\sum_{l=0}^{m}\left|\prod_{k=0}^{\omega-1} E_k\right|^l\right)\\
&< \left(\left|\prod_{k=0}^{\omega-1} E_k\right| \sum_{l=0}^{\omega-1} \frac{1}{\left|\prod_{k=0}^{l} E_{c_{k+1}}\right|}\right)\left(\sum_{l=0}^{\infty}\left|\prod_{k=0}^{\omega-1} E_k\right|^l\right)\\
&= \frac{\left|\prod_{k=0}^{\omega-1} E_k\right|}{1 - \left|\prod_{k=0}^{\omega-1} E_k\right|} \sum_{l=0}^{\omega-1} \frac{1}{\left|\prod_{k=0}^{l} E_{c_{k+1}}\right|},
\end{aligned}
$$

where $c_k = k - \left\lfloor \frac{k}{\omega} \right\rfloor \omega$ for $k \in \mathbb{N}_0$. Similarly, we get

$$
|e_p(t_n, t_0)| \sum_{j=0}^{n} \frac{1}{|e_p(t_{j+1}, t_0)|} < \frac{\left|\prod_{k=0}^{\omega-1} E_k\right|}{1 - \left|\prod_{k=0}^{\omega-1} E_k\right|} \sum_{l=0}^{\omega-1} \frac{1}{\left|\prod_{k=0}^{l} E_{c_{k+l}}\right|}
$$

for the case $n \equiv l \pmod{\omega}$ with $l \in \{0, 1, 2, \cdots, \omega - 1\}$. Therefore, (2.1) is Ulam stable with Ulam stability constant

$$
\mathcal{U} = \left(\max_{j \in \{0,1,2,\cdots,\omega-1\}} \sum_{l=0}^{\omega-1} \frac{1}{\left|\prod_{k=0}^{l} E_{c_{k+j}}\right|}\right) \frac{\mu_{\sup}\left|\prod_{k=0}^{\omega-1} E_k\right|}{\left|\left|\prod_{k=0}^{\omega-1} E_k\right| - 1\right|}
$$

for $\prod_{k=0}^{\omega-1} E_k = \prod_{k=0}^{\omega-1}(1 + \mu_k p_k)$ and $\left|\prod_{k=0}^{\omega-1} E_k\right| > 1$. This ends case (ii). Overall, we have shown that (2.1) is Ulam stable if and only if $\left|\prod_{k=0}^{\omega-1} E_k\right| = \left|\prod_{k=0}^{\omega-1}(1 + \mu_k p_k)\right| \neq 1$, with Ulam constant given by (2.21), completing the proof. □

Theorem 2.6 *Let* $\mathbb{T} = \{t_n\}_{n=0}^{\infty}$ *be an isolated time scale. Let* $p = p(t)$ *be given by* (2.18) *for all* $t = t_n \in \mathbb{T}$*; videlicet,* p *is* ω*-periodic. If there exist real constants* $a > 0$ *and* $b \geq 0$ *such that* (2.13) *holds, then* (2.1) *is not Ulam stable.*

Proof The proof is a straightforward extension of the argument given in the proof of Theorem 2.3 and is omitted. □

2.5 Conclusion and Future Directions

In this chapter, we investigate the stability and instability of first-order dynamic equations on isolated time scales with 3-periodic coefficients, and then in the general case with ω-periodic coefficients. We find that the linear ω-periodic dynamic equation is Ulam stable if and only if the exponential function has modulus different from unity. If the graininess of the isolated time scale increases at least linearly to infinity, the ω-periodic dynamic equation is not Ulam stable. We include as examples the isolated time scales with single step size, two step sizes, the triangular numbers, and q-difference equations. In the future, we aim to investigate whether more sophisticated descriptions of the graininess can lead to more results.

Acknowledgments The second author was supported by JSPS KAKENHI Grant Number JP20K03668.

References

1. D.R. Anderson, M. Onitsuka, Hyers-Ulam stability for a discrete time scale with two step sizes. Appl. Math. Comput. **344–345**, 128–140 (2019)
2. D.R. Anderson, M. Onitsuka, Best constant for Hyers-Ulam stability of second-order h-difference equations with constant coefficients. Results Math. **74**, 151 (2019). https://doi.org/10.1007/s00025-019-1077-9
3. D.R. Anderson, M. Onitsuka, Best constant for Hyers-Ulam stability of two step sizes linear difference equations. J. Math. Anal. Appl. **496**(2), 124807 (2021). https://doi.org/10.1016/j.jmaa.2020.124807
4. D.R. Anderson, M. Onitsuka, Hyers-Ulam stability for Cayley quantum equations and its application to h-difference equations. Mediterranean J. Math. **18**, 168 (2021)
5. D.R. Anderson, M. Onitsuka, Ulam stability for nonautonomous quantum equations. J. Inequalities Appl. **2021**, 161 (2021). https://doi.org/10.1186/s13660-021-02699-4
6. D.R. Anderson, M. Onitsuka, Ulam stability and instability of first-order linear 1-periodic and 2-periodic dynamic equations on isolated time scales, in *Dynamic Calculus and Equations on Time Scales*, ed. by S.G. Georgiev (De Gruyter, Berlin, 2023). ISBN 978-3-11-118289-6
7. D.R. Anderson, M. Onitsuka, J.M. Rassias, Best constant for Ulam stability of first-order h-difference equations with periodic coefficient. J. Math. Anal. Appl. **491**, 124363 (2020). https://doi.org/10.1016/j.jmaa.2020.124363
8. S. András, A.R. Mészáros, Ulam–Hyers stability of dynamic equations on time scales via Picard operators. Appl. Math. Comput. **219**, 4853–4864 (2013)

9. M. Bohner, R. Chieochan, Floquet theory for q-difference equations. Sarajevo J. Math. **8**(21), 355–366 (2012), no. 2
10. M. Bohner, A. Peterson, *Dynamic Equations on Time Scales, An Introduction with Applications* (Birkhäuser, Boston, 2001)
11. M. Bohner, T. Cuchta, S. Streipert, Delay dynamic equations on isolated time scales and the relevance of one-periodic coefficients. Math. Meth. Appl. Sci. **45**(10), 5821–5838 (2022)
12. M. Bohner, J. Mesquita, S. Streipert, Periodicity on isolated time scales. Math. Nachr. **295**, 259–280 (2022). https://doi.org/10.1002/mana.201900360
13. N. Brillouët-Belluot, J. Brzdęk, K. Ciepliński, On some recent developments in Ulam's type stability. Abstract Appl. Anal. **2012**. Article ID 716936, 41p.
14. J. Brzdęk, P. Wójcik, On approximate solutions of some difference equations. Bull. Australian Math. Soc. **95**(3), 476–481 (2017)
15. J. Brzdęk, D. Popa, I. Raşa, B. Xu, Ulam stability of operators, in *Mathematical Analysis and its Applications* (Academic Press, Cambridge, 2018)
16. R. Fukutaka, M. Onitsuka, Best constant in Hyers–Ulam stability of first-order homogeneous linear differential equations with a periodic coefficient. J. Math. Anal. Appl. **473**(2), 1432–1446 (2019)
17. L. Hua, Y. Li, J. Feng, On Hyers–Ulam stability of dynamic integral equation on time scales. Mathematica Aeterna **4**(6), 559–571 (2014)
18. D.H. Hyers, On the stability of the linear functional equation. Proc. Nat. Acad. Sci. U.S.A. **27**, 222–224 (1941)
19. S.-M. Jung, Y.W. Nam, Hyers–Ulam stability of Pielou logistic difference equation. J. Nonlinear Sci. Appl. **10**, 3115–3122 (2017)
20. Y.W. Nam, Hyers–Ulam stability of hyperbolic Möbius difference equation. Filomat **32**(13), 4555–4575 (2018). https://doi.org/10.2298/FIL1813555N
21. Y.W. Nam, Hyers–Ulam stability of elliptic Möbius difference equation. Cogent Math. Stat. **5**(1), 1–9 (2018)
22. Y.W. Nam, Hyers–Ulam stability of loxodromic Möbius difference equation. Appl. Math. Comput. **356**(1), 119–136 (2019). https://doi.org/10.1016/j.amc.2019.03.033
23. M. Onitsuka, Influence of the stepsize on Hyers–Ulam stability of first-order homogeneous linear difference equations. Int. J. Differ. Equ. **12**(2), 281–302 (2017)
24. M. Onitsuka, Hyers–Ulam stability of second-order nonhomogeneous linear difference equations with a constant step size. J. Comput. Anal. Appl. **28**(1), 152–165 (2020)
25. D. Popa, Hyers–Ulam stability of the linear recurrence with constant coefficients. Adv. Differ. Equ. **2005**, 407076 (2005)
26. D. Popa, Hyers–Ulam–Rassias stability of a linear recurrence. J. Math. Anal. Appl. **309**, 591–597 (2005)
27. H. Rasouli, S. Abbaszadeh, M. Eshaghi, Approximately linear recurrences. J. Appl. Anal. **24**(1), 81–85 (2018)
28. Th.M. Rassias, On the stability of linear mapping in Banach spaces. Proc. Amer. Math. Soc. **72**, 297–300 (1978)
29. Y.H. Shen, The Ulam stability of first order linear dynamic equations on time scales. Results Math. **72**(4), 1881–1895 (2017)
30. Y.H. Shen, Y.J. Li, Hyers–Ulam stability of first order nonhomogeneous linear dynamic equations on time scales. Comm. Math. Res. **35**(2), 139–148 (2019). https://doi.org/10.13447/j.1674-5647.2019.02.05
31. S.M. Ulam, *A Collection of the Mathematical Problems* (Interscience, New York, 1960)

Chapter 3
Solutions of Dynamic Sturm-Liouville Conformable Initial and Boundary Value Problems

Fatma Ayça Çetinkaya, Tom Cuchta, and Nick Wintz

Abstract In this chapter, we provide the variation of parameters formula and Green's functions for the conformable delta Sturm-Liouville initial value and boundary value problems, respectively. Numerous examples are provided to demonstrate the theory.

Keywords Conformable · Time scale · Sturm-Liouville · Initial value problem · Boundary value problem · Green's function · Variation of parameters

2020 MSC Primary 34N05; Secondary 34B24, 34B27

3.1 Introduction

The conformable calculus was introduced in [1, 14] and later expanded to resemble a proportional derivative controller in [6]. Since then, conformable calculus has become increasingly popular in the literature, with some recent results in diverse

F. A. Çetinkaya
Department of Mathematics, Faculty of Science, Mersin University, Mersin, Turkey

Department of Mathematics, University of Tennessee at Chattanooga, Mersin, Türkiye
e-mail: faycacetinkaya@mersin.edu.tr; fatmaayca-cetinkaya@utc.edu

T. Cuchta (✉)
Department of Mathematics and Physics, Marshall University, Huntington, WV, USA
e-mail: cuchta@marshall.edu

N. Wintz
Department of Mathematics, Computer Sciences, and Informational Technology, Lindenwood University, Saint Charles, MO, USA
e-mail: nwintz@lindenwood.edu

P. Benevieri, J. G. Mesquita (eds.), *Functional Differential Equations and Dynamic Equations on Time Scales*, Advances in Continuum Mechanics 52,
https://doi.org/10.1007/978-3-031-83327-4_3

areas such as control theory [16], ocean engineering [18], and nonlinear reaction-diffusion problems [13]. Applications of conformable calculus to areas such as fluid mechanics [4, 17] and solid mechanics [12, 19] are common.

The theory was generalized to a large class of conformable operators in the time scales calculus in [7]. We build on the work [11] in studying properties of solutions of dynamic equations with respect to the conformable Sturm-Liouville operator

$$Lx(t) = \left(p(t)x^{\Delta_\alpha}\right)^{\Delta_\alpha} + q(t)x(\sigma(t)), \tag{3.1}$$

where x^{Δ_α} denotes one of a class of conformable dynamic delta derivatives on a time scale, formally introduced below. A similar operator

$$\tilde{L}x(t) = \left(p(t)x^{\Delta_\alpha}(t)\right)^{\nabla_\alpha} + q(t)x(t) \tag{3.2}$$

is studied in [5, Chapter 8], but it mixes forward Δ_α and backward ∇_α derivatives, nor does it include composition with σ on the second term. A different mixed dynamic conformable operator on time scales appears in the recent work [3], but it uses a different and more restricted theory of conformable calculus. Our operator (3.1) generalizes the primary operator studied in the classic monograph [9, Chapter 4], while (3.2) generalizes the theory in the successor monograph [10, Section 4.3].

3.2 Preliminaries and Definitions

We direct the readers to the monographs [9, 10] and the survey paper [2] for routine definitions and an introduction to the time scales calculus. The so-called simple useful formula is [9, Theorem 1.16(iv)]

$$f(\sigma(t)) = f(t) + \mu(t)f^{\Delta}(t). \tag{3.3}$$

The Leibniz rule for differentiation of a Δ-integral where both the integral and integrand depend on the same variable t is known [9, Theorem 1.117(i)] as

$$\left[\int_a^t f(t,s)\Delta s\right]^{\Delta} = f(\sigma(t),t) + \int_a^t f^{\Delta}(t,s)\Delta s. \tag{3.4}$$

We now define the fundamental functions that extend the classical time scales calculus to its conformable case, following [8]. If $\mathbb{T}$ is a time scale and $\alpha \in [0,1]$, then consider functions $\kappa_0, \kappa_1 : \mathbb{T} \times [0,1] \to [0,\infty)$ to be rd-continuous functions such that for all $t \in \mathbb{T}$,

$$\begin{cases} \lim_{\alpha\to 0^+} \kappa_1(t;\alpha) = 1, & \lim_{\alpha\to 0^+} \kappa_0(t;\alpha) = 0, \\ \lim_{\alpha\to 1^-} \kappa_1(t;\alpha) = 0, & \lim_{\alpha\to 1^-} \kappa_0(t;\alpha) = 1, \\ \kappa_1(t;\alpha) \neq 0 \text{ and } \kappa_0(t;\alpha) \neq 0 \text{ whenever } \alpha \in (0,1). \end{cases} \tag{3.5}$$

In practice, we suppress the parameter α. The conformable delta derivative of a function $f\colon \mathbb{T} \to \mathbb{R}$ is defined using κ_0 and κ_1 by

$$f^{\Delta_\alpha}(t) = \kappa_1(t)f(t) + \kappa_0(t)f^{\Delta}(t), \tag{3.6}$$

which rearranges to the useful formula

$$f^{\Delta}(t) = \frac{1}{\kappa_0(t)} f^{\Delta_\alpha}(t) - \frac{\kappa_1(t)}{\kappa_0(t)} f(t). \tag{3.7}$$

The product rule for the Δ_α derivative is [8, Theorem 3.10(iii)]

$$(fg)^{\Delta_\alpha} = fg^{\Delta_\alpha} + g^{\sigma} f^{\Delta_\alpha} - fg^{\sigma}\kappa_1. \tag{3.8}$$

Let $\alpha \in (0,1]$, $s,t \in \mathbb{T}$ with $s \le t$, and let function $p : \mathbb{T} \to \mathbb{R}$ be rd-continuous. If the function $t \mapsto \dfrac{p(t) - \kappa_1(t)}{\kappa_0(t)}$ is regressive and rd-continuous, then the conformable dynamic exponential function $E_p\colon \mathbb{T} \times \mathbb{T} \to \mathbb{C}$ is defined in terms of the classical dynamic exponential by

$$E_p(t, t_0; \alpha) = e_{\frac{p(\cdot) - \kappa_1(\cdot;\alpha)}{\kappa_0(\cdot;\alpha)}}(t, t_0). \tag{3.9}$$

The conformable dynamic exponential function is an eigenfunction of the conformable fractional derivative operator, i.e.,

$$E_p^{\Delta_\alpha}(t, s; \alpha) = p(t)E_p(t, s; \alpha). \tag{3.10}$$

The conformable fractional Δ-integral of f is defined by its relation to the standard Δ-integral on time scales as

$$\int_a^t f(\tau)E_0(\sigma(t), \tau)\Delta_\alpha \tau := \int_a^t \frac{f(\tau)E_0(t, \tau)}{\kappa_0(\tau)} \Delta\tau,$$

and an analog of the fundamental theorem of calculus is [8, Theorem 3.13(i)]

$$\left(\int_a^t f(\tau)E_0(\sigma(t), \tau)\Delta_\alpha \tau\right)^{\Delta_\alpha} = f(t)E_0(\sigma(t), t). \tag{3.11}$$

The conformable analog of [9, Theorem 1.75] is given by Çetinkaya and Cuchta [11, Proposition 3.1]

$$\int_t^{\sigma(t)} f(\tau)\Delta_\alpha \tau = \frac{f(t)\mu(t)}{\kappa_0(t) - \mu(t)\kappa_1(t)}. \tag{3.12}$$

3.3 Main Results

The following result is self-contained and may be useful for future work; it is an analog of [9, Theorem 1.117]. See also [8, Theorem 3.13(iv)] and the similar result for the nabla dynamic calculus [5, Theorem 1.7.10].

Theorem 3.1 *Let $f : \mathbb{T} \times \mathbb{T}^\kappa \to \mathbb{R}$ be continuous at (t, t) with $t \in \mathbb{T}^\kappa$, and let $f^\Delta(t, \cdot)$ be rd-continuous on $[a, \sigma(t)]$. If $g(t) = \int_a^t f(t, \tau)E_0(\sigma(t), \tau)\Delta_\alpha \tau$, then*

$$g^{\Delta_\alpha}(t) = \int_a^t \Big[f^{\Delta_\alpha}(t, \tau) - \kappa_1(\tau)f(\sigma(t), \tau)\Big]E_0(\sigma(t), \tau)\Delta_\alpha \tau + f(\sigma(t), t)\left(1 - \frac{\mu(t)\kappa_1(t)}{\kappa_0(t)}\right). \tag{3.13}$$

Proof Using (3.6), compute

$$\begin{aligned}
g^{\Delta_\alpha}(t) &= \kappa_1(t)\int_a^t f(t, \tau)E_0(\sigma(t), \tau)\Delta_\alpha \tau + \kappa_0(t)\left[\int_a^t \frac{f(t, \tau)E_0(t, \tau)}{\kappa_0(\tau)}\Delta \tau\right]^\Delta \\
&= \kappa_1(t)\int_a^t f(t, \tau)E_0(\sigma(t), \tau)\Delta_\alpha \tau \\
&\quad + \kappa_0(t)\left[\frac{f(\sigma(t), t)E_0(\sigma(t), t)}{\kappa_0(t)} + \int_a^t \frac{\big[f(t, \tau)E_0(t, \tau)\big]^\Delta}{\kappa_0(\tau)}\Delta \tau\right] \\
&= \kappa_1(t)\int_a^t f(t, \tau)E_0(\sigma(t), \tau)\Delta_\alpha \tau + \kappa_0(t)\left[\frac{f(\sigma(t), t)E_0(\sigma(t), t)}{\kappa_0(t)}\right. \\
&\quad \left. + \int_a^t \frac{f^\Delta(t, \tau)E_0(t, \tau) + f(\sigma(t), \tau)E_0^\Delta(t, \tau)}{\kappa_0(\tau)}\Delta \tau\right].
\end{aligned}$$

By (3.9), we observe that $E_0^\Delta = -\frac{\kappa_1}{\kappa_0}E_0$, so

$$
\begin{aligned}
g^{\Delta_\alpha}(t) &= \kappa_1(t)\int_a^t f(t,\tau)E_0(\sigma(t),\tau)\Delta_\alpha\tau + \kappa_0(t)\int_a^t f^{\Delta}(t,\tau)E_0(\sigma(t),\tau)\Delta_\alpha\tau \\
&\quad + \kappa_0(t)\int_a^t \frac{f(\sigma(t),t)\left[\frac{-\kappa_1(t)}{\kappa_0(t)}E_0(t,\tau)\right]}{\kappa_0(\tau)}\Delta\tau + f(\sigma(t),t)E_0(\sigma(t),t) \\
&= \int_a^t \underbrace{\left[\kappa_1(t)f(t,\tau) + \kappa_0(t)f^{\Delta}(t,\tau)\right]}_{=f^{\Delta_\alpha}(t,\tau)} E_0(\sigma(t),\tau)\Delta_\alpha\tau \\
&\quad - \kappa_1(t)\int_a^t \frac{f(\sigma(t),\tau)E_0(t,\tau)}{\kappa_0(\tau)}\Delta\tau + f(\sigma(t),t)E_0(\sigma(t),t) \\
&= \int_a^t f^{\Delta_\alpha}(t,\tau)E_0(\sigma(t),\tau)\Delta\tau - \kappa_1(t)\int_a^t f(\sigma(t),\tau)E_0(\sigma(t),\tau)\Delta_\alpha\tau \\
&\quad + f(\sigma(t),t)\left(1 - \frac{\mu(t)\kappa_1(t)}{\kappa_0(t)}\right) \\
&= \int_a^t \left[f^{\Delta_\alpha}(t,\tau) - \kappa_1(t)E_0(\sigma(t),\tau)\right]E_0(\sigma(t),\tau)\Delta_\alpha\tau \\
&\quad + f(\sigma(t),t)\left(1 - \frac{\mu(t)\kappa_1(t)}{\kappa_0(t)}\right),
\end{aligned}
$$

completing the proof. □

The following variant of the Leibniz differentiation rule will turn out to be the one we need for studying conformable Sturm-Liouville dynamic equations. It differs from (3.13) in that we are computing the conformable delta derivative of a classical time scales integral.

Theorem 3.2 *Let $g : \mathbb{T} \times \mathbb{T}^\kappa \to \mathbb{R}$ be continuous at (t,t) with $t \in \mathbb{T}^\kappa$, and let $g^{\Delta_\alpha}(t,\cdot)$ be rd-continuous on $[a,\sigma(t)]$. If $f(t) = \int_a^t g(t,s)\Delta s$, then*

$$
f^{\Delta_\alpha}(t) = \kappa_0(t)g(\sigma(t),t) + \int_a^t g^{\Delta_\alpha}(t,s)\Delta s.
$$

Proof First, use (3.4) and (3.7) to obtain

$$
\begin{aligned}
f^{\Delta}(t) &= \left(\int_a^t g(t,s)\Delta s\right)^{\Delta} \\
&= g(\sigma(t),t) + \int_a^t g^{\Delta}(t,s)\Delta s \\
&= g(\sigma(t),t) + \int_a^t \left[\frac{1}{\kappa_0(t)}g^{\Delta_\alpha}(t,s) - \frac{\kappa_1(t)}{\kappa_0(t)}g(t,s)\right]\Delta s \\
&= g(\sigma(t),t) + \frac{1}{\kappa_0(t)}\int_a^t \left[g^{\Delta_\alpha}(t,s) - \kappa_1(t)g(t,s)\right]\Delta s.
\end{aligned}
$$

Now (3.6) implies

$$f^{\Delta_\alpha}(t) = \kappa_1(t)\int_a^t g(t,s)\Delta s + \kappa_0(t)g(\sigma(t),t) + \int_a^t \left[g^{\Delta_\alpha}(t,s) - \kappa_1(t)g(t,s)\right]\Delta s$$
$$= \kappa_0(t)g(\sigma(t),t) + \int_a^t g^{\Delta_\alpha}(t,s)\Delta s,$$

completing the proof. □

Using the operator L, we have the relevant homogeneous equation

$$Ly(t) = 0 \tag{3.14}$$

and the initial value problem

$$Ly = h(t), \quad y(a) = 0, \quad y^{\Delta_\alpha}(a) = 0. \tag{3.15}$$

To establish the variation of constants formula, we first define a Cauchy function. A function $x\colon \mathbb{T} \times \mathbb{T}^{\kappa^2} \to \mathbb{R}$ is called a Cauchy function for (3.14) provided that the following three axioms hold: For all $s \in \mathbb{T}^{\kappa^2}$,

$$Lx(\cdot, s) = 0, \tag{CF1}$$

$$x(\sigma(s), s) = 0, \tag{CF2}$$

and

$$x^{\Delta_\alpha}(\sigma(s), s) = \frac{1}{p(\sigma(s))}. \tag{CF3}$$

One can observe similarities and difference between our definition of a Cauchy function and that of [5, Definition 8.5.2], which defines Cauchy functions for (3.2). Now we find a particular Cauchy function in the case when $q(t) = 0$.

Example 3.1 The function $x(t,s) = \displaystyle\int_{\sigma(s)}^t \frac{E_0(\tau,\sigma(s))}{p(\tau)}\Delta_\alpha \tau$ is a Cauchy function for (3.14) when $q(t) = 0$. To see why, first note that (CF2) holds since

$$x(\sigma(s), s) = \int_{\sigma(s)}^{\sigma(s)} \frac{E_0(\tau,\sigma(s))}{p(\tau)}\Delta_\alpha \tau = 0.$$

Now use (3.11) to compute

$$\begin{aligned}
x^{\Delta_\alpha}(t,s) &= \int_{\sigma(s)}^{t} \frac{E_0(\tau,\sigma(s))}{p(\tau)E_0(\sigma(t),\tau)} E_0(\sigma(t),\tau)\Delta_\alpha \tau \\
&= \left(\frac{E_0(t,\sigma(s))}{p(t)E_0(\sigma(t),t)}\right) E_0(\sigma(t),t) \\
&= \frac{E_0(t,\sigma(s))}{p(t)}.
\end{aligned}$$

From here, it follows that (CF3) holds since

$$x^{\Delta_\alpha}(\sigma(s),s) = \frac{E_0(\sigma(s),\sigma(s))}{p(\sigma(s))} = \frac{1}{p(\sigma(s))}.$$

Finally, multiply by $p(t)$ and use (3.10) to derive (CF1):

$$\left(p(t)x^{\Delta_\alpha}\right)^{\Delta_\alpha} = E_0^{\Delta_\alpha}(t,\sigma(s)) = 0.$$

With the idea of a Cauchy function, we prove the variation of constants formula for the initial value problem (3.15).

Theorem 3.3 *If $f \in C_{\mathrm{rd}}$ and $x(t,s)$ is a Cauchy function for* (3.14), *then*

$$y(t) = \int_a^t x(t,s)\frac{h(s)}{\kappa_0(s)}\Delta s \tag{3.16}$$

solves the initial value problem (3.15).

Proof It is immediately clear that $y(t)$ satisfies the initial condition $y(a) = 0$. Now using Theorem 3.2 and (CF2), compute

$$y^{\Delta_\alpha}(t) = \kappa_0(t)x(\sigma(t),t)\frac{h(t)}{\kappa_0(t)} + \int_a^t x^{\Delta_\alpha}(t,s)\frac{h(s)}{\kappa_0(s)}\Delta s = \int_a^t x^{\Delta_\alpha}(t,s)\frac{h(s)}{\kappa_0(s)}\Delta s. \tag{3.17}$$

The initial condition $y^{\Delta_\alpha}(a) = 0$ is satisfied by inspection. What remains to show is that $Ly = h$. Compute again using Theorem 3.2 and (CF3)

$$\begin{aligned}
\left(p(t)x^{\Delta_\alpha}(t,s)\right)^{\Delta_\alpha} &= \left(\int_a^t p(t)x^{\Delta_\alpha}(t,s)\frac{h(s)}{\kappa_0(s)}\Delta s\right)^{\Delta_\alpha} \\
&= \kappa_0(t)p(\sigma(t))x^{\Delta_\alpha}(\sigma(t),t)\frac{h(t)}{\kappa_0(t)} \\
&\quad + \int_a^t \left[p(t)x^{\Delta_\alpha}(t,s)\right]^{\Delta_\alpha}\frac{h(s)}{\kappa_0(s)}\Delta s \\
&= h(t) + \int_a^t \left[p(t)x^{\Delta_\alpha}(t,s)\right]^{\Delta_\alpha}\frac{h(s)}{\kappa_0(s)}\Delta s.
\end{aligned}$$

Also using (3.12) and (CF2), compute

$$\begin{aligned}
q(t)y(\sigma(t)) &= q(t)\int_a^{\sigma(t)} x(\sigma(t),s)\frac{h(s)}{\kappa_0(s)}\Delta s\\
&= q(t)\int_a^{t} x(\sigma(t),s)\frac{h(s)}{\kappa_0(s)}\Delta s + q(t)\int_t^{\sigma(t)} x(\sigma(t),s)\frac{h(s)}{\kappa_0(s)}\Delta s\\
&= q(t)\int_a^{t} x(\sigma(t),s)\frac{h(s)}{\kappa_0(s)}\Delta s + q(t)\frac{x(\sigma(t),t)\mu(t)}{\kappa_0(t)-\mu(t)\kappa_1(t)}\\
&= q(t)\int_a^{t} x(\sigma(t),s)\frac{h(s)}{\kappa_0(s)}\Delta s.
\end{aligned}$$

Thus by (CF1),

$$\begin{aligned}
\big(p(t)x^{\Delta_\alpha}(t,s)\big)^{\Delta_\alpha} &+q(t)x(\sigma(t))\\
&= h(t) + \int_a^t \left\{\big[p(t)x^{\Delta_\alpha}(t,s)\big]^{\Delta_\alpha} + q(t)x(\sigma(t),s)\right\}\frac{h(s)}{\kappa_0(s)}\Delta s\\
&= h(t) + \int_a^t Lx(t,s)\frac{h(s)}{\kappa_0(s)}\Delta\\
&= h(t),
\end{aligned}$$

completing the proof. □

The next theorem will generalize [9, Theorem 4.67] to the conformable dynamic equations case. For constants A and B and rd-continuous function h, we now consider the nonhomogeneous boundary value problems

$$Ly = 0, \quad \beta_1 y(a) - \beta_2 y^{\Delta_\alpha}(a) = \gamma_1 y(\sigma^2(b)) + \gamma_2 y^{\Delta_\alpha}(\sigma(b)) = 0, \tag{3.18}$$

and

$$Ly = h, \quad \beta_1 y(a) - \beta_2 y^{\Delta_\alpha}(a) = A, \quad \gamma y(\sigma^2(b)) + \gamma_2 y^{\Delta_\alpha}(\sigma(b)) = B. \tag{3.19}$$

Theorem 3.4 *Assume that the homogeneous two-point conformable fractional boundary value problem* (3.18) *has only the trivial solution. Then the nonhomogeneous two-point boundary value problem* (3.19) *has a unique solution.*

Proof Suppose u and v are linearly independent solutions of (3.14). Then a general solution of (3.14) is given by

$$y(t) = c_1 u(t) + c_2 v(t).$$

Then x satisfies the boundary conditions in (3.18) if and only if

$$c_1\Big[\beta_1 u(a) - \beta_2 u^{\Delta_\alpha}(a)\Big] + c_2\Big[\beta_1 v(a) - \beta_2 v^{\Delta_\alpha}(a)\Big] = 0$$

and

$$c_1\Big[\gamma_1 u(\sigma^2(b)) + \gamma_2 u^{\Delta_\alpha}(\sigma(b))\Big] + c_2\Big[\gamma_1 v(\sigma^2(b)) + \gamma_2 v^{\Delta_\alpha}(\sigma(b))\Big] = 0,$$

i.e., if and only if

$$M\begin{pmatrix} c_1 \\ c_2 \end{pmatrix} = 0, \quad M = \begin{pmatrix} \beta_1 u(a) - \beta_2 u^{\Delta_\alpha}(a) & \beta_1 v(a) - \beta_2 v^{\Delta_\alpha}(a) \\ \gamma_1 u(\sigma^2(b)) + \gamma_2 u^{\Delta_\alpha}(\sigma(b)) & \gamma_1 v(\sigma^2(b)) + \gamma_2 v^{\Delta_\alpha}(\sigma(b)) \end{pmatrix}. \tag{3.20}$$

Since (3.18) has only trivial solution, we get that $c_1 = c_2 = 0$ is the unique solution of (3.20). So,

$$\det M \neq 0 \tag{3.21}$$

holds. Now if w is a solution of the nonhomogeneous equation $Ly = h(t)$, then a general solution takes the form

$$y(t) = a_1 u(t) + a_2 v(t) + w(t).$$

The function y satisfies the boundary conditions in (3.19) if and only if

$$M\begin{pmatrix} a_1 \\ a_2 \end{pmatrix} = \begin{pmatrix} A - \Big[\beta_1 w(a) - \beta_2 w^{\Delta_\alpha}(a)\Big] \\ B - \Big[\gamma_1 w(\sigma^2(b)) + \gamma_2 w^{\Delta_\alpha}(\sigma(b))\Big] \end{pmatrix}. \tag{3.22}$$

As a consequence of (3.21), we get that (3.22) has a unique solution, implying that the nonhomogeneous two-point boundary value problem (3.19) has a unique solution, completing the proof. □

Now we want to define Green's function for a two-point boundary value problem. First, we define one more boundary value problem: Let $x(t,s)$ be a Cauchy function for (3.14), and for each $s \in [a,b] \cap \mathbb{T}$, consider the problem

$$\begin{aligned} &Lu(\cdot,s) = 0, \quad \beta_1 u(a,s) - \beta_2 u^{\Delta_\alpha}(a,s) = 0, \\ &\gamma_1 u(\sigma^2(b),s) + \gamma_2 u^{\Delta_\alpha}(\sigma(b),s) = -\gamma x(\sigma^2(b),s) - \gamma_2 x^{\Delta_\alpha}(\sigma(b),s) \end{aligned}. \tag{3.23}$$

It is clear from Theorem 3.4 that there is a unique solution to (3.23). We define Green's function for (3.18) by

$$G(t,s) = \begin{cases} u(t,s), & t \le s \\ v(t,s), & t \ge \sigma(s), \end{cases}$$

where $v(t,s) = u(t,s) + x(t,s)$. It follows that for all $s \in [a,b] \cap \mathbb{T}$, $v(\cdot,s)$ is a solution of (3.14) and satisfies the second boundary condition of (3.18), since by (3.23),

$$\begin{aligned}&\gamma_1 v(\sigma^2(b),s)+\gamma_2 v^{\Delta_\alpha}(\sigma(b),s)\\&=\gamma_1\Big[u(\sigma^2(b),s)+x(\sigma^2(b),s)\Big]+\gamma_2\Big[u^{\Delta_\alpha}(\sigma(b),s)+x^{\Delta_\alpha}(\sigma(b),s)\Big]\\&=\Big[\gamma_1 u(\sigma^2(b),s)+\gamma_2 u^{\Delta_\alpha}(\sigma(b),s)\Big]+\Big[\gamma_1 x(\sigma^2(b),s)+\gamma_2 x^{\Delta_\alpha}(\sigma(b),s)\Big]=0.\end{aligned} \tag{3.24}$$

Now we are prepared to generalize [9, Theorem 4.70] to the conformable case.

Theorem 3.5 *If* (3.18) *has only the trivial solution and* $y(t)=\int_a^{\sigma(b)} G(t,s)\frac{h(s)}{\kappa_0(s)}\Delta s$, *then* y *is the unique solution of* (3.18).

Proof First we express $x(t)$ in a convenient way:

$$\begin{aligned}y(t)&=\int_a^{\sigma(b)} G(t,s)\frac{h(s)}{\kappa_0(s)}\Delta s\\&=\int_a^t v(t,s)\frac{h(s)}{\kappa_0(s)}\Delta s+\int_a^{\sigma(b)} u(t,s)h(s)\Delta s\\&=\int_a^{\sigma(b)} u(t,s)\frac{h(s)}{\kappa_0(s)}\Delta s+\int_a^t x(t,s)\frac{h(s)}{\kappa_0(s)}\Delta s.\end{aligned} \tag{3.25}$$

Let $z(t)$ denote the second integral in the last line. The integral defining z is precisely (3.16), so z solves (3.15). Now since $Lu(t,s)=0$, compute

$$Ly(t)=\int_a^{\sigma(b)} Lu(t,s)\frac{h(s)}{\kappa_0(s)}\Delta s+Lz(t)=h(t).$$

It remains to show that y satisfies the boundary conditions of (3.18). First, it is clear that $y(a)=\int_a^{\sigma(b)} u(a,s)\frac{h(s)}{\kappa_0(s)}\Delta s$. Now use (3.17) to compute

$$y^{\Delta_\alpha}(t)=\int_a^{\sigma(b)} u^{\Delta_\alpha}(t,s)\frac{h(s)}{\kappa_0(s)}\Delta s+\int_a^t x^{\Delta_\alpha}(t,s)\frac{h(s)}{\kappa_0(s)}\Delta s,$$

from which it is clear that $y^{\Delta_\alpha}(a)=\int_a^{\sigma(b)} u^{\Delta_\alpha}(a,s)\frac{h(s)}{\kappa_0(s)}\Delta s$. Therefore, applying (3.23) yields the first boundary condition

$$\beta_1 y(a)-\beta_2 y^{\Delta_\alpha}(a)=\int_a^{\sigma(b)}\Big[\beta_1 u(a,s)-\beta_2 u^{\Delta_\alpha}(a,s)\Big]\frac{h(s)}{\kappa_0(s)}\Delta s=0.$$

Now we focus our attention on the second boundary condition of (3.18). First we rewrite y starting from (3.25), and using $u=v-y$ and routine properties of the delta integral, we get

$$
\begin{aligned}
y(t) &= \int_a^{\sigma(b)} u(t,s)\frac{h(s)}{\kappa_0(s)}\Delta s + \int_a^t x(t,s)\frac{h(s)}{\kappa_0(s)}\Delta s \\
&= \int_a^{\sigma(b)} v(t,s)\frac{h(s)}{\kappa_0(s)}\Delta s + \int_{\sigma(b)}^t x(t,s)\frac{h(s)}{\kappa_0(s)}\Delta s.
\end{aligned}
$$

□

If we let $w(t)$ denote the second integral at the bottom, then it follows from Theorem 3.3 that w solves the initial value problem

$$Lw = h, \quad w(\sigma(b)) = 0, \quad w^{\Delta_\alpha}(\sigma(b)) = 0.$$

By (3.3),

$$w(\sigma^2(b)) = w(\sigma(b)) + \mu(\sigma(b))w^{\Delta_\alpha}(\sigma(b)) = 0.$$

Finally, we compute the remaining boundary condition using (3.24)

$$
\begin{aligned}
&\gamma_1 y(\sigma^2(b)) + \gamma_2 y^{\Delta_\alpha}(\sigma(b)) \\
&\qquad = \int_a^{\sigma(b)} \left[\gamma_1 v(\sigma^2(b), s) + \gamma_2 v^{\Delta_\alpha}(\sigma(b), s)\right]\frac{h(s)}{\kappa_0(s)}\Delta s = 0,
\end{aligned}
$$

completing the proof that y solves (3.18). The fact that y is unique follows from Theorem 3.4.

3.4 Examples

Below we plot some solutions of the initial and boundary value problems we studied above on a variety of time scales. These plots are done using `the timescalecalculus python` package by rewriting the equation as a system of equations.

We consider the equation $Lx = f$ and rewrite it for numerical computation. Here we simplify and will assume that $\kappa_0(t) = \kappa_0 = \alpha$ and $\kappa_1(t) = \kappa_1 = 1 - \alpha$, so that the calculus is simplified and the κ_j obey (3.5). Now using (3.6), we rewrite

$$p(t)x^{\Delta_\alpha}(t) = \kappa_1 p(t)x(t) + \kappa_0 p(t)x^{\Delta}(t),$$

and hence by the product rule for the delta derivative,

$$
\begin{aligned}
\left(p(t)x^{\Delta_\alpha}\right)^{\Delta} &= \left(\kappa_1 p^{\Delta}(t)x(t) + \kappa_1 p^{\sigma}(t)x^{\Delta}(t)\right) + \left(\kappa_0 p^{\Delta}(t)x^{\Delta}(t) + \kappa_0 p^{\sigma}(t)x^{\Delta\Delta}(t)\right) \\
&= \kappa_1 p^{\Delta}(t)x(t) + \left(\kappa_1 p^{\sigma}(t) + \kappa_0 p^{\Delta}(t)\right)x^{\Delta}(t) + \kappa_0 p^{\sigma}(t)x^{\Delta\Delta}(t).
\end{aligned}
$$

Now again by (3.6) and suppressing functions' arguments, we have

$$\begin{aligned}\left(px^{\Delta_\alpha}\right)^{\Delta_\alpha} &= \kappa_1 p x^{\Delta_\alpha} + \kappa_0\left(px^{\Delta_\alpha}\right)^{\Delta}\\ &= \kappa_1 p\Big(\kappa_1 x + \kappa_0 x^{\Delta}\Big) + \kappa_0\kappa_1 p^{\Delta} x + \kappa_0\Big(\kappa_1 p^{\sigma} + \kappa_0 p^{\Delta}\Big)x^{\Delta} + \kappa_0^2 p^{\sigma} x^{\Delta\Delta}\\ &= \Big(\kappa_1^2 p + \kappa_0\kappa_1 p^{\Delta}\Big)x + \Big(\kappa_0\kappa_1 p + \kappa_0\kappa_1 p^{\sigma} + \kappa_0^2 p^{\Delta}\Big)x^{\Delta} + \kappa_0^2 p^{\sigma} x^{\Delta\Delta}.\end{aligned}$$

Thus the nonhomogeneous equation $Lx = h$ can be rewritten as

$$x^{\Delta\Delta} = \frac{h}{\kappa_0^2 p^{\sigma}} - \frac{\kappa_1^2 p + \kappa_0\kappa_1 p^{\Delta} + q}{\kappa_0^2 p^{\sigma}} x - \frac{\kappa_0\kappa_1 p + \kappa_0\kappa_1 p^{\sigma} + \kappa_0^2 p^{\Delta} + q\mu}{\kappa_0^2 p^{\sigma}} x^{\Delta}. \tag{3.26}$$

Let $\begin{bmatrix} x_1 \\ x_2 \end{bmatrix} = \begin{bmatrix} x \\ x^{\Delta} \end{bmatrix}$, so (3.26) yields

$$\begin{bmatrix} x_1 \\ x_2 \end{bmatrix}^{\Delta} = \begin{bmatrix} x_2 \\ \dfrac{h}{\kappa_0^2 p^{\sigma}} - \dfrac{\kappa_1^2 p + \kappa_0\kappa_1 p^{\sigma} + q}{\kappa_0^2 p^{\sigma}} x_1 - \dfrac{\kappa_0\kappa_1 p + \kappa_0\kappa_1 p^{\sigma} + \kappa_0^2 p^{\Delta} + q\mu}{\kappa_0^2 p^{\sigma}} x_2 \end{bmatrix}. \tag{3.27}$$

We use (3.27) to plot[1] the solutions of initial and boundary value problems.

Example 3.2 Consider this time scale conformable analog of a model for the temperature distribution T in a tapered fin [15, Example 3.7].

$$\left(-tT^{\Delta_\alpha}(t)\right)^{\Delta_\alpha} + 5.00156T(\sigma(t)) = 373.117, \quad T(0) = 250, \quad T(4) = 250. \tag{3.28}$$

We assume that the temperature of the fin is uniformly distributed throughout its width, while the heat transfer of interest is along its surface in a two-dimensional direction. In this setting, the temperature of the object is monitored by a sensor with discrete, continuous, or hybrid measurements. Here, the sensors are read sequentially based on their location on the fin. Let

$$\mathbb{T} = \{0, 0.5, 1, 1.5, 2.0, 2.1, 2.2, 2.2, 2.3, 2.4, 2.5, 2.6, 2.7, 2.8, 2.9, 3.0, 3.5, 4.0\}$$

denote the positions of the sensors along the fin. We now present plots of the solution of (3.28) across various values of α.

[1] The code for our plots can be found on GitHub at https://github.com/tomcuchta/timescalecalculus/tree/master/examples/conformablebvp.

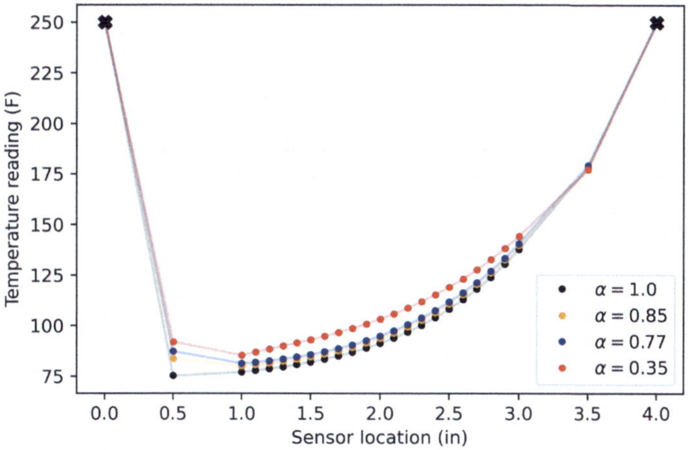

Based on the time scale selected, the temperature readings have the same basic shape regardless of our choice of α.

Example 3.3 Let $\mathbb{T} = \{0, 1, \dots, 15\}$ be a purely discrete time scale, and consider the boundary value problem

$$Lx = 0, \quad x(0) = 10, \quad x(15) = -10,$$

with $p(t) = \dfrac{-7}{t+5}$ and $q(t) = \dfrac{1}{t+1}$.

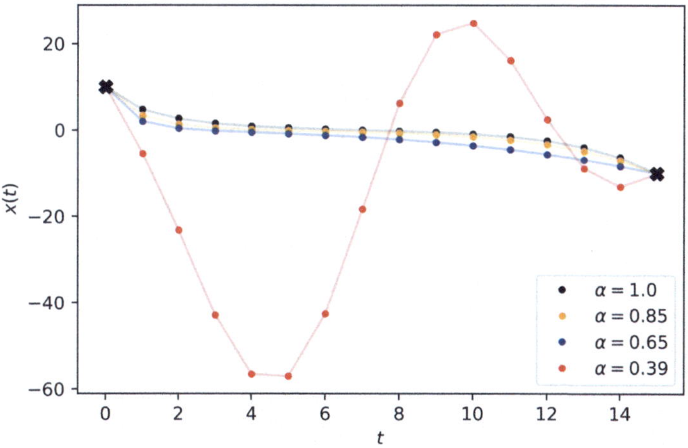

Example 3.4 Consider the boundary value problem

$$Lx = 0, \quad x(0) = 0, \quad x(3) = 15,$$

where $p(t) = 0.5$ and $q(t) = \dfrac{1}{\sqrt{t}+0.1}$ for two different time scales. The first is the time scale $\mathbb{T} = [0, 1] \cup [2, 3]$, and the second is the time scale $\mathbb{T} = [0, 1] \cup \{1.1, 1.2, \ldots, 1.9\} \cup [2, 3]$. We see from these examples that the solutions have different qualitative behavior on the two time scales, verifying that the structure of the domain is a crucial factor for the behavior of solutions. The behavior is dependent not only on the selection of the time scale but also on the value of α.

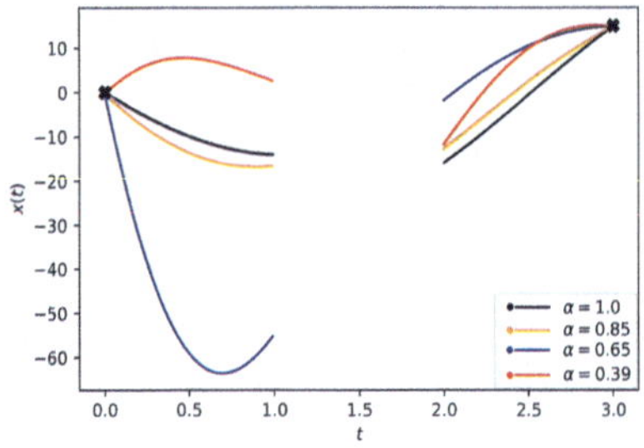

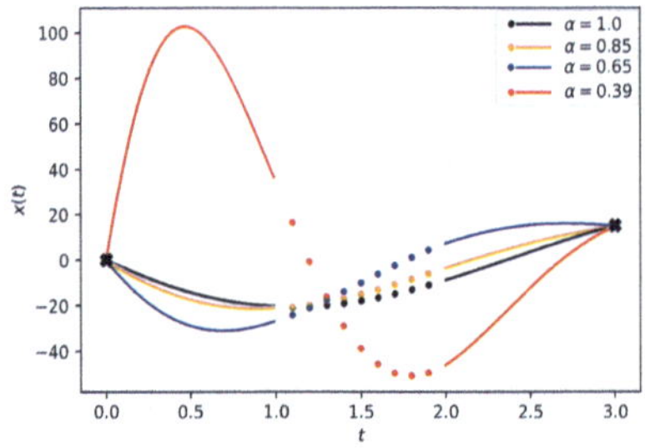

Example 3.5 Consider the following analog of the continuous axial deformation u of an elastic bar with length L given by [15, Example 3.6]:

$$\left(\left(\frac{t}{10} - 6\right)\left(u' - 0.00072\right)\right)' = 10000. \tag{3.29}$$

Here we take the same time scale as we used in Example 3.2. One way to find a time scale analog of (3.29) is to replace each derivative operator with a Δ_α operator as

$$\left(\left(\frac{t}{10} - 6\right)\left(u^{\Delta_\alpha} - 0.00072\right)\right)^{\Delta\alpha} = 10000. \tag{3.30}$$

We will take the time scale to be

$$\mathbb{T} = \{0, 1, 2, 3, 3.5, 4, 4.5, 5, 5.5, 6, 6.5, 7, 8, 9, 10, 12, 14, 14.5, 15, 15.5, 16, 16.5, 17, 19, 21, 23, 23.5, 24, 24.5, 25, 25.5, 26, 28, 29, 29.5, 29.6, 30\}.$$

Since (3.30) is not in the form (3.1), we will use (3.8) and (3.7) to put it into that form. Now let $g_1(t) = u^{\Delta_\alpha}(t) - 0.00072$ and $g_2(t) = \frac{t}{10} - 6$ so that

$$g_1^{\Delta_\alpha}(t) = u^{\Delta_\alpha^2}(t) - \kappa_1(t)0.00072 = u^{\Delta_\alpha^2}(t) - (1-\alpha)0.00072,$$

and

$$g_2^{\Delta_\alpha}(t) = \kappa_1(t)\left(\frac{t}{10} - 6\right) + \kappa_0(t)\frac{1}{10} = (1-\alpha)\left(\frac{t}{10} - 6\right) + \frac{\alpha}{10}.$$

Thus using (3.8), the left-hand side of (3.30) becomes

$$\begin{aligned}\left(u^{\Delta_\alpha}(t) - 0.00072\right)&\left((1-\alpha)\left(\frac{t}{10} - 6\right) + \frac{\alpha}{10}\right)\\ &+\left(\frac{\sigma(t)}{10} - 6\right)\left(u^{\Delta_\alpha^2}(t) - (1-\alpha)0.00072\right)\\ &\qquad - (1-\alpha)\left(u^{\Delta_\alpha} - 0.00072\right)\left(\frac{\sigma(t)}{10} - 6\right).\end{aligned} \tag{3.31}$$

Since

$$\begin{aligned}\left(\left(\frac{t}{10} - 6\right)u^{\Delta_\alpha}\right)^{\Delta_\alpha} &= \left(\frac{\sigma(t)}{10} - 6\right)u^{\Delta_\alpha^2}\\ &+\left((1-\alpha)\left(\frac{t}{10} - 6\right) + \frac{\alpha}{10}\right)u^{\Delta_\alpha} - (1-\alpha)u^{\Delta_\alpha}\left(\frac{\sigma(t)}{10} - 6\right),\end{aligned}$$

we have that (3.30) simplifies to

$$\begin{aligned}&\left(\left(\frac{t}{10} - 6\right)u^{\Delta_\alpha}\right)^{\Delta_\alpha} + 0.00072\left(\alpha\left(\frac{\sigma(t)}{10} - 6\right) - (1-\alpha)\left(\frac{t}{10} - 6\right) - \frac{\alpha}{10}\right)\\ &= 10000.\end{aligned}$$

We now plot the solution under the assumptions of $u(0) = 0$ and $u(30) = 0.03741$.

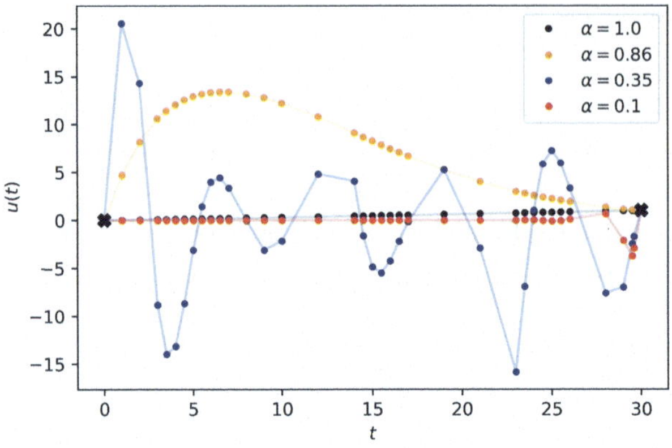

3.5 Conclusion

We have provided fundamental results for the future study of Sturm-Liouville theory in the delta conformable calculus on time scales. We proved two new Leibniz formulas for conformable differentiation of classical and conformable delta integrals, and then we proved variation of parameters and Green's function for delta conformable Sturm-Liouville initial value and boundary value problems. We also considered two dynamic BVPs relating to heat transfer of a fin and axial bending of a rod. Future work in this direction includes exploration of eigenvalue comparisons for these problems, extension to fourth order to explore conformable beam equations, introduction of delay, and exploration of physical applications of time scales calculus.

References

1. T. Abdeljawad, On conformable fractional calculus. J. Comput. Appl. Math. **279**, 57–66 (2015)
2. R.P. Agarwal, M. Bohner, D. O'Regan, A. Peterson, Dynamic equations on time scales: a survey. J. Comput. Appl. Math. **141**(1–2), 1–26 (2006)
3. B.P. Allahverdiev, H. Tuna, Conformable fractional Sturm–Liouville problems on time scales. Math. Methods Appl. Sci. **45**(4), 2299–2314 (2022)
4. M. Al-Smadi, O. Abu Arqub, S. Hadid, An attractive analytical technique for coupled system of fractional partial differential equations in shallow water waves with conformable derivative. Commun. Theor. Phys. **72**(8), 085001 (2020)
5. D.R. Anderson, S.G. Georgiev, *Conformable Dynamic Equations on Time Scales* (Chapman and Hall/CRC, Boca Raton, 2020)
6. D.R. Anderson, D.J. Ulness, Newly defined conformable derivatives. Adv. Dyn. Syst. Appl. **10**(2), 109–137 (2015)
7. B. Bayour, D.F.M. Torres, Existence of solution to a local fractional nonlinear differential equation. J. Comput. Appl. Math. **312**, 127–133 (2016)

8. B. Bayour, A. Hammoudi, D.F.M. Torres, A truly conformable calculus on time scales. Glob. Stoch. Anal. **5**(1), 1–14 (2018)
9. M. Bohner, A. Peterson, *Dynamic Equations on Time Scales: An Introduction with Applications* (Birkhäuser Boston Inc., Boston, 2001)
10. M. Bohner, A. Peterson, *Advances in Dynamic Equations on Time Scales* (Birkhäuser Boston Inc., Boston, 2003)
11. F.A. Çetinkaya, T. Cuchta, Sturm–Liouville and Riccati conformable dynamic equations. Adv. Dyn. Syst. Appl. **15**(1), 1–13 (2020)
12. O.O. Fadodun, O.P. Layeni, A.P. Akinola, Fractional wave propagation in radially vibrating non-classical cylinder. Earthquakes Struct. **13**(5), 465–471 (2017)
13. A.A. Hamou, Z. Hammouch, E. Azroul, P. Agarwal, Monotone iterative technique for solving finite difference systems of time fractional parabolic equations with initial/periodic conditions. Appl. Numer. Math. **181**, 561–593 (2022)
14. R. Khalil, M.A. Horani, A. Yousef, M. Sababheh, A new definition of fractional derivative. J. Comput. Appl. Math. **264**(65), 65–70 (2014)
15. P.K. Kythe, D. Wei, *An Introduction to Linear and Nonlinear Finite Element Analysis* (Birkhäuser Boston Inc., Boston, 2004)
16. L. Sadek, B. Abouzaid, E.M. Sadek, H.T. Alaoui, Controllability, observability and fractional linear-quadratic problem for fractional linear systems with conformable fractional derivatives and some applications. Int. J. Dyn. Control **11**, 214–228 (2023)
17. M. Usman, W. Alhejaili, M. Hamid, N. Khan, Fractional analysis of Jeffrey fluid over a vertical plate with time-dependent conductivity and diffusivity: a low-cost spectral approach. J. Comput. Sci. **63**, 101769 (2022)
18. U. Younas, J. Ren, Diversity of wave structures to the conformable fractional dynamical model. J. Ocean Eng. Sci. **8**(5), 559–572 (2023)
19. L. Zhang, H. Zhou, X. Wang, T. Deng, C. Chen, H. Zhang, T. Nagel, Modeling the visco-elastoplastic behavior of deep coal based on conformable derivative. Mech. Time-Dependent Mat. **28**, 1–21 (2023)

Chapter 4
Dynamic Hermite-Hadamard Inequality for r-Convex Symmetric Functions on Time Scales and Their Applications

Ahmed A. El-Deeb and Sandra Pinelas

Abstract In this chapter, we prove some new inequalities of Hermite-Hadamard-type for r-convex functions on an arbitrary time scale using the diamond-α dynamic integrals, which is defined as a linear combination of the delta and nabla integrals. These inequalities extend and improve some known dynamic inequalities on time scales and unify and extend some continuous inequalities and their corresponding discrete analogs.

Keywords Hermite-Hadamard inequality · r-Convex function · Diamond-α calculus · Time scales · Dynamic inequalities

4.1 Introduction

A well-celebrated inequality for the class of convex functions is the classical Hermite-Hadamard inequality. The inequality gives us an estimate, from below and from above, of the mean value of a convex function $f : [a, b] \to \mathbb{R}$:

$$f\left(\frac{a+b}{2}\right) \leq \frac{1}{b-a}\int_a^b f(t)dt \leq \frac{f(a)+f(b)}{2}. \tag{4.1}$$

A. A. El-Deeb (✉)
Department of Mathematics, Faculty of Science, Al-Azhar University, Nasr City, Cairo, Egypt
e-mail: ahmedeldeeb@azhar.edu.eg

S. Pinelas
Departamento de Ciências Exatas e Engenharia, Amadora, Portugal

Departamento de Matemática, Universidade de Aveiro, Aveiro, Portugal

P. Benevieri, J. G. Mesquita (eds.), *Functional Differential Equations and Dynamic Equations on Time Scales*, Advances in Continuum Mechanics 52,
https://doi.org/10.1007/978-3-031-83327-4_4

For more details see [19, pp. 50–51]. In [12], L. Fejér proved a weighted version of inequality (4.1). Precisely, if f is as above and $\hat{p} : [a, b] \to \mathbb{R}$ is a nonnegative integrable function, symmetric with respect to the middle point $(a+b)/2$, then

$$f\left(\frac{a+b}{2}\right)\int_a^b \hat{p}(t)dt \leq \int_a^b \hat{p}(t)f(t)dt \leq \frac{f(a)+f(b)}{2}\int_a^b \hat{p}(t)dt. \tag{4.2}$$

However, Choquet's theory provides the best understanding of the Hermite-Hadamard inequality in the framework of positive Radon measures on arbitrary compact convex sets. See [19, pp. 192-195] for details.

A. M. Fink [13] introduced an important remark concerning by noticing the possibility to consider certain signed measures.

Let $\hat{\mu}$ be a real Borel measure on $[a, b]$, with $\hat{\mu}([a, b]) > 0$. We say that $\hat{\mu}$ is a Hermite-Hadamard measure if for every convex function $f : [a, b] \to \mathbb{R}$ the following inequalities hold:

$$f(x_{\hat{\mu}}) \leq \frac{1}{\hat{\mu}([a,b])}\int_a^b f(x)d\hat{\mu}(x) \tag{4.3}$$

$$\leq \frac{b-x_{\hat{\mu}}}{b-a}f(a) + \frac{x_{\hat{\mu}-a}}{b-a}f(b), \tag{4.4}$$

where

$$x_{\hat{\mu}} = \frac{1}{\hat{\mu}([a,b])}\int_a^b x d\hat{\mu}(x)$$

is the barycenter of $\hat{\mu}$.

A. M. Fink [13] gave a complete characterization of the real-valued Borel measures $\hat{\mu}$ for which (4.3) works for every convex function, precisely the fulfillment of the following condition of the end-positivity:

$$\int_a^t (t-x)d\hat{\mu}(x) \geq 0 \quad and \int_a^t (t-x)d\hat{\mu}(x) \geq 0, \quad for \quad every \quad t \in [a,b].$$

In the same paper, A. M. Fink proved a sufficient condition for (4.3), but his arguments can be modified to get a complete characterization of the measures for which (4.3) works:

Theorem 4.1 *Let $\hat{\mu}$ be a real Borel measure on $[a, b]$ with $\hat{\mu}([a, b]) > 0$ and having the barycenter $x_{\hat{\mu}}$. Then the inequality*

$$\frac{1}{\hat{\mu}([a,b])}\int_a^b f(x)d\hat{\mu}(x) \leq \frac{b-x_{\hat{\mu}}}{b-a}f(a) + \frac{x_{\hat{\mu}-a}}{b-a}f(b)$$

works for all continuous convex functions $f : [a, b] \to \mathbb{R}$ if and only if

$$\frac{b-t}{b-a}\int_a^t (x-a)d\hat{\mu}(x) + \frac{t-a}{b-a}\int_t^b (b-x)d\hat{\mu}(x), \quad for \quad all \quad t \in [a,b].$$

Using the above results we can easily check that $d\hat{\mu} = (x^2 - \lambda)dx$ is a Hermite-Hadamard measure for all $\lambda \leq 1/16$.

P. Czinder and Z. Páles [9] (see also [8]) have extended the Hermite-Hadamard inequality in another direction, by considering functions that mix the up and down convexity:

Theorem 4.2 *Suppose that I is an interval and $f : I \to \mathbb{R}$ is a function symmetric with respect to an element $m \in \bar{I}$, that is,*

$$f(x) + f(2m - x) = 2f(m) \quad for \quad all \quad x \in I \cap (-\infty, m]. \tag{4.5}$$

If f is convex over the interval $I \cap (-\infty, m]$ and concave on the interval $I \cap [m, \infty)$, then for every interval $[a, b] \subset I$ with $(a+b)/2 \geq m$, the following inequalities hold true:

$$f\left(\frac{a+b}{2}\right) \geq \frac{1}{b-a}\int_a^b f(t)dt \geq \frac{f(a)+f(b)}{2}. \tag{4.6}$$

If $(a+b)/2 \leq m$, then the inequalities (4.6) *should be reversed.*

An appropriate version is valid for functions that are concave over the interval $I \cap (-\infty, m]$ and convex over $I \cap [m, \infty)$.

Notice that the inequality (4.1) can be derived from (4.6), by choosing m as one of the endpoints of I.

Florea and Niculescu [14] proved that Theorem 4.2 works actually in the general framework of Hermite-Hadamard measures.

Theorem 4.3 *Suppose that $f : I \to \mathbb{R}$ verifies the symmetry condition* (4.5) *and is convex over the interval $I \cap (-\infty, m]$ and concave over the interval $I \cap [m, \infty)$.*

If $(a+b)/2 \geq m$ and $\hat{\mu}$ is a Hermite-Hadamard measure on each of the intervals $[a, 2m - a]$ and $[2m - a, b]$ and is invariant with respect to the map $\bar{T}(x) = 2m - x$ on $[a, 2m - a]$, then

$$f(x_{\hat{\mu}}) \geq \frac{1}{\hat{\mu}([b,a])}\int_a^b f(t)d\hat{\mu} \geq \frac{b - x_{\hat{\mu}}}{b-a}f(a) + \frac{x_{\hat{\mu}} - a}{b-a}f(b). \tag{4.7}$$

If $(a+b)/2 \leq m$, then the inequalities (4.7) *work in a reverse way, provided μ is a Hermite-Hadamard measure on each of the intervals $[a, 2m - b]$ and $[2m - b, b]$ and is invariant with respect to the map $\bar{T}(x) = 2m - x$ on $[2m - b, b]$.*

The function $f : I \to \mathbb{R}$ is said to be convex on I if for all $x, y \in I$ and $t \in [0, 1]$, one has the inequality:

$$f(tx + (1-t)y) \leq tf(t) + (1-t)f(y).$$

A function $f : I \to (0, \infty)$ is said to be log-convex if $\log(f)$ is convex or, equivalently, if for all $x, y \in I$ and $t \in [0, 1]$ one has the inequality:

$$f(tx + (1-t)y) \le [f(t)]^t [f(y)]^{1-t}. \tag{4.8}$$

We note that if f and g are convex functions and g is monotonic nondecreasing, then $f \circ g$ is convex. Moreover, since $f = \exp(\log f)$, it follows that a log-convex function is convex, but the converse is not true [23, p. 7]. This fact is obvious from (4.8) as by the arithmetic-geometric mean inequality, we have

$$[f(t)]^t [f(y)]^{1-t} \le tf(x) + (1-t)f(y),$$

for all $x, y \in I$ and $t \in [0, 1]$.

In [22] the author generalized Hermite-Hadamard inequality to an r-convex positive function, which is defined on an interval $[a, b]$. A positive function f is called r-convex on $[a, b]$, if for each $x, y \in [a, b]$ and $t \in [a, b]$,

$$f(tx + (1-t)y) \le \begin{cases} \left[tf^r(x) + (1-t)f^r(y) \right]^{1/r}, & r \neq 0 \\ [f(x)]^t [f(y)]^{1-t}, & r = 0. \end{cases}$$

It is obvious that 0-convex functions are simply log-convex functions and r-convex functions are ordinary convex functions. One should note that if f is r-convex in $[a, b]$, then f^r is a convex function ($r > 0$).

In [18] and [25] the authors introduced some refinements of the Hadamard inequality for r-convex functions. Bessenyei in [2] proved Hermite-Hadamard inequality for generalized 3-convex functions. Also in [18] the authors proved that if f is r-convex in the interval $[a, b]$ and $0 \le r \le 1$, then

$$\frac{1}{b-a} \int_a^b f(x)dx \le \frac{r}{r+1} \left[f^r(a) + f^r(b) \right]^{1/r}. \tag{4.9}$$

In [26] the authors proved that if f is r-convex in $[a, b]$ and $r \ge 1$, then

$$\frac{1}{b-a} \int_a^b f(x)dx \le \left[\frac{1}{2} [f^r(a) + f^r(b)] \right]^{1/r}. \tag{4.10}$$

In the same paper, the authors proved the following inequality for r-convex functions and $r \ge 0$, which may be considered as an extension and refinement of (4.9) and (4.10):

$$\frac{1}{b-a} \int_a^b f(x)dx \le \frac{r}{r+1} \frac{f^{r+1}(b) - f^{r+1}(a)}{f^r(b) - f^r(a)}.$$

In 2001, by using (4.1), Dragomir [11] proved the following Hadamard's type result for functions, defined on a rectangle, that are convex on the coordinates.

Theorem 4.4 *Let $f : [a, b] \times [c, d] \to \mathbb{R}$ be such that the partial mappings*

$$f_y : [a, b] \to \mathbb{R},\ f_y(u) = f(u, y) \quad \text{and} \quad f_x : [c, d] \to \mathbb{R},\ f_x(v) = f(x, v),$$

defined for all $y \in [c, d]$ and $x \in [a, b]$, be r-convex with $r \geq 1$. Then the following inequalities hold:

$$\begin{aligned} & f\left(\frac{a+b}{2}, \frac{c+d}{2}\right) \\ & \leq \frac{1}{2}\left[\frac{1}{b-a}\int_a^b f\left(x, \frac{c+d}{2}\right)dx + \frac{1}{d-c}\int_c^d f\left(\frac{a+b}{2}, y\right)dy\right] \\ & \leq \frac{1}{(b-a)(d-c)}\int_a^b\int_c^d f(x, y)dxdy \\ & \leq \frac{1}{4(b-a)}\int_a^b \Big[f(x, c) + f(x, d)\Big]dx + \frac{1}{4(d-c)}\int_c^d \Big[f(a, y) + f(d, y)\Big]dy \\ & \leq \frac{f(a, c) + f(a, d) + f(b, c) + f(b, d)}{4}. \end{aligned} \tag{4.11}$$

The above inequalities are sharp.

Recently, the theory of time scales, which has been initiated by Stefan Hilger in his PhD thesis [15] in order to unify discrete and continuous analysis, has gained a lot of attention (see Sect. 4.2 for a brief overview). During the past decade, a considerable number of dynamic inequalities have been given by many authors who were motivated by some applications. The general idea is to prove a result for a dynamic inequality where the domain of the unknown function is a so-called time scale $\mathbb{T}$, which is an arbitrary nonempty closed subset of $\mathbb{R}$. The three most popular time scales are $\mathbb{T} = \mathbb{R}$, $\mathbb{T} = \mathbb{Z}$, and $\mathbb{T} = \overline{q^{\mathbb{Z}}} = \{q^z : z \in \mathbb{Z}\} \cup \{0\}$, where $q > 1$. The books [6] and [7] organize and summarize much of time scales calculus.

In 2008, Dinu [10] introduced a time scale version of the Hermite-Hadamard inequality by proving the following theorem:

Theorem 4.5 *Let $\mathbb{T}$ be a time scale and $a, b \in \mathbb{T}$. Let $f : [a, b] \to \mathbb{R}$ be a continuous convex function. Then*

$$f(x_\alpha) \leq \frac{1}{b-a}\int_a^b f(t) \diamond_\alpha t \leq \frac{b - x_\alpha}{b-a} f(a) + \frac{x_\alpha - a}{b-a} f(b), \tag{4.12}$$

where

$$x_\alpha = \frac{1}{b-a}\int_a^b t \diamond_\alpha t. \tag{4.13}$$

Also, Dinu [10] computed the following integral:

$$\int_a^b t \diamond_{1/2} t = \frac{b^2 + a^2}{2}. \tag{4.14}$$

In the same paper, the author extended (4.12) by establishing a weighted Hermite-Hadamard inequality for time scales:

Theorem 4.6 *Let $\mathbb{T}$ be a time scale and $a, b \in \mathbb{T}$. Let $f : [a, b] \to \mathbb{R}$ be a continuous convex function. Let $\omega : \mathbb{T} \to \mathbb{R}$ be a continuous function such that $\omega(t) \geq 0$ for all $t \in \mathbb{T}$, and $\int_a^b \omega(t) \diamond t > 0$. Then the following inequalities hold:*

$$f(x_{\omega,\alpha}) \leq \frac{1}{\int_a^b \omega(t) \diamond_\alpha t} \int_a^b \omega(t) f(t) \diamond_\alpha t \leq \frac{b - x_{\omega,\alpha}}{b-a} f(a) + \frac{x_{\omega,\alpha} - a}{b-a} f(b), \tag{4.15}$$

where

$$x_{\omega,\alpha} = \frac{1}{\int_a^b \omega(t) \diamond_\alpha t} \int_a^b t\omega(t) \diamond_\alpha t. \tag{4.16}$$

Definition 4.1 ([10]) Let $\mathbb{T}$ be a time scale, $a, b \in \mathbb{T}$, $\omega : \mathbb{T} \to \mathbb{R}_+$ be a positive weight, and $\alpha \in [0, 1]$. One says that a function $f : [a, b] \to \mathbb{R}$ is (ω, α)-symmetric on $[a, b]_\mathbb{T}$ if the following conditions are satisfied:

(i)

$$\frac{b - x_{\omega,\alpha}}{b-a} f(a) + \frac{x_{\omega,\alpha} - a}{b-a} f(b) = f(x_{\omega,\alpha}); \tag{4.17}$$

(ii)

$$\int_a^b f(t)\omega(t) \diamond_\alpha t = f(x_{\omega,\alpha}) \int_a^b \omega(t) \diamond_\alpha t. \tag{4.18}$$

Here, $x_{\omega,\alpha} = \int_a^b t\omega(t)\diamond_\alpha t / \int_a^b \omega(t)\diamond_\alpha t$.

By using Definition 4.1, Dinu [10] gave an extension of Theorem 4.3, for time scale, using functions not necessarily symmetric in the usual sense.

Theorem 4.7 *Let $\mathbb{T}$ be a time scale, $a \leq c \leq b \in \mathbb{T}$, $\omega : \mathbb{T} \to \mathbb{R}_+$ be a positive weight, and $\alpha \in [0, 1]$. Let $p = \int_a^c t\omega(t)\diamond_\alpha t / \int_a^c \omega(t)\diamond_\alpha t$ and $q = \int_c^b t\omega(t)\diamond_\alpha t / \int_c^b \omega(t)\diamond_\alpha t$.*

(i) If the function $f : [a, b] \to \mathbb{R}$ is (ω, α)-symmetric on $[a, c]_\mathbb{T}$ and convex on $[p, b]$, then

$$f(x_{\omega,\alpha}) \leq \frac{1}{\int_a^b \omega(t) \diamond_\alpha t} \int_a^b \omega(t) f(t) \diamond_\alpha t \leq \frac{b - x_{\omega,\alpha}}{b-a} f^r(a) + \frac{x_{\omega,\alpha} - a}{b-a} f^r(b). \tag{4.19}$$

If f is concave on $[p, b]$, then the inequalities in (4.19) *are reversed.*

(ii) If the function $f : [a, b] \to \mathbb{R}$ is (ω, α)-symmetric on $[a, c]_{\mathbb{T}}$ and concave on $[a, q]$, then one has (4.19).

If f is concave on $[a, q]$, then the inequalities in (4.19) *are reversed.*

In this chapter, we prove and extend some dynamic Hermite-Hadamard inequalities involving r-convex functions on time scales via the combined diamond-α dynamics. Our results extend and generalize the mentioned results of Florea and Niculescu in [14], Dinu in [10], Gholamreza et al. in [26], and Dragomir in [11]. The present chapter is arranged as follows: In Sect. 4.2, we give a brief background of the theory of time scales. In Sect. 4.3, we formulate and prove our main results.

4.2 Preliminaries on Time Scales

First, we recall some time scales essentials and some universal symbols used in the present chapter. From now on, $\mathbb{R}$ and $\mathbb{Z}$ denote the set of all real numbers and the set of all integers, respectively.

A time scale $\mathbb{T}$ is an arbitrary nonempty closed subset of $\mathbb{R}$. We suppose throughout the book that $\mathbb{T}$ has the topology that it inherits from the standard topology on $\mathbb{R}$. The forward jump operator $\sigma : \mathbb{T} \to \mathbb{T}$ is defined for any $\tau \in \mathbb{T}$ by

$$\sigma(\tau) := \inf\{s \in \mathbb{T} : s > \tau\},$$

and the backward jump operator $\rho : \mathbb{T} :\to \mathbb{T}$ is defined for any $\tau \in \mathbb{T}$ by

$$\rho(\tau) := \sup\{s \in \mathbb{T} : s < \tau\}.$$

In the previous two definitions, we have $\inf \emptyset = \sup \mathbb{T}$ (i.e., if τ is the minimum of $\mathbb{T}$, then $\sigma(\tau) = \tau$) and $\sup \emptyset = \inf \mathbb{T}$ (i.e., if τ is the maximum of $\mathbb{T}$, then $\rho(\tau) = \tau$), where $\emptyset$ is the empty set.

A point $\tau \in \mathbb{T}$ with $\inf \mathbb{T} < \tau < \sup \mathbb{T}$ is said to be right-scattered if $\sigma(\tau) > \tau$, right-dense if $\sigma(\tau) = \tau$, left-scattered if $\rho(\tau) < \tau$, and left-dense if $\rho(\tau) = \tau$. Points that are simultaneously right-dense and left-dense are called dense points, whereas points that are simultaneously right-scattered and left-scattered are called isolated points.

We define the forward graininess function $\mu : \mathbb{T} \to [0, \infty)$ for any $\tau \in \mathbb{T}$ by $\mu(\tau) := \sigma(\tau) - \tau$, and the backward graininess function $\nu : \mathbb{T} \to [0, \infty)$ is defined for any $\tau \in \mathbb{T}$ by $\nu(\tau) := \tau - \rho(\tau)$.

Let $u\colon \mathbb{T} \to \mathbb{R}$ be a function. Then the function $u^\sigma : \mathbb{T} \to \mathbb{R}$ is defined by $u^\sigma(\tau) = u(\sigma(\tau)), \forall \tau \in \mathbb{T}$, that is, $u^\sigma = u \circ \sigma$. In a similar manner, the function $u^\rho : \mathbb{T} \to \mathbb{R}$ is defined by $u^\rho(\tau) = u(\rho(\tau)), \forall \tau \in \mathbb{T}$, that is, $u^\rho = u \circ \rho$.

The sets $\mathbb{T}^\kappa$, $\mathbb{T}_\kappa$, and $\mathbb{T}^\kappa_\kappa$ are defined as follows: If $\mathbb{T}$ has a left-scattered maximum τ_1, then $\mathbb{T}^\kappa = \mathbb{T} - \{\tau_1\}$; otherwise $\mathbb{T}^\kappa = \mathbb{T}$. If $\mathbb{T}$ has a right-scattered minimum τ_2, then $\mathbb{T}^\kappa = \mathbb{T} - \{\tau_2\}$; otherwise $\mathbb{T}_\kappa = \mathbb{T}$. Finally, we have $\mathbb{T}^\kappa_\kappa = \mathbb{T}^\kappa \cap \mathbb{T}_\kappa$.

The interval $[a, b]$ in $\mathbb{T}$ is defined by

$$[a, b]_\mathbb{T} = \{\tau \in \mathbb{T} : a \le \tau \le b\}.$$

Open intervals and half-closed interval are defined similarly.

Suppose $u\colon \mathbb{T} \to \mathbb{R}$ is a function and $\tau \in \mathbb{T}^\kappa$. Then we say that $u^\Delta(\tau) \in \mathbb{R}$ is the delta derivative of u at τ if for any $\varepsilon > 0$ there exists a neighborhood U of τ such that for all $s \in U$

$$\left|[u(\sigma(\tau)) - u(s)] - u^\Delta(\tau)[\sigma(\tau) - s]\right| \le \varepsilon|\sigma(\tau) - s|.$$

Furthermore, u is said to be delta differentiable on $\mathbb{T}^\kappa$ if it is delta differentiable at each $\tau \in \mathbb{T}^\kappa$.

Similarly, we say that $u^\nabla(\tau) \in \mathbb{R}$ is the nabla derivative of u at τ if for any $\varepsilon > 0$ there exists a neighborhood V of τ such that for all $s \in V$

$$\left|[u(\rho(\tau)) - u(s)] - u^\nabla(\tau)[\rho(\tau) - s]\right| \le \varepsilon|\rho(\tau) - s|.$$

Moreover, u is said to be nabla differentiable on $\mathbb{T}_\kappa$ if it is nabla differentiable at each $\tau \in \mathbb{T}_\kappa$.

A function $u\colon \mathbb{T} \to \mathbb{R}$ is called right-dense continuous (rd-continuous) if u is continuous at all the right-dense points in $\mathbb{T}$ and its left-sided limits exist at all left-dense points in $\mathbb{T}$. On the other hand, u is called left-dense continuous (ld-continuous) if u is continuous at all the left-dense points in $\mathbb{T}$ and its right-sided limits exist at all right-dense points in $\mathbb{T}$.

A function $F\colon \mathbb{T} \to \mathbb{R}$ is said to be a delta antiderivative of $u\colon \mathbb{T} \to \mathbb{R}$ if $F^\Delta = u(\tau)$ for all $\tau \in \mathbb{T}^\kappa$. In this case, the definite integral of u is defined by

$$\int_a^b u(\tau)\Delta\tau = F(b) - F(a) \qquad for\ all \quad a, b \in \mathbb{T}.$$

A function $G\colon \mathbb{T} \to \mathbb{R}$ is said to be a nabla antiderivative of $v\colon \mathbb{T} \to \mathbb{R}$ if $G^\nabla = v(\tau)$ for all $\tau \in \mathbb{T}_\kappa$. In this case, the definite integral of v is defined by

$$\int_a^b v(\tau)\nabla\tau = G(b) - G(a) \qquad for\ all \quad a, b \in \mathbb{T}.$$

We will need the following important relations between calculus on time scales $\mathbb{T}$ and continuous calculus on $\mathbb{R}$ and discrete calculus on $\mathbb{Z}$. Note that:

(i) If $\mathbb{T} = \mathbb{R}$, then

$$\sigma(\tau) = \rho(\tau) = \tau, \quad \mu(\tau) = \nu(\tau) = 0, \quad u^{\Delta}(\tau) = u^{\nabla}(\tau) = u'(\tau),$$
$$\int_a^b u(\tau)\Delta\tau = \int_a^b u(\tau)\nabla\tau = \int_a^b u(\tau)d\tau. \tag{4.20}$$

(ii) If $\mathbb{T} = \mathbb{Z}$, then

$$\sigma(\tau) = \tau + 1, \quad \rho(\tau) = \tau - 1, \quad \mu(\tau) = \nu(\tau) = 1,$$
$$u^{\Delta}(\tau) = \Delta u(\tau), \quad u^{\nabla}(\tau) = \nabla u(\tau),$$
$$\int_a^b u(\tau)\Delta\tau = \sum_{\tau=a}^{b-1} u(\tau), \quad \int_a^b u(\tau)\nabla\tau = \sum_{\tau=a+1}^{b} u(\tau), \tag{4.21}$$

where Δ and ∇ are the forward and backward difference operators, respectively.

Now, we briefly introduce the diamond-α dynamic derivative and the diamond-α dynamic integral, and we refer the reader to [24] for a comprehensive development of the calculus of the diamond-α dynamic derivative and the diamond-α dynamic integration.

Let $\mathbb{T}$ be a time scale and $f(t)$ be differentiable on $\mathbb{T}$ in the Δ and ∇ senses. For $t \in T$, we define the diamond-α dynamic derivative $f^{\diamond_\alpha}(t)$ by

$$f^{\diamond_\alpha}(t) = \alpha f^{\Delta}(t) + (1-\alpha)f^{\nabla}(t), \qquad 0 \le \alpha \le 1. \tag{4.22}$$

Thus f is diamond-α differentiable if and only if f is Δ and ∇ differentiable. The diamond-α derivative reduces to the standard Δ derivative for $\alpha = 1$ or the standard ∇ derivative for $\alpha = 0$. On the other hand, it represents a "weighted dynamic derivative" for $\alpha \in (0, 1)$. Furthermore, the combined dynamic derivative offers a centralized derivative formula on any uniformly discrete time scale $\mathbb{T}$ when $\alpha = 1/2$.

Let $f, g\colon \mathbb{T} \to \mathbb{R}$ be diamond-α differentiable at $t \in \mathbb{T}$. Then:

(i) $f + g : \mathbb{T} \to \mathbb{R}$ is diamond-α differentiable at $t \in T$ with

$$(f+g)^{\diamond_\alpha}(t) = f^{\diamond_\alpha}(t) + g^{\diamond_\alpha}(t). \tag{4.23}$$

(ii) For any constant c, $cf : \mathbb{T} \to \mathbb{R}$ is diamond-α differentiable at $t \in T$ with

$$(cf)^{\diamond_\alpha}(t) = cf^{\diamond_\alpha}(t). \tag{4.24}$$

(ii) $fg : \mathbb{T} \to \mathbb{R}$ is diamond-α differentiable at $t \in T$ with

$$(fg)^{\diamond_\alpha}(t) = f^{\diamond_\alpha}(t)g(t) + \alpha f^{\sigma}(t)g^{\Delta}(t) + (1-\alpha)f^{\rho}(t)g^{\nabla}(t). \tag{4.25}$$

Let $a, t \in \mathbb{T}$, and $h\colon \mathbb{T} \to \mathbb{R}$. Then the diamond-$\alpha$ integral from a to t of h is defined by

$$\int_a^t h(\tau) \diamond_\alpha \tau = \alpha \int_a^t h(t)\Delta\tau + (1-\alpha)\int_a^t h(\tau)\nabla\tau, \qquad 0 \le \alpha \le 1. \tag{4.26}$$

We may notice that the $\diamond_\alpha$ integral is a combined Δ and ∇ integral. In general, we do not have

$$\left(\int_a^t h(\tau) \diamond_\alpha \tau\right)^{\diamond_\alpha} = h(t), \qquad t \in \mathbb{T}. \tag{4.27}$$

Theorem 4.8 *Let $a, b, t \in \mathbb{T}$, $c \in \mathbb{R}$. Let f and g be two continuous functions on $[a, b]$. Then:*

(1) $\int_a^t [f(\tau) + g(\tau)] \diamond_\alpha \tau = \int_a^t f(\tau) \diamond_\alpha \tau + \int_a^t g(\tau) \diamond_\alpha \tau.$
(2) $\int_a^t cf(\tau) \diamond_\alpha \tau = c\int_a^t f(\tau) \diamond_\alpha \tau.$
(3) $\int_a^t f(\tau) \diamond_\alpha \tau = \int_a^b f(\tau) \diamond_\alpha \tau + \int_b^t f(\tau) \diamond_\alpha \tau.$
(4) $\int_a^t f(\tau) \diamond_\alpha \tau = -\int_t^a f(\tau) \diamond_\alpha \tau.$
(5) If $f(t) \le g(t)$ for all $\tau \in [a, b]_{\mathbb{T}}$, then $\int_a^b f(\tau) \diamond_\alpha \tau \le \int_a^b g(\tau) \diamond_\alpha \tau.$

Let u be a $\diamond-\alpha$ differentiable function on $[a, b]_{\mathbb{T}}$. Then u is increasing if $u^{\diamond_\alpha}(\tau) > 0$, nondecreasing if $u^{\diamond_\alpha}(\tau) \ge 0$, decreasing if $u^{\diamond_\alpha}(\tau) < 0$, and nonincreasing if $u^{\diamond_\alpha}(\tau) \le 0$ on $[a, b]_{\mathbb{T}}$.

The two-variable time scales calculus and multiple integration on time scales were introduced in [3, 4] (see also [5]). Let $\mathbb{T}_1$ and $\mathbb{T}_2$ be two time scales, and put $\mathbb{T}_1 \times \mathbb{T}_2 = \{(t, s) : t \in \mathbb{T}_1, s \in \mathbb{T}_2\}$, which is a complete metric space with the metric d defined by

$$d((t, s), (\acute{t}, \acute{s})) = \sqrt{(t-\acute{t})^2 + (s-\acute{s})^2}, \quad \forall (t, s), (\acute{t}, \acute{s}) \in \mathbb{T}_1 \times \mathbb{T}_2.$$

To make this work self-content, we now recall some basic definitions of the diamond-α partial dynamic calculus on time scales [17].

Suppose $a < b$ are points in $\mathbb{T}_1$, $c < d$ are points in $\mathbb{T}_2$, $[a, b]$ is the closed bounded interval in $\mathbb{T}_1$, and $[c, d]$ is the closed bounded interval in $\mathbb{T}_2$. We introduce a "rectangle" in $\mathbb{T}_1 \times \mathbb{T}_2$ by

$$R = [a, b] \times [c, d] = \{(t, s) : t \in [a, b], s \in [c, d]\}.$$

Let $\{t_0, t_1, \ldots, t_n\} \subset [a, b]$, where $a = t_0 < t_1 < \cdots < t_n = b$, and $s_0, s_1, \ldots, s_k \subset [c, d]$, where $c = s_0 < s_1 < \cdots < s_k = d$. The numbers n and k may be arbitrary positive integers. We call the collection of intervals

$$P_1 = \{[t_{i-1}, t_i) : 1 \le i \le n\}$$

a $\diamond_\alpha$-partition of $[a, b)$ and denote the set of all $\diamond_\alpha$-partition of $[a, b)$ by $\mathcal{P}([a, b])$. Similarly, the collection of intervals

$$P_2 = \{[s_{j-1}, s_j) : 1 \le s \le k\}$$

is called a $\diamond_\alpha$-partition of $[c, d)$ and denote the set of all $\diamond_\alpha$-partition of $[c, d)$ by $\mathcal{P}([c, d])$. Let

$$R_{ij} = [t_{i-1,t_i}) \times [s_{j-1}, s_j), \quad \text{where} \quad 1 \le i \le n, 1 \le j \le k.$$

We call the collection $P = \{R_{ij} : 1 \le i \le n, 1 \le j \le k\}$ a $\diamond_\alpha$-partition of R, generated by the $\diamond_\alpha$-partition of P_1 and P_2 of $[a, b)$ and $[c, d)$, respectively, and write $P = P_1 \times P_2$. The rectangles P_{ij}, $1 \le i \le n, 1 \le j \le k$, are called the subrectangles of the partition P. The set of all $\diamond_\alpha$-partition of R is denoted by $\mathcal{P}(R)$.

Let $f : R \to \mathbb{R}$ be a bounded function. We set

$$M = \sup\{f(t, s) : (t, s) \in R\} \quad \text{and} \quad m = \inf\{f(t, s) : (t, s) \in R\},$$

and for $1 \le i \le n, 1 \le j \le k$, we set

$$M_{ij} = \sup\{f(t, s) : (t, s) \in R_{ij}\} \quad \text{and} \quad m_{ij} = \inf\{f(t, s) : (t, s) \in R_{ij}\}.$$

The upper Darboux $\diamond_\alpha$-integral of $U(f)$ of f over R and the lower Darboux $\diamond_\alpha$-integral $L(f)$ of R are defined, respectively, by

$$U(f) = \inf\{U(f, P) : P \in \mathcal{P}(R)\}$$

and

$$L(f) = \sup\{L(f, P) : P \in \mathcal{P}(R)\},$$

where

$$U(f, P) = \sum_{i=1}^{n} \sum_{j=1}^{k} M_{ij}(t_i - t_{i-1})(s_j - s_{j-1}),$$

and

$$L(f, P) = \sum_{i=1}^{n} \sum_{j=1}^{k} m_{ij}(t_i - t_{i-1})(s_j - s_{j-1}).$$

Definition 4.2 ([21]) We say that f is $\diamond_\alpha$-integrable over R provided $L(f) = U(f)$. In this case, we write $\int_R f(t, s) \diamond_\alpha t \diamond_\alpha s$ for this common value. We call this integral the Darboux $\diamond_\alpha$-integral.

Theorem 4.9 ([21]) *If $L(f, P) = U(f, P)$ for some $\diamond_\alpha$-partition of R, then the function f is $\diamond_\alpha$-integrable over R and*

$$\int_R f(t,s) \diamond_\alpha t \diamond_\alpha s = L(f, P) = U(f, P).$$

In the proofs of our main results, we will need the following Jensen inequality for convex functions and Minkowski inequality on time scales.

Theorem 4.10 ([16], Integral Jensen Inequality) *Assume $u \in C([a, b], (c, d))$ and $v \in C([a, b], \mathbb{R})$ with*

$$\int_a^b |v(t)|dt > 0.$$

Moreover, suppose $F \in C((c, d), \mathbb{R})$ is a convex function. Then

$$F\left(\frac{\int_a^b u(t)|v(t)|dt}{\int_a^b |v(t)|dt}\right) \le \frac{\int_a^b F(u(t))|v(t)|dt}{\int_a^b |v(t)|dt}.$$

Theorem 4.11 ([1], Dynamic Jensen Inequality) *Assume $u \in C_{\mathrm{rd}}([a, b]_{\mathbb{T}}, (c, d))$ and $v \in C_{\mathrm{rd}}([a, b]_{\mathbb{T}}, \mathbb{R})$ such that*

$$\int_a^b |v(\tau)| \diamond_\alpha \tau > 0.$$

Moreover, suppose $F \in C((c, d), \mathbb{R})$ is a convex function. Then

$$F\left(\frac{\int_a^b u(\tau)|v(\tau)| \diamond_\alpha \tau}{\int_a^b |v(\tau)| \diamond_\alpha \tau}\right) \le \frac{\int_a^b F(u(\tau))|v(\tau)| \diamond_\alpha \tau}{\int_a^b |v(\tau)| \diamond_\alpha \tau}. \tag{4.28}$$

Theorem 4.12 ([1], Dynamic Minkowski Inequality) *Let $u, v \in C_{\mathrm{rd}}\big([a, b]_{\mathbb{T}}, \mathbb{R}\big)$ and $p > 1$. Then*

$$\left[\int_a^b \left|u(t) + v(t)\right|^p \Delta t\right]^{\frac{1}{p}} \le \left[\int_a^b \left|u(t)\right|^p \Delta t\right]^{\frac{1}{p}} + \left[\int_a^b \left|v(t)\right|^p \Delta t\right]^{\frac{1}{p}}. \tag{4.29}$$

Also, we will need the following algebraic inequalities in the proof of our results:

$$a^\lambda + b^\lambda \le (a+b)^\lambda \le 2^{\lambda-1}\big(a^\lambda + b^\lambda\big), \quad \text{if} \quad a, b \ge 0, \lambda \ge 1; \tag{4.30}$$

$$a^\lambda + b^\lambda \ge (a+b)^\lambda \ge 2^{\lambda-1}\big(a^\lambda + b^\lambda\big), \quad \text{if} \quad a, b \ge 0, 0 \le \lambda \le 1;$$

$$a^\lambda + b^\lambda \ge 2^{1-\lambda}\big(a^\lambda + b^\lambda\big), \quad \text{if} \quad a, b > 0, \lambda < 0.$$

Now, we are ready to state and prove the main results in this book.

4.3 Main Results

In this section, we will state and prove the main results and investigate some dynamic Hermite-Hadamard-type inequalities with r-convex functions on time scales.

Theorem 4.13 *Let $\mathbb{T}$ be a time scale and $a, b \in \mathbb{T}$. Let $f: [a, b] \to \mathbb{R}$ be a continuous* r*-convex function and $r \geq 1$. Then the following inequalities hold:*

$$f(x_\alpha) \leq \frac{1}{b-a}\int_a^b f(t) \diamond_\alpha t \leq \left[\frac{b-x_\alpha}{b-a} f^r(a) + \frac{x_\alpha - a}{b-a} f^r(b)\right]^{1/r}, \tag{4.31}$$

where $x_\alpha = \frac{1}{b-a}\int_a^b t \diamond_\alpha t$.

Proof Since $r \geq 1$, by Jensen inequality (4.28), we have

$$\left(\frac{1}{b-a}\int_a^b f(t) \diamond_\alpha t\right)^r \leq \frac{1}{b-a}\int_a^b f^r(t) \diamond_\alpha t. \tag{4.32}$$

By convexity of f^r and the right side of (4.12), we obtain

$$\frac{1}{b-a}\int_a^b f^r(t) \diamond_\alpha t \leq \frac{b-x_\alpha}{b-a} f^r(a) + \frac{x_\alpha - a}{b-a} f^r(b). \tag{4.33}$$

Then, from (4.32) and (4.33), we get

$$\frac{1}{b-a}\int_a^b f(t) \diamond_\alpha t \leq \left[\frac{b-x_\alpha}{b-a} f^r(a) + \frac{x_\alpha - a}{b-a} f^r(b)\right]^{1/r}. \tag{4.34}$$

This is the proof of the right-side inequality in (4.31).

Now, we will prove the left-hand side inequality in (4.31). Using Jensen inequality (4.28), by taking $u(t) = t$ and $v(t) = 1$, we get

$$f\left(\frac{\int_a^b t \diamond_\alpha t}{b-a}\right) \leq \left(\frac{\int_a^b f(t) \diamond_\alpha t}{b-a}\right). \tag{4.35}$$

Hence, from (4.35) and the definition of x_α, we get

$$f(x_\alpha) \leq \frac{1}{b-a}\int_a^b f(t) \diamond_\alpha t. \tag{4.36}$$

Then, from (4.34) and (4.36), we get the desired inequality (4.31). This completes the proof. □

Remark 4.1 Let $f: [a, b] \to \infty$ be a 1-convex function. Then the inequality in Theorem 4.13 reduces to the inequality in Theorem 4.5.

Remark 4.2 In Theorem 4.13, if $\mathbb{T} = \mathbb{R}$ and $f: [a, b] \to \infty$ is a 1-convex function, then we get the classical Hermite-Hadamard inequality (4.1).

Corollary 4.1 *In Theorem 4.13, if we take* $\alpha = 1$*, then we get the delta version of inequalities* (4.31) *as follows:*

$$f(x_1) \leq \frac{1}{b-a} \int_a^b f(t)\Delta t \leq \left[\frac{b-x_1}{b-a} f^r(a) + \frac{x_1-a}{b-a} f^r(b)\right]^{1/r},$$

where

$$x_1 = \frac{1}{b-a} \int_a^b t\Delta t.$$

Corollary 4.2 *In Theorem 4.13, if we take* $\alpha = 1/2$*, then we get the diamond-*$1/2$ *version of inequality* (4.31) *as follows:*

$$f\left(\frac{a+b}{2}\right) \leq \frac{1}{b-a} \int_a^b f(t) \diamond_{1/2} t \leq \left[\frac{f^r(a) + f^r(b)}{2}\right]^{1/r}. \tag{4.37}$$

Proof Applying (4.14) in the definition of x_α when $\alpha = 1/2$, we get

$$x_{1/2} = \frac{b+a}{2}. \tag{4.38}$$

Applying (4.38) in (4.31), we get the desired inequalities (4.37). This completes the proof. □

Corollary 4.3 *In Theorem 4.13, if we take* $\alpha = 0$*, then we get the nabla version of inequality* (4.31) *as follows:*

$$f(x_0) \leq \frac{1}{b-a} \int_a^b f(t)\nabla t \leq \left[\frac{b-x_0}{b-a} f^r(a) + \frac{x_0-a}{b-a} f^r(b)\right]^{1/r},$$

where

$$x_0 = \frac{1}{b-a} \int_a^b t\nabla t.$$

Now, we extend (4.2), (4.12), (4.15), and (4.31), by establishing a weighted version of the Hermite-Hadamard inequality on time scales.

Theorem 4.14 *Let* $\mathbb{T}$ *be a time scale and* $a, b \in \mathbb{T}$*. Let* $f: [a, b] \to \mathbb{R}$ *be a continuous* r*-convex function with* $r \geq 1$*. Let* $\omega: \mathbb{T} \to \mathbb{R}$ *be a continuous function such that* $\omega(t) \geq 0$ *for all* $t \in \mathbb{T}$*, and* $\int_a^b \omega(t) \diamond t > 0$*. Then the following inequalities hold:*

$$f(x_{\omega,\alpha}) \leq \frac{1}{\int_a^b \omega(t) \diamond_\alpha t} \int_a^b \omega(t) f(t) \diamond_\alpha t \leq \left[\frac{b - x_{\omega,\alpha}}{b-a} f^r(a) + \frac{x_{\omega,\alpha} - a}{b-a} f^r(b) \right]^{1/r}, \tag{4.39}$$

where $x_{\omega,\alpha} = \frac{1}{\int_a^b \omega(t) \diamond_\alpha t} \int_a^b t\omega(t) \diamond_\alpha t.$

Proof By using Theorem 4.6, we observe that the proof uses the same arguments as the proof of Theorem 4.13, and hence we omit the details. □

Remark 4.3 If we take $\omega(t) = 1$, then the inequality in Theorem 4.14 reduces to the inequality in Theorem 4.13.

Remark 4.4 Let $f : [a, b] \to [0, \infty)$ be a 1-convex function. Then the inequality in Theorem 4.14 reduces to the inequality in Theorem 4.6.

Corollary 4.4 *In Theorem 4.14, if we take* $\alpha = 1$, *then we get the nabla version of inequality* (4.39) *as follows:*

$$f(x_{\omega,1}) \leq \frac{1}{\int_a^b \omega(t) \Delta t} \int_a^b \omega(t) f(t) \Delta t \leq \left[\frac{b - x_{\omega,1}}{b-a} f^r(a) + \frac{x_{\omega,1} - a}{b-a} f^r(b) \right]^{1/r},$$

where

$$x_{\omega,1} = \frac{1}{\int_a^b \omega(t) \Delta t} \int_a^b t\omega(t) \Delta t.$$

Corollary 4.5 *In Theorem 4.14, if we take* $\alpha = 1/2$, *then we get the nabla version of inequality* (4.39) *as follows:*

$$f(x_{\omega,1/2}) \leq \frac{1}{\int_a^b \omega(t) \diamond_{1/2} t} \int_a^b \omega(t) f(t) \diamond_{1/2} t$$
$$\leq \left[\frac{b - x_{\omega,1/2}}{b-a} f^r(a) + \frac{x_{\omega,1/2} - a}{b-a} f^r(b) \right]^{1/r},$$

where

$$x_{\omega,1/2} = \frac{1}{\int_a^b \omega(t) \diamond_{1/2} t} \int_a^b t\omega(t) \diamond_{1/2} t.$$

Corollary 4.6 *In Theorem 4.14, if we take* $\alpha = 0$, *then we get the nabla version of inequality* (4.39) *as follows:*

$$f(x_{\omega,0}) \leq \frac{1}{\int_a^b \omega(t) \nabla t} \int_a^b \omega(t) f(t) \nabla t \leq \left[\frac{b - x_{\omega,0}}{b-a} f^r(a) + \frac{x_{\omega,0} - a}{b-a} f^r(b) \right]^{1/r},$$

where

$$x_{\omega,0} = \frac{1}{\int_a^b \omega(t)\nabla t} \int_a^b t\omega(t)\nabla t.$$

In the following theorem, we extend Theorem 4.3 and Theorem 4.7 by proving the following inequality for time scales using r-convex functions not necessarily symmetric in the usual sense:

Theorem 4.15 *Let $\mathbb{T}$ be a time scale, $a \leq c \leq b \in \mathbb{T}$, $\omega : \mathbb{T} \to \mathbb{R}_+$ be a positive weight, and $\alpha \in [0, 1]$. Let $p = \int_a^c t\omega(t)\diamond_\alpha t / \int_a^c \omega(t)\diamond_\alpha t$ and $q = \int_c^b t\omega(t)\diamond_\alpha t / \int_c^b \omega(t)\diamond_\alpha t$.*

(i) If the function $f : [a, b] \to \mathbb{R}$ is (ω, α)-symmetric on $[a, c]_{\mathbb{T}}$ and r-convex on $[p, b]$, then

$$\begin{aligned} f(x_{\omega,\alpha}) &\leq \frac{1}{\int_a^b \omega(t) \diamond_\alpha t} \int_a^b \omega(t) f(t) \diamond_\alpha t \\ &\leq \left[\frac{b - x_{\omega,\alpha}}{b-a} f^r(a) + \frac{x_{\omega,\alpha} - a}{b-a} f^r(b)\right]^{1/r}. \end{aligned} \tag{4.40}$$

If f is r-concave on $[p, b]$, then the inequalities in (4.40) are reversed.

(ii) If the function $f : [a, b] \to \mathbb{R}$ is (ω, α)-symmetric on $[a, c]_{\mathbb{T}}$ and r-concave on $[a, q]$, then one has (4.40).

If f is r-concave on $[a, q]$, then the inequalities in (4.40) are reversed.

Proof Suppose that f is (ω, α)-symmetric on $[a, c]_{\mathbb{T}}$ and r-convex on $[p, b]$. We will prove the left-hand side inequality in (4.40), we notice that

$$\begin{aligned} \int_a^b f(t)\omega(t) \diamond_\alpha t &= \int_a^c f(t)\omega(t) \diamond_\alpha t + \int_c^b f(t)\omega(t) \diamond_\alpha t \\ &= f(p) \int_a^c \omega(t) \diamond_\alpha t + \int_c^b f(t)\omega(t) \diamond_\alpha t. \end{aligned} \tag{4.41}$$

Using the (ω, α)-symmetric of the function f. Since f is r-convex on $[p, b]$ and $c \geq p$, then, using the left-hand side inequality in (4.39), we get $\int_c^b f(t)\omega(t) \diamond_\alpha t \geq f(q) \int_c^b \omega(t) \diamond_\alpha t$, and so

$$\begin{aligned} \frac{1}{\int_a^b \omega(t) \diamond_\alpha t} \int_a^b f(t)\omega(t) \diamond_\alpha t &\geq \frac{\int_a^c \omega(t) \diamond_\alpha t}{\int_a^b \omega(t) \diamond_\alpha t} f(p) + \frac{\int_c^b \omega(t) \diamond_\alpha t}{\int_a^b \omega(t) \diamond_\alpha t} f(q) \\ &\geq \left(\frac{\int_a^c \omega(t) \diamond_\alpha t}{\int_a^b \omega(t) \diamond_\alpha t} f^r(p) + \frac{\int_c^b \omega(t) \diamond_\alpha t}{\int_a^b \omega(t) \diamond_\alpha t} f^r(q)\right)^{1/r} \end{aligned}$$

$$\geq f\left(\frac{\int_a^c \omega(t)\diamond_\alpha t}{\int_a^b \omega(t)\diamond_\alpha t}p+\frac{\int_c^b \omega(t)\diamond_\alpha t}{\int_a^b \omega(t)\diamond_\alpha t}q\right)$$

$$= f\left(\frac{\int_a^b t\omega(t)\diamond_\alpha t}{\int_a^b \omega(t)\diamond_\alpha t}\right)$$

$$= f(x_{\omega,\alpha}), \tag{4.42}$$

where we used the definitions of p and q, combined with the convexity of f on $[p, b]$.

Now, we prove the right-hand side inequality in (4.40). Since $r \geq 1$, by Jensen inequality (4.28), we get

$$\left(\frac{1}{b-a}\int_a^b \omega(t)f(t)\diamond_\alpha t\right)^r \leq \frac{1}{b-a}\int_a^b \omega(t)f^r(t)\diamond_\alpha t. \tag{4.43}$$

By convexity of f^r and the right-hand side of (4.19), we obtain

$$\frac{1}{\int_a^b \omega(t)\diamond_\alpha t}\int_a^b \omega(t)f^r(t)\diamond_\alpha t \leq \frac{b-x_{\omega,\alpha}}{b-a}f^r(a)+\frac{x_{\omega,\alpha}-a}{b-a}f^r(b). \tag{4.44}$$

Then, from (4.43) and (4.44), we get

$$\frac{1}{\int_a^b \omega(t)\diamond_\alpha t}\int_a^b \omega(t)f(t)\diamond_\alpha t \leq \left[\frac{b-x_{\omega,\alpha}}{b-a}f^r(a)+\frac{x_{\omega,\alpha}-a}{b-a}f^r(b)\right]^{1/r}. \tag{4.45}$$

This is the proof of the right-hand side inequality in (4.40). From (4.42) and (4.45), we get the desired inequality (4.40). This completes the proof. □

Remark 4.5 As a particular case, in Theorem 4.15, if we take $c = a$ or $c = b$, we get Theorem 4.14.

Remark 4.6 Let $f\colon [a, b] \to [0, \infty)$ be a 1-convex function. Then the inequality in Theorem 4.15 reduces to the inequality in Theorem 4.40.

In the following, we give extensions and generalizations for the inequalities obtained in [20] due to Nwaeze.

Theorem 4.16 *Let $\mathbb{T}_1$ and $\mathbb{T}_2$ be two time scales and $a, b, x \in \mathbb{T}_1$, $c, d, y \in \mathbb{T}_2$ with $a < b$, $c < d$ and $f : [a, b] \times [c, d] \to \mathbb{R}$ be such that the partial mappings*

$$f_y : [a, b] \to \mathbb{R},\ f_y(u) = f(u, y) \quad \text{and} \quad f_x : [c, d] \to \mathbb{R},\ f_x(v) = f(x, v),$$

defined for all $y \in [c, d]$ and $x \in [a, b]$, are continuous and r-convex with $r \geq 1$. Then the following inequalities hold:

$$\frac{1}{2}\left[\frac{1}{b-a}\int_a^b f(x, s_\alpha) \diamond_\alpha x + \frac{1}{d-c}\int_c^d f(t_\alpha, y) \diamond_\alpha y\right]$$
$$\leq \frac{1}{(b-a)(d-c)}\int_a^b\int_c^d f(x, y) \diamond_\alpha x \diamond_\alpha y$$
$$\leq \frac{1}{2(b-a)(d-c)^{\frac{1}{r}}}\left\{\left(\int_a^b (d-s_\alpha)^{\frac{1}{r}} f(x, c) \diamond_\alpha x\right)^r\right.$$
$$\left.+\left(\int_a^b (s_\alpha - c)^{\frac{1}{r}} f(x, d) \diamond_\alpha x\right)^r\right\}^{1/r}$$
$$+\frac{1}{2(b-a)^{\frac{1}{r}}(d-c)}\left\{\left(\int_c^d (b-t_\alpha)^{\frac{1}{r}} f(a, y) \diamond_\alpha y\right)^r\right.$$
$$\left.+\left(\int_c^d (t_\alpha - a)^{\frac{1}{r}} f(b, y) \diamond_\alpha y\right)^r\right\}^{1/r}, \tag{4.46}$$

where $t_\alpha = \frac{1}{b-a}\int_a^b t \diamond_\alpha t$, *and* $s_\alpha = \frac{1}{d-c}\int_c^d s \diamond_\alpha s$.

Proof Applying Theorem 4.13 to the function f_y, we obtain

$$f_y(t_\alpha) \leq \frac{1}{b-a}\int_a^b f_y(x) \diamond_\alpha x \leq \left[\frac{b-t_\alpha}{b-a} f_y^r(a) + \frac{t_\alpha - a}{b-a} f_y^r(b)\right]^{1/r},$$

for all $y \in [c, d]$; that is,

$$f(t_\alpha, y) \leq \frac{1}{b-a}\int_a^b f(x, y) \diamond_\alpha x \leq \left[\frac{b-t_\alpha}{b-a} f^r(a, y) + \frac{t_\alpha - a}{b-a} f^r(b, y)\right]^{1/r}. \tag{4.47}$$

Taking the diamond-α integral on both sides of (4.47) over the interval $[c, d]$, applying items 1 and 5 of Theorem 4.8, and dividing by $d - c$, we get

$$\frac{1}{d-c}\int_c^d f(t_\alpha, y) \diamond_\alpha y \leq \frac{1}{(b-a)(d-c)}\int_a^b\int_c^d f(x, y) \diamond_\alpha x \diamond_\alpha y$$
$$\leq \frac{1}{d-c}\int_c^d \left[\frac{b-t_\alpha}{b-a} f^r(a, y)\right.$$
$$\left.+\frac{t_\alpha - a}{b-a} f^r(b, y)\right]^{1/r} \diamond_\alpha y. \tag{4.48}$$

Applying Minkowski's inequality (4.29) on the right-hand side of (4.48), we have

$$\frac{1}{d-c}\int_c^d \left[\frac{b-t_\alpha}{b-a} f^r(a, y) + \frac{t_\alpha - a}{b-a} f^r(b, y)\right]^{1/r} \diamond_\alpha y$$

$$
\begin{aligned}
&\leq \frac{1}{d-c}\left\{\left(\int_c^d \left[\frac{b-t_\alpha}{b-a} f^r(a,y)\right]^{1/r} \diamond_\alpha y\right)^r\right.\\
&\left.+\left(\int_c^d \left[\frac{t_\alpha-a}{b-a} f^r(b,y)\right]^{1/r} \diamond_\alpha y\right)^r\right\}^{1/r}\\
&\leq \frac{1}{d-c}\left\{\left(\int_c^d \left(\frac{b-t_\alpha}{b-a}\right)^{\frac{1}{r}} f(a,y) \diamond_\alpha y\right)^r\right.\\
&\left.+\left(\int_c^d \left(\frac{t_\alpha-a}{b-a}\right)^{\frac{1}{r}} f(b,y) \diamond_\alpha y\right)^r\right\}^{1/r}\\
&\leq \frac{1}{d-c}\left\{\frac{1}{b-a}\left(\int_c^d (b-t_\alpha)^{\frac{1}{r}} f(a,y) \diamond_\alpha y\right)^r\right.\\
&\left.+\frac{1}{b-a}\left(\int_c^d (t_\alpha-a)^{\frac{1}{r}} f(b,y) \diamond_\alpha y\right)^r\right\}^{1/r}\\
&\leq \frac{1}{(b-a)^{\frac{1}{r}}(d-c)}\left\{\left(\int_c^d (b-t_\alpha)^{\frac{1}{r}} f(a,y) \diamond_\alpha y\right)^r\right.\\
&\left.+\left(\int_c^d (t_\alpha-a)^{\frac{1}{r}} f(b,y) \diamond_\alpha y\right)^r\right\}^{1/r}. \qquad (4.49)
\end{aligned}
$$

From (4.48) and (4.49), we have

$$
\begin{aligned}
\frac{1}{d-c}\int_c^d f(t_\alpha,y)\diamond_\alpha y &\leq \frac{1}{(b-a)(d-c)}\int_a^b\int_c^d f(x,y)\diamond_\alpha x \diamond_\alpha y\\
&\leq \frac{1}{(b-a)^{\frac{1}{r}}(d-c)}\left\{\left(\int_c^d (b-t_\alpha)^{\frac{1}{r}} f(a,y) \diamond_\alpha y\right)^r\right.\\
&\left.+\left(\int_c^d (t_\alpha-a)^{\frac{1}{r}} f(b,y) \diamond_\alpha y\right)^r\right\}^{1/r}. \qquad (4.50)
\end{aligned}
$$

Similarly, applying Theorem 4.13 to f_x, we get

$$
f_x(t_\alpha) \leq \frac{1}{d-c}\int_c^d f_y(x)\diamond_\alpha y \leq \left[\frac{d-s_\alpha}{d-c} f_x^r(c) + \frac{s_\alpha-c}{d-c} f_x^r(d)\right]^{1/r},
$$

for all $y \in [a,b]$; that is,

$$
f(x,s_\alpha) \leq \frac{1}{d-c}\int_c^d f(x,y)\diamond_\alpha y \leq \left[\frac{d-s_\alpha}{d-c} f^r(x,c) + \frac{s_\alpha-c}{d-c} f^r(x,d)\right]^{1/r}. \qquad (4.51)
$$

Taking the diamond-α integral on both sides of (4.47) over the interval $[a, b]$, applying items 1 and 5 of Theorem 4.8, and dividing by $b - a$, we get

$$\frac{1}{b-a}\int_a^b f(x, s_\alpha)\diamond_\alpha x \leq \frac{1}{(b-a)(d-c)}\int_a^b\int_c^d f(x, y)\diamond_\alpha x\diamond_\alpha y$$
$$\leq \frac{1}{b-a}\int_a^b\left[\frac{d-s_\alpha}{d-c}f^r(x, c)\right.$$
$$\left.+\frac{s_\alpha - c}{b-a}f^r(x, d)\right]^{1/r}\diamond_\alpha y. \tag{4.52}$$

Similarly, applying Minkowski's inequality (4.29) on the right-hand side of (4.52) and similar to the above argument, we have

$$\frac{1}{b-a}\int_a^b f(x, s_\alpha)\diamond_\alpha x \leq \frac{1}{(b-a)(d-c)}\int_a^b\int_c^d f(x, y)\diamond_\alpha x\diamond_\alpha y$$
$$\leq \frac{1}{(b-a)(d-c)^{\frac{1}{r}}}\left\{\left(\int_a^b (d-s_\alpha)^{\frac{1}{r}} f(x, c)\diamond_\alpha x\right)^r\right.$$
$$\left.+\left(\int_a^b (s_\alpha - c)^{\frac{1}{r}} f(x, d)\diamond_\alpha x\right)^r\right\}^{1/r}. \tag{4.53}$$

Adding relations (4.48) and (4.52), we get

$$\frac{1}{b-a}\int_a^b f(x, s_\alpha)\diamond_\alpha x + \frac{1}{d-c}\int_c^d f(t_\alpha, y)\diamond_\alpha y$$
$$\leq \frac{2}{(b-a)(d-c)}\int_a^b\int_c^d f(x, y)\diamond_\alpha x\diamond_\alpha y$$
$$\leq \frac{1}{(b-a)(d-c)^{\frac{1}{r}}}\left\{\left(\int_a^b (d-s_\alpha)^{\frac{1}{r}} f(x, c)\diamond_\alpha x\right)^r\right.$$
$$\left.+\left(\int_a^b (s_\alpha - c)^{\frac{1}{r}} f(x, d)\diamond_\alpha x\right)^r\right\}^{1/r}$$
$$+\frac{1}{(b-a)^{\frac{1}{r}}(d-c)}\left\{\left(\int_c^d (b-t_\alpha)^{\frac{1}{r}} f(a, y)\diamond_\alpha y\right)^r\right.$$
$$\left.+\left(\int_c^d (t_\alpha - a)^{\frac{1}{r}} f(b, y)\diamond_\alpha y\right)^r\right\}^{1/r}. \tag{4.54}$$

Hence we get the desired inequality (4.46) from (4.54). This completes the proof.

□

Remark 4.7 Let $f : [a, b] \to (0, \infty)$ be a 1-convex function. Then the inequality in Theorem 4.16 reduces to the inequality in [20, Theorem 3.1].

Corollary 4.7 *In Theorem 4.16, if we take* $\alpha = 1$*, then we get the delta version of inequalities* (4.46) *as follows:*

$$\begin{aligned}
&\frac{1}{2}\left[\frac{1}{b-a}\int_a^b f(x, s_\alpha)\Delta x + \frac{1}{d-c}\int_c^d f(t_\alpha, y)\Delta y\right] \\
&\le \frac{1}{(b-a)(d-c)}\int_a^b\int_c^d f(x, y)\Delta x\Delta y \\
&\le \frac{1}{2(b-a)(d-c)^{\frac{1}{r}}}\left\{\left(\int_a^b (d-s_\alpha)^{\frac{1}{r}} f(x, c)\Delta x\right)^r\right. \\
&\left.+\left(\int_a^b (s_\alpha - c)^{\frac{1}{r}} f(x, d)\Delta x\right)^r\right\}^{1/r} \\
&+\frac{1}{2(b-a)^{\frac{1}{r}}(d-c)}\left\{\left(\int_c^d (b-t_\alpha)^{\frac{1}{r}} f(a, y)\Delta y\right)^r\right. \\
&\left.+\left(\int_c^d (t_\alpha - a)^{\frac{1}{r}} f(b, y)\Delta y\right)^r\right\}^{1/r}, \qquad (4.55)
\end{aligned}$$

where $t_1 = \frac{1}{b-a}\int_a^b t\Delta t$*, and* $s_1 = \frac{1}{d-c}\int_c^d s\Delta s$.

Corollary 4.8 *In Theorem 4.16, if we take* $\alpha = 1/2$*, then we get the diamond-1/2 version of inequality* (4.46) *as follows:*

$$\begin{aligned}
&\frac{1}{2}\left[\frac{1}{b-a}\int_a^b f\left(x, \frac{c+d}{2}\right)\diamond_{1/2} x + \frac{1}{d-c}\int_c^d f\left(\frac{a+b}{2}, y\right)\diamond_{1/2} y\right] \\
&\le \frac{1}{(b-a)(d-c)}\int_a^b\int_c^d f(x, y)\diamond_{1/2} x \diamond_{1/2} y \\
&\le \frac{1}{2^{\frac{1+r}{r}}(b-a)}\left\{\left(\int_a^b f(x, c)\diamond_{1/2} x\right)^r + \left(\int_a^b f(x, d)\diamond_{1/2} x\right)^r\right\}^{1/r} \\
&+\frac{1}{2^{\frac{1+r}{r}}(d-c)}\left\{\left(\int_c^d f(a, y)\diamond_{1/2} y\right)^r + \left(\int_c^d f(b, y)\diamond_{1/2} y\right)^r\right\}^{1/r}. \qquad (4.56)
\end{aligned}$$

Proof Applying (4.14) in the definition of s_α and t_α when $\alpha = 1/2$, we get

$$s_{1/2} = \frac{c+d}{2} \qquad (4.57)$$

and

$$t_{1/2} = \frac{a+b}{2}. \tag{4.58}$$

Applying (4.57) and (4.58) in (4.46), we get the desired inequalities (4.56). This completes the proof. □

Corollary 4.9 *In Theorem 4.16, if we take* $\alpha = 0$*, then we get the nabla version of inequality* (4.46) *as follows:*

$$\begin{aligned}
&\frac{1}{2}\left[\frac{1}{b-a}\int_a^b f(x,s_\alpha)\nabla x + \frac{1}{d-c}\int_c^d f(t_\alpha,y)\nabla y\right]\\
&\leq \frac{1}{(b-a)(d-c)}\int_a^b\int_c^d f(x,y)\nabla x\nabla y\\
&\leq \frac{1}{2(b-a)(d-c)^{\frac{1}{r}}}\left\{\left(\int_a^b (d-s_\alpha)^{\frac{1}{r}} f(x,c)\nabla x\right)^r\right.\\
&\left.+\left(\int_a^b (s_\alpha-c)^{\frac{1}{r}} f(x,d)\nabla x\right)^r\right\}^{1/r}\\
&+\frac{1}{2(b-a)^{\frac{1}{r}}(d-c)}\left\{\left(\int_c^d (b-t_\alpha)^{\frac{1}{r}} f(a,y)\nabla y\right)^r\right.\\
&\left.+\left(\int_c^d (t_\alpha-a)^{\frac{1}{r}} f(b,y)\nabla y\right)^r\right\}^{1/r},
\end{aligned} \tag{4.59}$$

where $t_0 = \frac{1}{b-a}\int_a^b t\nabla t$, *and* $s_0 = \frac{1}{d-c}\int_c^d s\nabla s$.

Theorem 4.17 *Let* $\mathbb{T}_1$ *and* $\mathbb{T}_2$ *be two time scales and* $a, b, x \in \mathbb{T}_1$, $c, d, y \in \mathbb{T}_2$ *with* $a < b$, $c < d$ *and* $f : [a,b]\times[c,d] \to \mathbb{R}$ *be such that the partial mappings*

$$f_y : [a,b] \to \mathbb{R},\ f_y(u) = f(u,y) \quad \text{and} \quad f_x : [c,d] \to \mathbb{R},\ f_x(v) = f(x,v),$$

defined for all $y \in [c,d]$ *and* $x \in [a,b]$, *are continuous and* r*-convex with* $r \geq 1$. *Then the following inequalities hold:*

$$\begin{aligned}
&f\left(t_\alpha, \frac{c+d}{2}\right) + f\left(\frac{a+b}{2}, s_\alpha\right)\\
&\leq \frac{1}{b-a}\int_a^b f\left(x,\frac{c+d}{2}\right)\diamond_\alpha x + \frac{1}{d-c}\int_c^d f\left(\frac{a+b}{2},y\right)\diamond_\alpha y
\end{aligned}$$

$$\leq \left[\frac{b-t_\alpha}{b-a} f^r\left(a, \frac{c+d}{2}\right) + \frac{t_\alpha - a}{b-a} f^r\left(b, \frac{c+d}{2}\right)\right]^{1/r}$$

$$+ \left[\frac{d-s_\alpha}{d-c} f^r\left(\frac{a+b}{2}, c\right) + \frac{s_\alpha - c}{d-c} f^r\left(\frac{a+b}{2}, d\right)\right]^{1/r}. \tag{4.60}$$

Proof Since the inequalities in (4.47) hold for all $y \in [c, d]$ and by using the assumption of our theorem, we therefore have that for $y = \frac{c+d}{2}$, the following inequality holds:

$$f\left(t_\alpha, \frac{c+d}{2}\right) \leq \frac{1}{b-a}\int_a^b f\left(x, \frac{c+d}{2}\right) \diamond_\alpha \tag{4.61}$$

$$\leq \left[\frac{b-t_\alpha}{b-a} f^r\left(a, \frac{c+d}{2}\right) + \frac{t_\alpha - a}{b-a} f^r\left(b, \frac{c+d}{2}\right)\right]^{1/r}.$$

Using a similar argument, we get from (4.51)

$$f\left(\frac{a+b}{2}, s_\alpha\right) \leq \frac{1}{d-c}\int_c^d f\left(\frac{a+b}{2}, y\right) \diamond_\alpha y \tag{4.62}$$

$$\leq \left[\frac{d-s_\alpha}{d-c} f^r\left(\frac{a+b}{2}, c\right) + \frac{s_\alpha - c}{d-c} f^r\left(\frac{a+b}{2}, d\right)\right]^{1/r}.$$

By adding (4.61) and (4.62), we get the desired inequality (4.60). This completes the proof. □

Remark 4.8 Let $f\colon [a, b] \to (0, \infty)$ be a 1-convex function. Then the inequality in Theorem 4.17 reduces to the inequality in [20, Theorem 3.5].

Corollary 4.10 *In Theorem 4.17, if we take $\alpha = 1$, then we get the delta version of inequalities* (4.60) *as follows:*

$$f\left(t_1, \frac{c+d}{2}\right) + f\left(\frac{a+b}{2}, s_1\right)$$

$$\leq \frac{1}{b-a}\int_a^b f\left(x, \frac{c+d}{2}\right)\Delta x + \frac{1}{d-c}\int_c^d f\left(\frac{a+b}{2}, y\right)\Delta y$$

$$\leq \left[\frac{b-t_1}{b-a} f^r\left(a, \frac{c+d}{2}\right) + \frac{t_1 - a}{b-a} f^r\left(b, \frac{c+d}{2}\right)\right]^{1/r}$$

$$+ \left[\frac{d-s_1}{d-c} f^r\left(\frac{a+b}{2}, c\right) + \frac{s_1 - c}{d-c} f^r\left(\frac{a+b}{2}, d\right)\right]^{1/r},$$

where $t_1 = \frac{1}{b-a}\int_a^b t\Delta t$, and $s_1 = \frac{1}{d-c}\int_c^d s\Delta s$.

Corollary 4.11 *In Theorem 4.17, if we take* $\alpha = 1/2$*, then we get the diamond-*$1/2$ *version of inequalities* (4.60) *as follows:*

$$\begin{aligned} f\left(\frac{a+b}{2}, \frac{c+d}{2}\right) &\leq \frac{1}{2(b-a)} \int_a^b f\left(x, \frac{c+d}{2}\right) \diamond_{1/2} x \\ &+ \frac{1}{2(d-c)} \int_c^d f\left(\frac{a+b}{2}, y\right) \diamond_{1/2} y \\ &\leq \frac{1}{2^{\frac{1+r}{r}}} \left[f^r\left(a, \frac{c+d}{2}\right) + f^r\left(b, \frac{c+d}{2}\right) \right]^{1/r} \\ &+ \frac{1}{2^{\frac{1+r}{r}}} \left[f^r\left(\frac{a+b}{2}, c\right) + f^r\left(\frac{a+b}{2}, d\right) \right]^{1/r}. \end{aligned} \tag{4.63}$$

Proof Applying (4.14) in the definition of s_α and t_α when $\alpha = 1/2$, we get

$$s_{1/2} = \frac{c+d}{2} \tag{4.64}$$

and

$$t_{1/2} = \frac{a+b}{2}. \tag{4.65}$$

Applying (4.64) and (4.65) in (4.60), we get the desired inequalities (4.63). This completes the proof. □

Corollary 4.12 *In Theorem 4.17, if we take* $\alpha = 0$*, then we get the nabla version of inequalities* (4.60) *as follows:*

$$\begin{aligned} & f\left(t_0, \frac{c+d}{2}\right) + f\left(\frac{a+b}{2}, s_0\right) \\ &\leq \frac{1}{b-a} \int_a^b f\left(x, \frac{c+d}{2}\right) \nabla x + \frac{1}{d-c} \int_c^d f\left(\frac{a+b}{2}, y\right) \nabla y \\ &\leq \left[\frac{b-t_0}{b-a} f^r\left(a, \frac{c+d}{2}\right) + \frac{t_0-a}{b-a} f^r\left(b, \frac{c+d}{2}\right) \right]^{1/r} \\ &+ \left[\frac{d-s_0}{d-c} f^r\left(\frac{a+b}{2}, c\right) + \frac{s_0-c}{d-c} f^r\left(\frac{a+b}{2}, d\right) \right]^{1/r}, \end{aligned}$$

where $t_0 = \frac{1}{b-a} \int_a^b t \nabla t$*, and* $s_0 = \frac{1}{d-c} \int_c^d s \nabla s$*.*

Theorem 4.18 *Let* $\mathbb{T}_1$ *and* $\mathbb{T}_2$ *be two time scales and* $a, b, x \in \mathbb{T}_1$*,* $c, d, y \in \mathbb{T}_2$ *with* $a < b$*,* $c < d$ *and* $f : [a, b] \times [c, d] \to \mathbb{R}$ *be such that the partial mappings*

$$f_y : [a, b] \to \mathbb{R},\ f_y(u) = f(u, y) \quad \text{and} \quad f_x : [c, d] \to \mathbb{R},\ f_x(v) = f(x, v),$$

defined for all $y \in [c, d]$ *and* $x \in [a, b]$, *be continuous and* r*-convex with* $r \geq 1$. *Then the following inequalities hold:*

$$\frac{1}{b-a}\int_a^b [f(x,c)+f(x,d)] \diamond_\alpha x + \frac{1}{d-c}\int_c^d [f(a,y)+f(b,y)] \diamond_\alpha y$$
$$\leq A_1^{1/r} f(a,c) + A_2^{1/r} f(a,d) + A_3^{1/r} f(b,c) + A_4^{1/r} f(b,d), \tag{4.66}$$

where

$$A_1 = \frac{b-t_\alpha}{b-a} + \frac{d-t_\alpha}{d-c},$$
$$A_2 = \frac{b-t_\alpha}{b-a} + \frac{s_\alpha-c}{d-c},$$
$$A_3 = \frac{t_\alpha-a}{b-a} + \frac{d-s_\alpha}{d-c},$$
$$A_4 = \frac{t_\alpha-a}{b-a} + \frac{s_\alpha-c}{d-c}.$$

Proof From (4.47) and by using the inequality (4.30), since $r \geq 1 \Rightarrow 0 < \frac{1}{r} \leq 1$, we get

$$\frac{1}{b-a}\int_a^b f(x,c) \diamond_\alpha x \leq \left[\frac{b-t_\alpha}{b-a} f^r(a,c) + \frac{t_\alpha-a}{b-a} f^r(b,c)\right]^{1/r} \tag{4.67}$$
$$\leq \left(\frac{b-t_\alpha}{b-a}\right)^{1/r} f(a,c) + \left(\frac{t_\alpha-a}{b-a}\right)^{1/r} f(b,c),$$

$$\frac{1}{b-a}\int_a^b f(x,d) \diamond_\alpha x \leq \left[\frac{b-t_\alpha}{b-a} f^r(a,d) + \frac{t_\alpha-a}{b-a} f^r(b,d)\right]^{1/r} \tag{4.68}$$
$$\leq \left(\frac{b-t_\alpha}{b-a}\right)^{1/r} f(a,d) + \left(\frac{t_\alpha-a}{b-a}\right)^{1/r} f(b,d).$$

Also, from (4.51) and by using the inequality (4.30), since $r \geq 1 \Rightarrow 0 < \frac{1}{r} \leq 1$, we get

$$\frac{1}{d-c}\int_c^d f(a,y) \diamond_\alpha y \leq \left[\frac{d-s_\alpha}{d-c} f^r(a,c) + \frac{s_\alpha-c}{d-c} f^r(a,d)\right]^{1/r} \tag{4.69}$$
$$\leq \left(\frac{d-s_\alpha}{d-c}\right)^{1/r} f(a,c) + \left(\frac{s_\alpha-c}{d-c}\right)^{1/r} f(a,d),$$

$$\frac{1}{d-c}\int_c^d f(b,y)\diamond_\alpha y \le \left[\frac{d-s_\alpha}{d-c}f^r(b,c)+\frac{s_\alpha-c}{d-c}f^r(b,d)\right]^{1/r} \tag{4.70}$$
$$\le \left(\frac{d-s_\alpha}{d-c}\right)^{1/r} f(b,c)+\left(\frac{s_\alpha-c}{d-c}\right)^{1/r} f(b,d).$$

Thus, from (4.67), (4.68), (4.69), and (4.70) by addition, we get the desired inequality (4.66). This completes the proof. □

Corollary 4.13 *In Theorem 4.18, if we take* $\alpha = 1$*, then we get the delta version of inequalities* (4.60) *as follows:*

$$\frac{1}{b-a}\int_a^b [f(x,c)+f(x,d)]\Delta x+\frac{1}{d-c}\int_c^d [f(a,y)+f(b,y)]\Delta y$$
$$\le \Upsilon_1^{1/r} f(a,c)+\Upsilon_2^{1/r} f(a,d)+\Upsilon_3^{1/r} f(b,c)+\Upsilon_4^{1/r} f(b,d), \tag{4.71}$$

where

$$\Upsilon_1 = \frac{b-t_1}{b-a}+\frac{d-t_1}{d-c},$$
$$\Upsilon_2 = \frac{b-t_1}{b-a}+\frac{s_1-c}{d-c},$$
$$\Upsilon_3 = \frac{t_1-a}{b-a}+\frac{d-s_1}{d-c},$$
$$\Upsilon_4 = \frac{t_1-a}{b-a}+\frac{s_1-c}{d-c},$$

and $t_1 = \frac{1}{b-a}\int_a^b t\Delta t$, *and* $s_1 = \frac{1}{d-c}\int_c^d s\Delta s$.

Corollary 4.14 *In Theorem 4.18, if we take* $\alpha = 1/2$*, then we get the delta version of inequalities* (4.60) *as follows:*

$$\frac{1}{b-a}\int_a^b [f(x,c)+f(x,d)]\diamond_{1/2} x+\frac{1}{d-c}\int_c^d [f(a,y)+f(b,y)]\diamond_{1/2} y$$
$$\le f(a,c)+f(a,d)+f(b,c)+f(b,d). \tag{4.72}$$

Proof Applying (4.14) in the definition of s_α and t_α when $\alpha = 1/2$, we get

$$s_{1/2} = \frac{c+d}{2} \tag{4.73}$$

and

$$t_{1/2} = \frac{a+b}{2}. \tag{4.74}$$

Applying (4.73) and (4.74) in (4.60), $A_1 = A_2 = A_3 = A_4 = 1$, and therefore, we get the desired inequalities (4.63). This completes the proof. □

Corollary 4.15 *In Theorem 4.18, if we take $\alpha = 0$, then we get the delta version of inequalities* (4.60) *as follows:*

$$\frac{1}{b-a}\int_a^b [f(x,c)+f(x,d)]\nabla x + \frac{1}{d-c}\int_c^d [f(a,y)+f(b,y)]\nabla y$$
$$\leq \hat{\Upsilon}_1^{1/r} f(a,c) + \hat{\Upsilon}_2^{1/r} f(a,d) + \hat{\Upsilon}_3^{1/r} f(b,c) + \hat{\Upsilon}_4^{1/r} f(b,d),$$

where

$$\hat{\Upsilon}_1 = \frac{b-t_0}{b-a} + \frac{d-t_0}{d-c},$$
$$\hat{\Upsilon}_2 = \frac{b-t_0}{b-a} + \frac{s_0-c}{d-c},$$
$$\hat{\Upsilon}_3 = \frac{t_0-a}{b-a} + \frac{d-s_0}{d-c},$$
$$\hat{\Upsilon}_4 = \frac{t_0-a}{b-a} + \frac{s_0-c}{d-c},$$

and $t_0 = \frac{1}{b-a}\int_a^b t\nabla t$, *and* $s_0 = \frac{1}{d-c}\int_c^d s\nabla s$.

Remark 4.9 Let $f : [a, b] \to (0, \infty)$ be a 1-convex function. Then the inequalities in Theorem 4.18 and Corollary 4.13 reduce to the inequalities in [20, Theorem 3.5 and Corollary 3.10], respectively.

Remark 4.10 Let $f : [a, b] \to (0, \infty)$ be a 1-convex function. If we take $\mathbb{T}_1 = \mathbb{T}_2 = \mathbb{R}$ in Corollaries (4.11), (4.10), and (4.13) and combine the resultant inequalities, we get [11, Theorem 1.1] due to Dragomir.

4.4 An Application

In this section, we present some applications of Corollary 4.13. We consider the following example.

Example 4.1 Let $\mathbb{T}_1 = \mathbb{T}_2 = \mathbb{Z}$ and $f : [0, 2] \times [1, 3] \to \mathbb{R}$ be r-convex and continuous on the coordinates. Then the following inequality holds:

$$f(0,2) + f(1,1) + f(1,3) + f(2,2) \leq f(0,1) + f(0,3) + f(2,1) + f(2,3).$$

Justification: The left-hand side of Corollary 4.13 gives

$$\frac{1}{2}\int_{0}^{2}[f(x,1)+f(x,3)]\Delta x+\frac{1}{2}\int_{1}^{3}[f(0,y)+f(2,y)]\Delta y$$

$$\leq \frac{1}{2}\sum_{x=0}^{1}[f(x,1)+f(x,3)]+\frac{1}{2}\sum_{y=1}^{2}[f(0,y)+f(2,y)]$$

$$= f(0,1)+\frac{1}{2}\Big[f(0,2)+f(0,3)+f(1,1)$$

$$+f(1,3)+f(2,1)+f(2,2)\Big]. \tag{4.75}$$

By a simple computation, we get $t_1 = \frac{1}{2}$ and $s_1 = \frac{3}{2}$.

$$\hat{\Upsilon}_1 = \left(\frac{3}{2}\right)^{1/r}, \hat{\Upsilon}_2 = 1 = \hat{\Upsilon}_3, \hat{\Upsilon}_4 = \left(\frac{1}{2}\right)^{1/r}.$$

Hence, the right-hand side of Corollary 4.13 amounts to

$$\left(\frac{3}{2}\right)^{1/r} f(0,1)+f(0,3)+f(2,1)+\left(\frac{1}{2}\right)^{1/r} f(2,3). \tag{4.76}$$

We obtain the desired inequality by combining (4.75) and (4.76) in the spirit of Corollary 4.13.

References

1. R.P. Agarwal, D. O'Regan, S. Saker, *Dynamic Inequalities on Time Scale* (Springer, Berlin, 2014)
2. M. Bessenyei, Hermite-Hadamard-type inequalities for generalized 3-convex functions. Publ. Math. (Debr.) **65**(1–2), 223–232 (2004)
3. M. Bohner, G.S. Guseinov, Partial differentiation on time scales. Dyn. Syst. Appl. **13**(3–4), 351–379 (2004)
4. M. Bohner, G.S. Guseinov, Multiple integration on time scales. Dyn. Syst. Appl. **13**(3–4), 579–606 (2005)
5. M. Bohner, G.S. Guseinov, Double integral calculus of variations on time scales. Comput. Math. Appl. **54**, 45–57 (2007)
6. M. Bohner, A. Peterson, *Dynamic Equations on Time Scales: An Introduction with Applications* (Birkhäuser Boston, Inc., Boston, 2001), x+358p.
7. M. Bohner, A. Peterson, *Advances in Dynamic Equations on Time Scales* (Birkhäuser Boston, Inc., Boston, 2003), xii+348p.
8. P. Czinder, A weighted Hermite-Hadamard-type inequality for convex symmetric function. Publ. Math. Debrecen **68**, 215–224 (2006)

9. P. Czinder, Z. Páles, An extension of the Hermite-Hadamard inequality and an application for Gini and Stolarsky means. JIPAM **5**(42), 1–8 (2004)
10. C. Dinu, Hermite-Hadamard inequality on time scales. J. Inequal. Appl. (2008), Art. ID 287947, 24pp.
11. S.S. Dragomir, On Hadamard's inequality for convex functions on the coordinates in a rectangle from the plane. Taiwan. J. Math. **4**, 775–788 (2001)
12. L. Fejér, Uber die Foirierrehen. Math. Naturwiss. Anz. Ungar. Akad. Wiss. II **24**, 360–390 (1906)
13. A.M. Fink, A best possible Hadamard inequality. Math. Inequal. Appl. **1**, 223-230 (1998)
14. A. Florea, C.P. Niculescu, A Hermite-Hadamard inequality for convex-concave symmetric functions. Bull. Math. Soci. Sci. Math. Roumanie **2**, 149–156 (2007)
15. S. Hilger, *Ein maßkettenkalkül mit anwendung auf zentrumsmannigfaltigkeiten* (Universität Würzburg, Würzburg,1988)
16. J.L.W.V. Jensen, Sur les fonctions convexes et les inégalités entre les valeurs moyennes. Acta Math. **30**(1), 175–193 (1906)
17. W.J. Liu, A. Tuna, Y. Jiang, Diamond-α weighted Ostrowski type and Gruss type inequalities on time scales. Appl. Math. Comput. **270**, 251–260 (2015)
18. N. Ngoc, N.V. Vinh, P.T. Hien, Integral inequalities of Hadamard type for r-convex functions. Math. Naturwiss. Anz. Ungar. Akad. Wiss. **35**(4), 172–178 (2009)
19. C.P. Niculescu, L.-E. Persson, *Convex Functions and their Applications. A Contemporary Approach* (Springer, New York, 2006)
20. E.R. Nwaeze, Time scale version of the Hermite–Hadamard inequality for functions convex on the coordinates (2017). Preprint. arXiv:1706.07995, 13p.
21. U.M. Özkan, B. Kaymakc, Basics of diamond-α partial dynamic calculus on time scales. Math. Comput. Model. **50**, 1253–1261 (2009)
22. M. Pearce, J. Peccaric, P. Josip, V. Simic, Stolarsky means and Hadamard's inequality. J. Math. Anal. Appl. **220**(1), 99–109 (1998)
23. J.E. Pecaric, F. Proschan, Y.L. Tong, *Convex Functions, Partial Orderings and Statistical Applications* (Academic Press, Cambridge, 1991)
24. Q. Sheng, M. Fadag, J. Henderson, J.M. Davis, An exploration of combined dynamic derivatives on time scales and their applications. Nonlinear Anal. Real World Appl. **7**(3), 395–413 (2006)
25. G. Yang, Refinements of Hadamard inequality for r-convex functions. Indian J. Pure Appl. Math. **10**(32), 1571–1579 (2001)
26. G. Zabandan, A. Bodaghi, A. Kılıçman, The Hermite-Hadamard inequality for r-convex functions. J. Ineq. Appl. **1**, 215p. (2012)

Chapter 5
Existence of Solutions and Continuous Dependence on Parameters for a Class of Measure Differential Equations

Claudio A. Gallegos, Hernán Henríquez, and Jaqueline Godoy Mesquita

Abstract This chapter is devoted to the study of the qualitative properties for a special class of measure differential equations (MDEs, for short). We establish several results on the existence of global solutions, including the existence of regulated, continuous, differentiable, and S-asymptotically ω-periodic solutions. Furthermore, we present a result on continuous dependence of solutions in terms of parameters. Our results are based on extensions of Krasnoselskii's fixed point theorem.

Keywords Measure differential equations · Kurzweil-Stieltjes integral · Existence of solutions · S-asymptotically ω-periodic functions · Continuous dependence on parameters

2020 MSC 45M99, 34A06, 34A38

5.1 Introduction

This work is dedicated to the study of qualitative properties for a class of measure differential equation—abbreviated, MDEs—of type

C. A. Gallegos
Departamento de Matemáticas, Universidad de Chile, UCH, Santiago, Chile

H. Henríquez
Departamento de Matemática, Universidad de Santiago de Chile, Santiago, Chile
e-mail: hernan.henriquez@usach.cl

J. G. Mesquita (✉)
Department of Mathematics, University of Brasília, Brasilia, Brazil
e-mail: jgmesquita@unb.br

P. Benevieri, J. G. Mesquita (eds.), *Functional Differential Equations and Dynamic Equations on Time Scales*, Advances in Continuum Mechanics 52,
https://doi.org/10.1007/978-3-031-83327-4_5

$$x(t) = x_0 - h(t_0, x(t_0)) + h(t, x(t)) + \int_{t_0}^{t} f(s, x(s))\mathrm{d}g(s), \quad t \in J, \tag{5.1}$$

$$x(t_0) = x_0 \in \mathbb{R}^n, \tag{5.2}$$

where $J = [t_0, t_0 + a]$ with $t_0 \in \mathbb{R}$ and $a > 0$. In this problem, $f, h\colon J \times \mathbb{R}^n \to \mathbb{R}^n$ and $g\colon J \to \mathbb{R}$ denote functions whose properties will be described later, and the integral on the right-hand side is considered in the sense of Kurzweil-Stieltjes.

Specifically, our main concern in this work is the study of existence of solutions for problem MDE (5.1) and (5.2). We analyze this equation by imposing certain technical conditions over the function f, most of which are related to the concept of integration in the sense of Kurzweil-Stieltjes, see conditions (A1), (A2), and (A2*) in Sect. 5.3. Further, we consider the function h satisfying a nonlinear contraction assumption, see conditions (A3) and (A3*) in Sect. 5.3. The abovementioned conditions allow us to guarantee the existence of solutions for problem MDE (5.1) and (5.2), see Theorems 5.6 and 5.7. Another goal in this chapter is to establish special classes of solutions for problem MDE (5.1) and (5.2), including continuous and differentiable solutions, see Propositions 5.1 and 5.2 in Sect. 5.4. Additionally, we study solutions defined on unbounded intervals in order to establish S-asymptotically ω-periodic solutions for problem MDE (5.1) and (5.2), see Theorems 5.8 and 5.9. We emphasize that this concept of periodicity has yet to be considered in the framework of measure differential equations. Besides of the aforementioned existence results, we establish a theorem of continuous dependence on parameters for problem MDE (5.1) and (5.2).

Some of the earliest contributors in the theory of MDEs were W. W. Schmaedeke, R. R. Sharma, and P. C. Das (see [10, 37, 39]). Over the years, several authors have developed and motivated the qualitative theory of these equations and principally focused on the existence of solutions, stability theory, and applications to other types of systems exhibiting discontinuous behavior, see, for instance, [9, 13, 14, 17, 18, 21, 30, 32, 36]. One of the interests in studying this type of equations relies on its formulation, which can include two types of problems widely studied in the specialized literature nowadays, such as is the case of dynamic equations on time scales and impulsive equations.

In relation to dynamic equations, the notion of time scales has been intensively developed in dynamical systems and related topics. Some of the reasons are the capacity of this theory to unify the discrete and continuous analysis and include some other developments such as quantum analysis and its applications to the modeling of strongly nonlinear systems. We refer to [1, 6, 7, 24]. Recently, an interesting theoretical point of view was considered in the time scales setting. In [40], A. Slavík gave a first connection between measure differential equations and dynamic equations on time scales. Henceforth, this innovating approach has been useful to develop the theory of dynamic equations on time scales from a general perspective, allowing to consider more general conditions than the usual ones. We refer to [16–19, 21, 22] for recent advances.

Let us consider a time scale $\mathbb{T}$, i.e., a nonempty closed subset of $\mathbb{R}$, and let Δ be the operator delta derivative. The analog of the problem MDE (5.1) and (5.2) in the framework of time scales can be formulated by

$$[x(t) - h(t, x(t))]^{\Delta} = f(t, x(t)), \quad t \in J \cap \mathbb{T}, \tag{5.3}$$

for which, under appropriate assumptions (see [41]), its integral form is given by

$$x(t) = x_0 - h(t_0, x(t_0)) + h(t, x(t)) + \int_{t_0}^{t} f(t, x(t))\Delta s, \quad t \in J \cap \mathbb{T}. \tag{5.4}$$

As previously mentioned, due to the development carried out in [40], we can insert equation (5.4) in the context of measure differential equations of type (5.1). In consequence, the qualitative theory developed in this chapter can be applied to dynamic equations of type (5.4). For instance, Theorems 5.6 and 5.7 improve the existence results founded in [41], and also it is possible to provide new existence results of S-asymptotically ω-periodic solutions for (5.4).

Furthermore, it is also possible to include as a special case of measure differential equations the so-called *impulsive differential equations*. Indeed, it is a known fact that measure differential equations of type (5.1) encompass the following hybrid impulsive system:

$$[x(t) - h(t, x(t))]' = f(t, x(t)), \quad t \neq \tau_j, \tag{5.5}$$

$$\Delta^{+}x(\tau_j) = I_j(x(\tau_j)), \quad j \in \{1, \ldots, m\}, \tag{5.6}$$

where $f, h\colon J \times \mathbb{R}^n \to \mathbb{R}^n$, $I_j\colon \mathbb{R}^n \to \mathbb{R}^n$ for each $j \in \{1, \ldots, m\}$, $\{\tau_j\}_{j\in\{1,\ldots,m\}}$ is an increasing real sequence, and $\Delta^{+}x(\tau_j) = x(\tau_j^{+}) - x(\tau_j)$.

In the classical theory of ordinary differential equations, Eq. (5.5) is called hybrid differential equation with linear perturbation of second type. We refer the reader to [12, 31, 41], and the references therein for details on this topic.

Under appropriate assumptions, we can rewrite (5.5) and (5.6) in the following integral form:

$$x(t) = x(t_0) + h(t, x(t)) - h(t_0, x(t_0)) + \int_{t_0}^{t} f(s, x(s))\,\mathrm{d}s + \sum_{\substack{j\in\{1,\ldots,m\},\\ t_0 \leqslant \tau_j < t}} I_j(x(\tau_j)), \tag{5.7}$$

where the integral on the right-hand side is considered in the Kurzweil sense. Note that Eq. (5.7) is a special case of the measure differential equation (5.1). For more details about this aspect, we mention [17, 33]. Moreover, using the correspondence found in [17], it is possible to regard measure differential equations of type (5.1) to study the hybrid impulsive dynamic equation on time scales given by

$$x(t) = x(t_0) + h(t, x(t)) - h(t_0, x(t_0)) + \int_{t_0}^{t} f(s, x(s))\Delta s + \sum_{\substack{j\in\{1,\ldots,m\},\\ t_0\leqslant\tau_j<t}} I_j(x(\tau_j)), \tag{5.8}$$

where the integral on the right-hand side of (5.8) represents the Δ-integral on the time scale in the sense of Kurzweil. Consequently, the qualitative theory obtained in this chapter can also be translated to these hybrid impulsive cases. The equations given by (5.7) and (5.8) have many important applications, since they can describe models that have abrupt state changes. This type of behavior can be found in many different situations and phenomena such as bursting rhythm models in medicine, frequency modulated systems, and ingestion of medicine, among others. We refer to [2–5, 29], and the references therein.

We point out that measure differential equations play an important role for models in continuum mechanics, since they encompass impulsive differential equations and dynamic equations on time scales. It is a known fact that in Continuum Mechanics, systems often experience sudden changes or discontinuities due to external forces or interactions and equations with impulses allow us to model these events accurately by incorporating sudden changes in the system's behavior at specific time points. This is crucial for understanding and predicting the response of materials and structures to sudden impacts, collisions, or other discontinuous phenomena.

Equations with impulses are also important for designing control strategies in Continuum Mechanics, particularly for systems that require rapid responses to external disturbances or inputs. By incorporating impulse-like control actions into the mathematical model, we can develop effective control algorithms that can quickly stabilize or manipulate the system in real time, ensuring desired performance and stability criteria are met.

This chapter is organized in five sections. In Sect. 5.2, a brief summary of the Kurzweil-Stieltjes integral and properties that we will use later is included. In Sect. 5.3, we provide existence results for problem MDE (5.1) and (5.2). In Sect. 5.4, we investigate the existence of solutions with special properties as continuity, differentiability, and the existence of S-asymptotically ω-periodic solutions. In Sect. 5.5, we provide a continuous dependence on parameters result for problem MDE (5.1) and (5.2).

5.2 Preliminaries

In this section, we present some basic concepts of Kurzweil-Stieltjes integration theory. For more details, the reader can see [33, 38]. Throughout this text, X will always denote a Banach space endowed with a norm $\|\cdot\|_X$, and $[a, b] \subset \mathbb{R}$ is a compact interval. In particular, for $X = \mathbb{R}^n$, we write $\|\cdot\|$ instead of $\|\cdot\|_{\mathbb{R}^n}$. In addition, we use the following notation: $\mathbb{R}^+ = (0, +\infty)$ and $\mathbb{R}_0^+ = [0, +\infty)$.

A function $\delta\colon [a, b] \to \mathbb{R}^+$ is called a *gauge* on $[a, b]$. If δ is a gauge on $[a, b]$, a *tagged partition* of the interval $[a, b]$ with subdivision points $a = s_0 \leqslant s_1 \leqslant \cdots \leqslant s_k = b$, and *tags* $\tau_i \in [s_{i-1}, s_i]$, $i = 1, \ldots, k$, is called *δ-fine* if

$$[s_{i-1}, s_i] \subset (\tau_i - \delta(\tau_i), \tau_i + \delta(\tau_i)), \quad i = 1, \ldots, k.$$

It is important to mention that given a gauge $\delta\colon [a, b] \to \mathbb{R}^+$, it is always possible to obtain a δ-fine tagged partition of the interval $[a, b]$, see [33, Cousin's lemma 6.1.3].

Definition 5.1 ([33, Definition 6.1.2]) We say that a function $f\colon [a, b] \to \mathbb{R}^n$ is *Kurzweil-Stieltjes integrable* on $[a, b]$ with respect to a function $g\colon [a, b] \to \mathbb{R}$ if there is a vector $\mathcal{I} \in \mathbb{R}^n$ such that for every $\varepsilon > 0$, there exists a gauge $\delta\colon [a, b] \to \mathbb{R}^+$ such that

$$\left\| \sum_{i=1}^{k} f(\tau_i)\,(g(s_i) - g(s_{i-1})) - \mathcal{I} \right\| < \varepsilon$$

for all δ-fine tagged partition of $[a, b]$. In this case, $\mathcal{I}$ is called the *Kurzweil-Stieltjes integral of f with respect to g over* $[a, b]$, and it will be denoted by $\int_a^b f(s)\mathrm{d}g(s)$ or just by $\int_a^b f\mathrm{d}g$.

As it should be expected, the Kurzweil-Stieltjes integral satisfies the usual properties of linearity, integrability on subintervals, and additivity with respect to adjacent intervals, among other usual properties in integration theory (see [33]).

When g is the identity function, R. Henstock in 1961 (see [27]) introduced another concept of integral for real-valued functions which is equivalent to the integral in the sense of Kurzweil. In the technical literature, due to its equivalence with the integral of Kurzweil, this integral is known as *Henstock-Kurzweil integral* or *gauge integral* (see [28]). The relationship between different integration concepts can be summarized in the following chain of inclusions:

$$\mathcal{R}([a, b], \mathbb{R}) \subset \mathcal{L}_1([a, b], \mathbb{R}) \subset H([a, b], \mathbb{R}) = \mathcal{K}([a, b], \mathbb{R}),$$

where $\mathcal{R}([a, b], \mathbb{R})$ denotes the space of Riemann integrable functions, $\mathcal{L}_1([a, b], \mathbb{R})$ stands for the space of Lebesgue integrable functions, $H([a, b], \mathbb{R})$ denotes the space of Henstock integrable functions, and $\mathcal{K}([a, b], \mathbb{R})$ denotes the space of Kurzweil integrable functions.

To complete these general comments on the Kurzweil-Stieltjes integral, we point out that the Kurzweil-Stieltjes definition allows us to integrate a broad class of functions. For instance, as it was explained previously, if $g(t) \equiv t$, then we obtain the Henstock-Kurzweil integral, allowing us to integrate highly oscillating functions. On the other hand, the Kurzweil-Stieltjes integral allows us to integrate a function f with respect to g even when both of them have points of discontinuity in common, which we know that is not possible for the integral in the Riemann-Stieltjes sense. Therefore, for all these reasons, it seems most convenient to work with the Kurzweil-Stieltjes integral when continuity of both functions f and g is not required. The relation among other types of Stieltjes integrals and the Kurzweil-Stieltjes can be found in [33].

In the rest of this section, we develop in detail some fundamental technical aspects to justify the results presented in the following sections. Initially, we recall the concept of regulated function which plays an important role in this work.

Definition 5.2 A function $f\colon [a,b] \to X$ is called *regulated* if the limits below exist

$$\lim_{s\to t^-} f(s) = f(t^-) \text{ for } t \in (a,b] \text{ and } \lim_{s\to t^+} f(s) = f(t^+) \text{ for } t \in [a,b).$$

The space consisting of all regulated functions $f\colon [a,b] \to X$ will be denoted by $G([a,b],X)$, and it is a Banach space under the usual norm of the uniform convergence $\|f\|_\infty = \sup_{a\leqslant t\leqslant b} \|f(t)\|_X$.

In addition, we recall that the vector space consisting of all functions $f\colon [a,b] \to X$ of bounded variation on $[a,b]$, denoted by $BV([a,b],X)$, is a Banach space endowed with the norm

$$\|f\|_{BV} = \|f(a)\|_X + \mathrm{var}_a^b(f), \quad f \in BV([a,b],X),$$

where $\mathrm{var}_a^b(f)$ denotes the variation of the function f on $[a,b]$. In the case $X=\mathbb{R}$, we abbreviate the notation by writing $BV([a,b])$ instead of $BV([a,b],\mathbb{R})$. It is well known that $BV([a,b],X) \subset G([a,b],X)$.

The next result yields sufficient conditions to ensure the existence of the Kurzweil-Stieltjes integral.

Theorem 5.1 ([38, Corollary 1.34]) *Let $f\colon [a,b] \to \mathbb{R}^n$ be a regulated function, and let $g\colon [a,b] \to \mathbb{R}$ be a bounded variation function. Then the integral $\int_a^b f\mathrm{d}g$ exists, and*

$$\left\| \int_a^b f(s)\mathrm{d}g(s) \right\| \leqslant \|f\|_\infty \mathrm{var}_a^b(g).$$

The following results establish important properties of the Kurzweil-Stieltjes integral.

Theorem 5.2 ([33, Corollary 6.5.4]) *Let $f\colon [a,b] \to \mathbb{R}^n$ and $g\colon [a,b] \to \mathbb{R}$ be a pair of functions such that g is regulated, and the integral $\int_a^b f\mathrm{d}g$ exists. Then the function $p\colon [a,b] \to \mathbb{R}^n$ given by*

$$p(t) = \int_a^t f(s)\mathrm{d}g(s), \quad t \in [a,b]$$

is regulated and satisfies

$$p(t^+) = p(t) + f(t)\Delta^+ g(t), \quad t \in [a, b),$$

$$p(t^-) = p(t) - f(t)\Delta^- g(t), \quad t \in (a, b],$$

where $\Delta^+ g(t) = g(t^+) - g(t)$ *and* $\Delta^- g(t) = g(t) - g(t^-)$.

Remark 5.1 In the case that $g: [a, b] \to \mathbb{R}$ is a regulated function, the Kurzweil-Stieltjes integral $\int_a^b f \mathrm{d}g$ exists for $f \in BV([a, b], \mathbb{R}^n)$, see [33, Theorem 6.3.11].

The subsequent definition will allow us to appropriately manage sets of regulated functions.

Definition 5.3 ([20]) A set $\mathcal{F} \subset G([a, b], \mathbb{R}^n)$ is said to be *equiregulated* if for every $\varepsilon > 0$ the following conditions hold:

(i) For every $\tau_0 \in (a, b]$, there exists $\delta > 0$ such that $\|x(t) - x(\tau_0^-)\| < \varepsilon$ for all $\tau_0 - \delta < t < \tau_0$ and all $x \in \mathcal{F}$.
(ii) For every $\tau_0 \in [a, b)$, there exists $\delta > 0$ such that $\|x(s) - x(\tau_0^+)\| < \varepsilon$ for all $\tau_0 < s < \tau_0 + \delta$ and all $x \in \mathcal{F}$.

Next we present a type of Arzelà-Ascoli Theorem to characterize the compact sets in the space of regulated functions.

Lemma 5.1 ([20, Corollary 2.4]) *A set* $\mathcal{F} \subset G([a, b], \mathbb{R}^n)$ *is relatively compact if and only if for every* $t \in [a, b]$ *the set* $\{x(t) : x \in \mathcal{F}\}$ *is bounded in* $\mathbb{R}^n$ *and* $\mathcal{F}$ *is equiregulated.*

We finish this section with a generalization of Lebesgue's Dominated Convergence Theorem.

Theorem 5.3 ([33, Theorem 6.8.10]) *Let* $g \in BV([a, b])$ *and let* $f, f_k : [a, b] \to \mathbb{R}^n$, $k \in \mathbb{N}$, *be functions such that the integral* $\int_a^b f_k \mathrm{d}g$ *exists for every* $k \in \mathbb{N}$ *and* $\lim_{k \to +\infty} f_k(t) = f(t)$ *for each* $t \in [a, b]$. *Assume that there exists a positive constant* $K > 0$ *such that for every subdivision* $a = \sigma_0 < \sigma_1 < \ldots < \sigma_l = b$ *of the interval* $[a, b]$ *and every finite subset* $\{m_1, m_2, \ldots, m_l\}$ *of* $\mathbb{N}$, *the inequality*

$$\left\| \sum_{j=1}^{l} \int_{\sigma_{j-1}}^{\sigma_j} f_{m_j}(s) \mathrm{d}g(s) \right\| < K$$

holds. Then the integral $\int_a^b f \mathrm{d}g$ *exists, and*

$$\lim_{k \to +\infty} \int_a^b f_k \mathrm{d}g = \int_a^b f \mathrm{d}g.$$

5.3 Existence Results

In this section, we present some results on the existence of solutions for problem MDE (5.1) and (5.2). In order to state our results, we will make use of some additional concepts and a variant of Krasnoselskii's fixed point theorem.

Definition 5.4 A continuous nondecreasing function $\varphi\colon \mathbb{R}_0^+ \to \mathbb{R}_0^+$ is said to be a $\mathfrak{D}$-*function* if $\varphi(0) = 0$ and $\varphi(s) > 0$ for $s > 0$.

Definition 5.5 An operator $T : X \to X$ is called *nonlinear contraction* if there is a $\mathfrak{D}$-function φ satisfying $\varphi(t) < t$ for $t > 0$, and

$$\|Tx - Tz\|_X \leqslant \varphi(\|x - z\|_X),$$

for all $x, z \in X$.

Definition 5.5 is taken from [34]. Originally, the concept of nonlinear contraction appears in [8], in which the function φ is considered upper semicontinuous from the right and satisfying $\varphi(t) < t$ for $t > 0$.

Next, we recall some known concepts of compactness for nonlinear operators. See [35] for details.

Definition 5.6 An operator $T\colon X \to X$ is said to be:

- *Compact* if $T(X)$ is a relatively compact subset of X.
- *Totally bounded* if $T(E)$ is a relatively compact subset of X for any bounded subset $E \subset X$.
- *Completely continuous* if it is continuous and totally bounded.

We next recall an extension of Krasnoselskii's fixed point theorem, which is essential to establish our results in this section. This extension was achieved by using an improvement of the Banach contraction mapping principle. The reader can see [8] for a discussion about this issue.

Theorem 5.4 ([34, Theorem 1]) *Let S be a closed, convex, and bounded subset of a Banach space X, and let $A\colon S \to X$, $B\colon S \to X$ be two operators which satisfy the following conditions:*

(a) *A is a nonlinear contraction.*
(b) *B is completely continuous.*
(c) *$Au + Bv \in S$ for all $u, v \in S$.*

Then the operator equation $Au + Bu = u$ has a solution in S.

In the rest of this section, we assume that $J = [t_0, t_0 + a]$ with $t_0 \in \mathbb{R}$ and $a > 0$, and the function $g\colon J \to \mathbb{R}$ is nondecreasing and left-continuous. To state our results, we introduce a set of additional conditions about $h, f\colon J \times \mathbb{R}^n \to \mathbb{R}^n$.

(A0) For every $t \in J$, the function $f(t, \cdot)$ is continuous.

(A1) For every $u \in \mathbb{R}^n$, the function $f(\cdot, u)$ is Kurzweil-Stieltjes integrable w.r.t. g on J.

(A2) There exists a function $M: J \to \mathbb{R}$ which is Kurzweil-Stieltjes integrable w.r.t. g such that

$$\left\| \int_c^d f(s, u) \mathrm{d}g(s) \right\| \leqslant \int_c^d M(s) \mathrm{d}g(s),$$

for all $u \in \mathbb{R}^n$ and $[c, d] \subseteq J$.

(A3) For every $v \in \mathbb{R}^n$, the function $h(\cdot, v)$ is regulated on J, and there exists a $\mathfrak{D}$-function $\varphi: \mathbb{R}_0^+ \to \mathbb{R}_0^+$ with $\varphi(t) < t$ such that

$$\|h(t, u) - h(t, v)\| \leqslant \varphi(\|u - v\|)$$

for all $u, v \in \mathbb{R}^n$, and $t \in J$.

In order to illustrate a situation in which the condition (A1) is satisfied, we will present a property to ensure that a function $s \mapsto f(s, x(s))$ is regulated on J for each $x \in G(J, \mathbb{R}^n)$. For a function $x: J \to \mathbb{R}^n$, we denote by $Im(x) = \{x(t) : t \in J\}$ the range set of x.

Theorem 5.5 *Let $x \in G(J, \mathbb{R}^n)$. Assume that $f: J \times \mathbb{R}^n \to \mathbb{R}^n$ is a function such that for each $v \in \overline{Im(x)}$, the function $f(\cdot, v)$ is regulated on J. Let $y(s): J \to \mathbb{R}^n$ the function given by $y(s)(t) = f(s, x(t))$. Assume that $\{y(s)(\cdot): s \in J\}$ is an equiregulated set of functions. Then the function $z: J \to \mathbb{R}^n$ given by $z(t) = f(t, x(t))$ is regulated.*

Proof We will start by showing that the function z has limits on the right. For this purpose, we will consider the set $J_1 = [t_0, t_0 + a)$, and we define the function $L: J_1 \to \mathbb{R}^n$ by

$$L(s) = \lim_{t \to s^+} f(t, x(s^+)).$$

Let $s_0 \in J_1$. We claim that $\lim\limits_{t \to s_0^+} z(t) = L(s_0)$. In fact, for $t \in J_1$, we can write

$$z(t) - L(s_0) = z(t) - f(t, x(s_0^+)) + f(t, x(s_0^+)) - L(s_0).$$

Since the functions $y(\xi) = f(\xi, x(\cdot))$ are equiregulated for $\xi \in J$, it follows from Definition 5.3 that for every $\varepsilon > 0$, we can choose $\delta_0 > 0$ such that

$$\|f(\xi, v) - f(\xi, x(s_0^+))\| < \varepsilon/2,$$

for all $\xi \in J$ and $v \in \mathbb{R}^n$ such that $\|v - x(s_0^+)\| < \delta_0$. In addition, since $x(\cdot)$ is a regulated function, there exists $\delta_1 > 0$ such that $\|x(t) - x(s_0^+)\| < \delta_0$ for $s_0 < t < s_0 + \delta_1$. Combining these assertions, we infer that

$$\|f(\xi, x(t)) - f(\xi, x(s_0^+))\| < \varepsilon/2$$

for all $\xi \in J$ and $t \in (s_0, s_0 + \delta_1)$. In particular, taking $\xi = t$, we obtain that

$$\|z(t) - f(t, x(s_0^+))\| = \|f(t, x(t)) - f(t, x(s_0^+))\| < \varepsilon/2.$$

Furthermore, since $\lim_{t \to s_0^+} f(t, x(s_0^+)) = L(s_0)$, there exists $\delta_2 > 0$ such that

$$\|f(t, x(s_0^+)) - L(s_0)\| < \varepsilon/2,$$

for all $s_0 < t < s_0 + \delta_2$. Therefore, for $\delta = \min\{\delta_1, \delta_2\}$ and gathering our previous estimations, we obtain

$$\|z(t) - L(s_0)\| < \varepsilon,$$

for all $s_0 < t < s_0 + \delta$. This shows that $\lim_{t \to s_0^+} f(t, x(t)) = L(s_0)$. Proceeding in similar way, we can prove that $\lim_{t \to s_0^-} f(t, x(t))$ exists for $s_0 \in (t_0, t_0 + a]$, which completes the proof of the theorem. □

Corollary 5.1 *Let $x \in G(J, \mathbb{R}^n)$. Assume that $f : J \times \mathbb{R}^n \to \mathbb{R}^n$ satisfies the conditions:*

(a) For every $v \in \overline{Im(x)}$, the function $f(\cdot, v)$ is regulated on J.
(b) The functions $f(t, \cdot)$ are continuous uniformly for $t \in J$.

Then the function $z : J \to \mathbb{R}^n$ given by $z(t) = f(t, x(t))$ is regulated.

Proof Applying Theorem 5.5 we only need to prove that the functions $f(s, x(\cdot))$ are equiregulated for $s \in J$. Let $\varepsilon > 0$. It follows from condition (b) that we can choose $\delta > 0$ such that

$$\|f(s, u) - f(s, v)\| < \varepsilon$$

for all $s \in J$ and $u, v \in \mathbb{R}^n$ such that $\|u - v\| < \delta$. Let $t \in [t_0, t_0 + a)$. There exists $\delta_1 > 0$ such that $\|x(t^+) - x(t')\| < \delta$ for all $t < t' < t + \delta_1$. Collecting these assertions, we deduce that

$$\|f(s, x(t^+)) - f(s, x(t'))\| < \varepsilon$$

for all $s \in J$ and all $t < t' < t + \delta_1$, which completes the proof. □

Remark 5.2 From Corollary 5.1, it is immediate to note that condition (A3) implies that for a fixed $x \in G(J, \mathbb{R}^n)$, the function $h(\cdot, x(\cdot))$ is regulated on J.

Returning to the more general context of integrable functions in the sense of Kurzweil-Stieltjes and in order to establish our results, we recall the following property.

Lemma 5.2 ([32, Lemma 3.1]) *Assume that the function f satisfies the conditions (A0)–(A2). Let $x\colon J \to \mathbb{R}^n$ be a regulated function. Then the function $y(\cdot)$ given by $y(t) = f(t, x(t))$ for $t \in J$ is Kurzweil-Stieltjes integrable w.r.t. g on J and*

$$\left\| \int_c^d f(s, x(s))\mathrm{d}g(s) \right\| \leqslant \int_c^d M(s)\mathrm{d}g(s), \tag{5.9}$$

for all interval $[c, d] \subseteq J$.

In what follows, let us consider the following concept of solution for the problem MDE (5.1) and (5.2).

Definition 5.7 A regulated function $x\colon J \to \mathbb{R}^n$ is said to be a *solution of problem MDE* (5.1) *and* (5.2) if the following conditions hold:

(i) The function $t \to h(t, x(t))$ is regulated on J.
(ii) The integral equation (5.1) is verified, where the integral on the right-hand side is understood in the Kurzweil-Stieltjes sense.

We are now in a position to establish a result on the existence of solutions for problem MDE (5.1) and (5.2).

Theorem 5.6 *Assume that hypotheses* (A0)–(A3) *hold. Assume further that*

$$\liminf_{r\to\infty} \frac{\varphi(r)}{r} < 1. \tag{5.10}$$

Then there exists a solution for problem MDE (5.1) *and* (5.2) *defined on J.*

Proof Using condition (5.10), we infer that there exists $N > 0$ large enough such that

$$\frac{\varphi(N)}{N} + \frac{\|x_0 - h(t_0, x_0)\| + H_0 + K_0}{N} < 1,$$

where $H_0 = \sup\{\|h(t, 0)\| : t \in J\}$ and $K_0 = \int_{t_0}^{t_0+a} M(s)\mathrm{d}g(s)$.

Let S be the subset of $G(J, \mathbb{R}^n)$ defined by

$$S := \{x \in G(J, \mathbb{R}^n) : x(t_0) = x_0, \ \|x\|_\infty \leqslant N\}.$$

Clearly, S is a nonempty, closed, convex, and bounded subset of $G(J, \mathbb{R}^n)$. We define the operators $A\colon S \to G(J, \mathbb{R}^n)$ and $B\colon S \to G(J, \mathbb{R}^n)$ by

$$Ax(t) = h(t, x(t)), \quad t \in J, \tag{5.11}$$

and

$$Bx(t) = x_0 - h(t_0, x_0) + \int_{t_0}^{t} f(s, x(s))\mathrm{d}g(s), \quad t \in J. \tag{5.12}$$

Using these definitions, the integral equation (5.1) can be transformed into the operator equation

$$Ax(t) + Bx(t) = x(t), \quad t \in J. \tag{5.13}$$

From (A3), it follows that the operator A is a nonlinear contraction. In fact, for every $t \in J$, we have

$$\|Ax(t) - Az(t)\| = \|h(t, x(t)) - h(t, z(t))\| \leqslant \varphi(\|x(t) - z(t)\|) \leqslant \varphi(\|x - z\|_\infty),$$

which implies that $\|Ax - Az\|_\infty \leqslant \varphi(\|x - z\|_\infty)$.

Next, we show that B is a completely continuous operator from S into $G(J, \mathbb{R}^n)$. First, we show that B is continuous on S. Let $(x_n)_{n\in\mathbb{N}}$ be a sequence in S converging to a function $x \in S$. Let $d = \{\sigma_0, \sigma_1, \ldots, \sigma_l\}$ be a subdivision of the interval $[t_0, t_0 + a]$. Applying Lemma 5.2, for every finite subset $\{m_1, m_2, \ldots, m_l\}$ of $\mathbb{N}$, we have

$$\begin{aligned}\left\|\sum_{j=1}^{l} \int_{\sigma_{j-1}}^{\sigma_j} f(s, x_{m_j}(s))\mathrm{d}g(s)\right\| &\leqslant \sum_{j=1}^{l} \int_{\sigma_{j-1}}^{\sigma_j} M(s)\mathrm{d}g(s)\\ &= \int_{t_0}^{t_0+a} M(s)\mathrm{d}g(s) = K_0 < \infty,\end{aligned}$$

which allows us to use the dominated convergence theorem for the Kurzweil-Stieltjes integral. Hence, combining condition (A0) and Theorem 5.3, we obtain

$$\begin{aligned}\lim_{n\to+\infty} Bx_n(t) &= \lim_{n\to+\infty}\left[x_0 - h(t_0, x_0) + \int_{t_0}^{t} f(s, x_n(s))\mathrm{d}g(s)\right]\\ &= x_0 - h(t_0, x_0) + \lim_{n\to+\infty}\int_{t_0}^{t} f(s, x_n(s))\mathrm{d}g(s)\\ &= x_0 - h(t_0, x_0) + \int_{t_0}^{t} f(s, x(s))\mathrm{d}g(s) = Bx(t),\end{aligned}$$

for all $t \in J$. This shows that B is a continuous map on S.

Next we prove that B is a compact operator on S. As a consequence of Lemma 5.1, we only need to show that $B(S)$ is a uniformly bounded and equiregulated set in $G(J, \mathbb{R}^n)$. In order to establish the first assertion, for an arbitrary $x \in S$ we estimate

$$\begin{aligned}\|Bx(t)\| &\leqslant \|x_0 - h(t_0, x_0)\| + \left\| \int_{t_0}^{t} f(s, x(s))\mathrm{d}g(s) \right\| \\ &\leqslant \|x_0 - h(t_0, x_0)\| + K_0,\end{aligned}$$

for all $t \in J$. Therefore, $\|Bx\|_\infty \leqslant \|x_0 - h(t_0, x_0)\| + K_0$, for all $x \in S$. This shows that B is uniformly bounded on S.

On the other hand, let $t_1, t_2 \in J$ with $t_1 < t_2$. For any $x \in S$, we have

$$\|Bx(t_2) - Bx(t_1)\| = \left\| \int_{t_1}^{t_2} f(s, x(s))\mathrm{d}g(s) \right\| \leqslant \int_{t_1}^{t_2} M(s)\mathrm{d}g(s) = p(t_2) - p(t_1),$$

where $p(\cdot)$ is the function given by $p(t) = \int_{t_0}^{t} M(s)\mathrm{d}g(s)$ for $t \in J$. Since by Theorem 5.2 the function p is regulated on J, the above estimates imply that $B(S)$ is an equiregulated set in $G(J, \mathbb{R}^n)$.

To finish this proof, we show that condition (c) of Theorem 5.4 holds. Let $u, v \in S$ be arbitrary. Clearly, $Au(t_0) + Bv(t_0) = h(t_0, u(t_0)) + x_0 - h(t_0, x_0) = x_0$. Also, we estimate

$$\begin{aligned}\|Au(t) + Bv(t)\| &\leqslant \|h(t, u(t))\| + \|x_0 - h(t_0, x_0)\| + \left\| \int_{t_0}^{t} f(s, v(s))\mathrm{d}g(s) \right\| \\ &\leqslant \|h(t, u(t)) - h(t, 0)\| + \|h(t, 0)\| + \|x_0 - h(t_0, x_0)\| + \\ &\quad + \int_{t_0}^{t} M(s)\mathrm{d}g(s) \\ &\leqslant \varphi(N) + \|x_0 - h(t_0, x_0)\| + H_0 + K_0 < N,\end{aligned}$$

for all $t \in J$. Consequently, $\|Au + Bv\|_\infty \leqslant N$, which in turn implies that $Au + Bv \in S$.

The preceding development shows that all the hypotheses of Theorem 5.4 are satisfied, which allows us to conclude that Eq. (5.13) has a solution in S. This shows that the problem MDE (5.1) and (5.2) has a solution defined on J. □

Next we will exhibit a simple application of Theorem 5.6. To present this result, we need a previous property on the Kurzweil-Stieltjes integration.

Lemma 5.3 *Assume that $g\colon [a, b] \to \mathbb{R}$ is a nondecreasing left-continuous function and that $\eta\colon [a, b] \to \mathbb{R}$ is a bounded Kurzweil-Stieltjes integrable w.r.t. g function. Let $u \in G([a, b])$. Then the function $u(\cdot)\eta(\cdot)$ is Kurzweil-Stieltjes integrable w.r.t. g.*

Proof By [33, Theorem 4.1.5] there exists a sequence $(u_n)_n$ of step functions which converges uniformly to u. It follows that each function $u_n(\cdot)\eta(\cdot)$ is Kurzweil-Stieltjes integrable w.r.t. g. Since the sequence $(u_n\eta)_n$ converges uniformly to $u\eta$, using [33, Theorem 6.3.8] we conclude that the function $u\eta$ is Kurzweil-Stieltjes integrable w.r.t. g. □

Example 5.1 Consider the following integral equation:

$$x(t) = \frac{1}{2}\sin^2(t)\ln(1+|x(t)|) + \int_0^t f(s,x(s))\mathrm{d}g(s), \ t \in [0,1], \tag{5.14}$$

where g is a nondecreasing left-continuous function, and $f\colon [0,1]\times\mathbb{R}\to\mathbb{R}$ is the function given by $f(s,z)=\eta(s)e^{\gamma\cos(z)}$ for all $(s,z)\in[0,1]\times\mathbb{R}$, where $\gamma>0$ is a constant and $\eta\colon [0,1]\to\mathbb{R}^+$ is a bounded and Kurzweil-Stieltjes integrable w.r.t. g function. It is clear that condition (A0) holds. Also, using Lemma 5.3, we obtain that condition (A1) holds. Furthermore,

$$\left|\int_c^d f(s,u)\mathrm{d}g(s)\right| = \left|\int_c^d \eta(s)e^{\gamma\cos u}\mathrm{d}g(s)\right| \leqslant e^{\gamma}\int_c^d \eta(s)\mathrm{d}g(s),$$

for all $u\in\mathbb{R}$ and $[c,d]\subseteq[0,1]$, which shows that condition (A2) is fulfilled with $M(s)=e^{\gamma}\eta(s)$ for $s\in[0,1]$.

In addition, let h be the function given by $h(t,u)=\frac{1}{2}\sin^2(t)\ln(1+|u|)$ defined on $[0,1]\times\mathbb{R}$. It is clear that h is continuous; hence the function $h(\cdot,v)$ is regulated on $[0,1]$. Moreover, for every $u,v\in\mathbb{R}$ we have

$$|h(t,u)-h(t,v)| \leqslant \frac{1}{2}\left|\ln\left(\frac{(1+|u|)}{(1+|v|)}\right)\right| \leqslant \varphi(|u-v|),$$

where $\varphi(t)=\frac{1}{2}\ln(1+t)$ for $t\geqslant 0$. Consequently, the function h satisfies the condition (A3).

Since conditions (A0)–(A3) are fulfilled and $\liminf_{r\to\infty}\frac{\varphi(r)}{r}=0$, it follows from Theorem 5.6 that the integral equation (5.14) has a solution defined on $[0,1]$.

Conditions (A2) and (A3) are very demanding since, said somewhat superficially, they tell us that the function f is uniformly bounded and that the function h uniformly satisfies a Lipschitz-type property of continuity. However, these conditions can be generalized as follows:

(A2*) There exists a function $M\colon J\times[0,+\infty)\to\mathbb{R}$ such that $M(\cdot,\alpha)$ is Kurzweil-Stieltjes integrable w.r.t. g for all $\alpha\geqslant 0$, the function $M(s,\cdot)$ is nondecreasing for all $s\in J$, and

$$\left\|\int_c^d f(s,u)\mathrm{d}g(s)\right\| \leqslant \int_c^d M(s,\|u\|)\mathrm{d}g(s),$$

for all $u\in\mathbb{R}^n$, and $[c,d]\subseteq J$.

(A3*) For every $v \in \mathbb{R}^n$, the function $h(\cdot, v)$ is regulated on J, and there exists a function $\varphi : \mathbb{R}_0^+ \times \mathbb{R}_0^+ \to \mathbb{R}_0^+$ such that for every $r \geqslant 0$, $\varphi(\cdot, r)$ is a $\mathfrak{D}$-function satisfying $\varphi(t, r) < t$ for all $t > 0$, and

$$\|h(t, u) - h(t, v)\| \leqslant \varphi(\|u - v\|, r)$$

for all $u, v \in \mathbb{R}^n$, $\|u\|, \|v\| \leqslant r$, and $t \in J$.

Theorem 5.7 *Assume that conditions* (A0), (A1), (A2*), *and* (A3*) *are fulfilled. Assume further that*

$$\liminf_{r\to\infty}\left(\frac{\varphi(r, r)}{r} + \frac{1}{r}\int_{t_0}^{t_0+a} M(s, r)\mathrm{d}g(s)\right) < 1. \tag{5.15}$$

Then there exists a solution for problem MDE (5.1)–(5.2) *defined on* J.

Proof Since the proof follows the same lines as the one made for Theorem 5.6, we will present here only a brief sketch of the proof.

(i) Using (5.15), we infer that there exists $N > 0$ large enough such that

$$\varphi(N, N) + H_0 + \|x_0 + h(t_0, x_0)\| + \int_{t_0}^{t_0+a} M(s, N)\mathrm{d}g(s) < N. \tag{5.16}$$

We define $S = \{x \in G(J, \mathbb{R}^n) : x(t_0) = x_0,\ \|x\|_\infty \leqslant N\}$.

(ii) As an immediate consequence of condition (A3*), we infer that the operator A defined by (5.11) is a nonlinear contraction with respect to the $\mathfrak{D}$-function $\varphi(\cdot, N)$ on S.

(iii) In this step, we show that the operator B defined by (5.12) is completely continuous. Let $(x_n)_{n\in\mathbb{N}}$ be a sequence in S that converges to x. Proceeding as in the proof of Theorem 5.6 and as a consequence of (A2*), Lemma 5.2, and Theorem 5.3, we obtain that $Bx_n(t) \to Bx(t)$ as $n \to \infty$ uniformly for $t \in J$, which shows that B is continuous. Turning to apply (A2*) and Lemma 5.2, we infer that

$$\left\|\int_{t_0}^{t} f(s, x(s))\mathrm{d}g(s)\right\| \leqslant \int_{t_0}^{t} M(s, N)\mathrm{d}g(s) \leqslant \int_{t_0}^{t_0+a} M(s, N)\mathrm{d}g(s)$$

for all $x \in S$. This implies that $B(S)$ is a bounded set. In addition, arguing as in the proof of Theorem 5.6, we derive that $B(S) \subset G(J, \mathbb{R}^n)$ is a relatively compact set.

(iv) Let u, v be arbitrary functions in S. Applying (A2*), (A3*), Lemma 5.2, and (5.16), we can estimate

$$\|Au(t) + Bv(t)\| \leqslant \|h(t, u(t))\| + \|x_0 - h(t_0, x_0)\| + \left\|\int_{t_0}^{t} f(s, v(s))\mathrm{d}g(s)\right\|$$

$$\leqslant \varphi(N, N) + \|h(t, 0)\| + \|x_0 - h(t_0, x_0)\| + \\ + \int_{t_0}^{t_0+a} M(s, N) \mathrm{d}g(s) \\ \leqslant N$$

for all $t \in J$, which implies that $Au + Bv \in S$.

As a consequence of steps (i)–(iv), we can affirm that all the hypotheses of Theorem 5.4 are satisfied, and thus this concludes the proof. □

5.4 Differentiability and Existence of S-Asymptotically ω-Periodic Solutions

In this section, our aim is to characterize the existence of special classes of solutions. In particular we study the existence of continuous, differentiable, and S-asymptotically ω-periodic solutions for problem MDE (5.1) and (5.2).

We begin with a result that guarantees the existence of continuous solutions.

Proposition 5.1 *Assume that conditions* (A1), (A2), *and* (A3*) *hold. Assume further that* g *is continuous, and for every* $v \in \mathbb{R}^n$, *the function* $h(\cdot, v)$ *is continuous. Then every solution* $x(\cdot)$ *of problem MDE* (5.1) *and* (5.2) *is a continuous function.*

Proof The proof of this statement is standard. For completeness of the text, we are going to include here an outline of it. Let $x(\cdot)$ be a solution of the problem MDE (5.1) and (5.2) with $\|x\|_\infty \leqslant r$. Suppose that $x(\cdot)$ is not continuous at $\bar{t}$. Then there must be $\varepsilon > 0$ and a sequence $(t_k)_k$ that converges to $\bar{t}$ and such that $\|x(t_k) - x(\bar{t})\| \geqslant \varepsilon$ for all $k \in \mathbb{N}$. It follows from expression (5.1) that

$$x(t_k)-x(\bar{t}) = h(t_k, x(t_k))-h(\bar{t}, x(\bar{t})) + \int_{t_0}^{t_k} f(s, x(s))\mathrm{d}g(s) - \int_{t_0}^{\bar{t}} f(s, x(s))\mathrm{d}g(s) \\ = h(t_k, x(t_k)) - h(t_k, x(\bar{t})) + h(t_k, x(\bar{t})) - h(\bar{t}, x(\bar{t})) + \\ + \int_{\bar{t}}^{t_k} f(s, x(s))\mathrm{d}g(s).$$

Without loss of generality, we assume that $t_k > \bar{t}$. From the previous decomposition, it follows that

$$\|x(t_k) - x(\bar{t})\| \leqslant \|h(t_k, x(t_k)) - h(\bar{t}, x(\bar{t}))\| + \\ + \left\| \int_{t_0}^{t_k} f(s, x(s))\mathrm{d}g(s) - \int_{t_0}^{\bar{t}} f(s, x(s))\mathrm{d}g(s) \right\|$$

$$\begin{aligned}&\leqslant \|h(t_k, x(t_k)) - h(t_k, x(\bar{t}))\| + \|h(t_k, x(\bar{t})) - h(\bar{t}, x(\bar{t}))\| + \\ &+ \left\| \int_{\bar{t}}^{t_k} f(s, x(s)) \mathrm{d}g(s) \right\| \\ &\leqslant \varphi(\|x(t_k) - x(\bar{t})\|, r) + \|h(t_k, x(\bar{t})) - h(\bar{t}, x(\bar{t}))\| + \\ &+ \left\| \int_{\bar{t}}^{t_k} f(s, x(s)) \mathrm{d}g(s) \right\|. \end{aligned} \tag{5.17}$$

Since $h(\cdot, x(\bar{t}))$ is continuous, then $\|h(t_k, x(\bar{t})) - h(\bar{t}, x(\bar{t}))\| \to 0$ as $k \to \infty$, and, using Theorem 5.2, $\int_{\bar{t}}^{t_k} f(s, x(s)) \mathrm{d}g(s) \to 0$ as $k \to \infty$. Since $\varphi([\varepsilon, 2r], r)$ is a compact set, we can choose $\delta > 0$ such that $\delta \leq t - \varphi(t, r)$ for all $t \in [\varepsilon, 2r]$. Using that $\|x(t_k) - x(\bar{t})\| \in [\varepsilon, 2r]$ and applying the estimate (5.17), we obtain

$$\|x(t_k) - x(\bar{t})\| \leqslant \|x(t_k) - x(\bar{t})\| - \delta + \|h(t_k, x(\bar{t})) - h(\bar{t}, x(\bar{t}))\| + \left\| \int_{\bar{t}}^{t_k} f(s, x(s)) \mathrm{d}g(s) \right\|,$$

which implies that

$$\delta \leqslant \|h(t_k, x(\bar{t})) - h(\bar{t}, x(\bar{t}))\| + \left\| \int_{\bar{t}}^{t_k} f(s, x(s)) \mathrm{d}g(s) \right\| \to 0, \ k \to \infty.$$

This contradiction shows that $x(\cdot)$ is continuous at $\bar{t}$. □

Next we study the differentiability of a solution of problem MDE (5.1) and (5.2). Before stating our result, we will introduce a notation. Assume that $h\colon [t_0, t_0 + a] \times \mathbb{R}^n \to \mathbb{R}^n$ is differentiable and $h(t, x) = col(h_1, \ldots, h_n)$. We denote $J_x h(p)$ to the Jacobian matrix

$$J_x h(p) = \begin{pmatrix} \frac{\partial h_1(p)}{\partial x_1} & \frac{\partial h_1(p)}{\partial x_2} & \cdots & \frac{\partial h_1(p)}{\partial x_n} \\ \frac{\partial h_2(p)}{\partial x_1} & \frac{\partial h_2(p)}{\partial x_2} & \cdots & \frac{\partial h_2(p)}{\partial x_n} \\ \vdots & \vdots & & \vdots \\ \frac{\partial h_n(p)}{\partial x_1} & \frac{\partial h_n(p)}{\partial x_2} & \cdots & \frac{\partial h_n(p)}{\partial x_n} \end{pmatrix}.$$

Furthermore,

$$h(t, x) = h(t_0, x_0) + \frac{\partial h(t_0, x_0)}{\partial t}(t - t_0) + J_x h(t_0, x_0)(x - x_0) + R(t, x, t_0, x_0),$$

where $\lim\limits_{t \to t_0, x \to x_0} \dfrac{R(t, x, t_0, x_0)}{\|(t, x) - (t_0, x_0)\|} = 0.$

On the other hand, for a function $u(\cdot)$ defined on an interval J, we denote by $u'_+(t)$—respectively, $u'_-(t)$—the right derivative of u at t—respectively, the left derivative—when the derivative exists.

Proposition 5.2 *Assume that conditions* (A1), (A2), *and* (A3*) *hold. Assume further that h is continuously differentiable, with $I - J_x h(t, x(t))$ invertible, and that f and g are continuous functions. Let $x(\cdot)$ be a solution of problem MDE* (5.1) *and* (5.2). *Then $x(\cdot)$ satisfies the following properties:*

(a) The function $x(\cdot)$ is differentiable a.e. for $t \in [t_0, t_0 + a]$.
(b) For each $t \in [t_0, t_0 + a)$ such that $g'_+(t)$ exists, there is also the right derivative $x'_+(t)$. Similarly, for each $t \in (t_0, t_0 + a]$ such that $g'_-(t)$ exists, there is also the left derivative $x'_-(t)$.
(c) If g is differentiable at t, then the function $x(\cdot)$ is also differentiable at t.
(d) If g is continuously differentiable at t, then the function $x(\cdot)$ is also continuously differentiable at t.

Proof It follows from Corollary 5.1 that $x(\cdot)$ is a continuous function. We begin the proof with a simple remark. Let $u(t) = \int_{t_0}^{t} f(s, x(s))\mathrm{d}g(s)$. Assume initially that the derivative $g'(t)$ exists. Then the derivative $u'(t)$ also exists and $u'(t) = f(t, x(t))g'(t)$. In fact, under our hypotheses, the Kurzweil-Stieltjes integral $\int_{t_0}^{t} f(s, x(s))\mathrm{d}g(s)$ coincides with the Riemann-Stieltjes integral of $f(s, x(s))$ w.r.t. g ([33, Theorem 6.2.12]), and the assertion is a well-known property for the Riemann-Stieltjes integral. It follows from (5.1) that

$$\frac{x(t+\Delta) - x(t)}{\Delta} = \frac{\partial h(t, x(t))}{\partial t} + J_x h(t, x(t)) \frac{x(t+\Delta) - x(t)}{\Delta} + \\ + \frac{R(t+\Delta, x(t+\Delta), t, x(t))}{\Delta} + \frac{u(t+\Delta) - u(t)}{\Delta},$$

which implies that

$$\frac{x(t+\Delta) - x(t)}{\Delta} = (I - J_x h(t, x(t)))^{-1} \left(\frac{\partial h(t, x(t))}{\partial t} + \frac{u(t+\Delta) - u(t)}{\Delta} \right. \\ \left. + \frac{R(t+\Delta, x(t+\Delta), t, x(t))}{\Delta} \right).$$

Hence, taking the limit above as $\Delta \to 0$ and using the previous remark, we obtain

$$x'(t) = (I - J_x h(t, x(t)))^{-1} \left(\frac{\partial h(t, x(t))}{\partial t} + f(t, x(t))g'(t) \right). \tag{5.18}$$

Since g is a nondecreasing function, the derivative $g'(t)$ exists a.e. Thus, the assertion (a) is an immediate consequence of the preceding development. Claims (b) and (c) are similarly demonstrated. Finally, statement (d) is a consequence of identity (5.18). □

We are next concerned with the existence of solutions on the interval $[0, \infty)$. Taking into account that our later objective is to establish the existence of approximately periodic solutions, initially we will restrict ourselves to investigate the existence of bounded solutions on $[0, \infty)$. We denote by $G_b([0, \infty), \mathbb{R}^n)$ the space consisting of all bounded regulated functions $u\colon [0, \infty) \to \mathbb{R}^n$ endowed with the norm of uniform convergence. We slightly modify condition (A1), as follows:

(A1*) For every $u \in \mathbb{R}^n$ and every $a > 0$, the function $f(\cdot, u)$ is Kurzweil-Stieltjes integrable w.r.t. g on the interval $[0, a]$.

Furthermore, from here on, the condition (A3*) should be understood as follows: For every $v \in \mathbb{R}^n$, the function $h(\cdot, v)$ belongs to $G_b([0, \infty), \mathbb{R}^n)$, and there exists a function $\varphi : \mathbb{R}_0^+ \times \mathbb{R}_0^+ \to \mathbb{R}_0^+$ such that for every $r \geqslant 0$, $\varphi(\cdot, r)$ is a $\mathfrak{D}$-function satisfying $\varphi(t, r) < t$ for all $t > 0$, and

$$\|h(t, u) - h(t, v)\| \leqslant \varphi(\|u - v\|, r)$$

for all $u, v \in \mathbb{R}^n$, $\|u\|, \|v\| \leqslant r$, and $t \geqslant 0$.

From now on, $\int_0^\infty M(s, N)\mathrm{d}g$ denotes the Kurzweil-Henstock-Stieltjes integral of $M(\cdot, N)$ with respect to g on the infinite interval $[0, \infty)$.

Theorem 5.8 *Assume that conditions (A0), (A1*), (A2*), and (A3*) hold. Suppose further that*

$$\liminf_{N\to\infty} \frac{1}{N} \left(\sup_{0\leqslant t<\infty, \|u\|\leqslant N} \|h(t, u)\| + \int_0^\infty M(s, N)\mathrm{d}g \right) < 1. \tag{5.19}$$

Then there exists a bounded solution for problem MDE (5.1) *and* (5.2) *on* $[0, \infty)$.

Proof It follows from Theorem 5.7 that problem MDE (5.1) and (5.2) has a solution on every interval of type $[t_0, t_0 + a]$. Using this observation, we will carry out an inductive construction to prove the existence of solution on $[0, \infty)$. Let us consider the intervals of the form $[k - 1, k]$ for $k \in \mathbb{N}$. Let $k = 1$. By Theorem 5.7, we know that there exists a solution, denoted by $x^1(\cdot)$, of problem MDE (5.1) and (5.2) defined on the interval $[0, 1]$. Suppose now that x^k for $k \in \mathbb{N}$ is a solution of problem MDE (5.1) and (5.2) on $[k-1, k]$. Proceeding as in the definition of $x^1(\cdot)$, let $x^{k+1}(\cdot)$ be a solution of the problem

$$x(t) = x(k) + h(t, x(t)) - h(k, x(k)) + \int_k^t f(s, x(s))\mathrm{d}g(s), \ k \leqslant t \leqslant k + 1,$$

$$x(k) = x^k(k) \in \mathbb{R}^n.$$

Let us define the function $x(t) = x^k(t)$ for $t \in [k-1, k]$. It is clear that $x(\cdot)$ is a solution of problem MDE (5.1) and (5.2) on $[0, \infty)$.

Assume now that $x(\cdot)$ is not bounded on $[0, \infty)$. Let $N(t) = \sup_{0 \leqslant s \leqslant t} \|x(s)\|$. It is clear that $N(t) \to \infty$ as $t \to \infty$. Hence, we can select $t \geqslant t_1$ such that $N(t) \geqslant N_1 > 0$. We can assume that $\frac{1}{N_1}\|x_0 - h(t_0, x_0)\| \leqslant \varepsilon/2$, for $\varepsilon > 0$ given. Since x is a solution, it follows that

$$\|x(t)\| \leqslant \|x_0 - h(t_0, x_0)\| + \|h(t, x(t))\| + \left\| \int_0^t f(s, x(s))\mathrm{d}g(s) \right\|$$

$$\leqslant \|x_0 - h(t_0, x_0)\| + \sup_{0 \leqslant \xi \leqslant t, \|u\| \leqslant N(t)} \|h(\xi, u)\| + \int_0^t M(s, N(t))\mathrm{d}g(s).$$

Hence, using (5.19) we obtain that

$$1 \leqslant \frac{1}{N(t)}\|x_0 - h(t_0, x_0)\| + \frac{1}{N(t)} \sup_{\xi < \infty, \|u\| \leqslant N(t)} \|h(\xi, u)\| +$$

$$+ \frac{1}{N(t)} \int_0^t M(s, N(t))\mathrm{d}g(s) < 1 - \varepsilon/2,$$

which is a contradiction. Thus, we complete the proof that $x(\cdot)$ is a bounded solution of problem MDE (5.1) and (5.2) on $[0, \infty)$. □

In the remaining of this section, we will study the existence of a type of solution that resembles a ω-periodic function. In [25, 26], it was introduced the concept of S-asymptotically ω-periodic functions for bounded continuous functions defined on $[0, \infty)$, and the authors studied the main properties of this class of functions. In what follows, we will extend the definition considered in those works to include regulated functions.

Definition 5.8 A bounded regulated function $x\colon [0, \infty) \to X$ is called *S-asymptotically ω-periodic* if there exists $\omega > 0$ such that

$$x(t + \omega) - x(t) \to 0, \ t \to \infty.$$

Before establishing the concepts necessary to study this topic, inspired by [25, Example 3.2], we exhibit an example to illustrate the concept of S-asymptotically ω-periodic regulated functions.

Example 5.2 Let $(a_n)_{n \in \mathbb{N}_0} \subset \mathbb{R}$ be a bounded sequence of real numbers consisting of numbers different from each other and such that $a_{n+1} - a_n \to 0$, as $n \to \infty$. Let $x\colon [0, \infty) \to \mathbb{R}$ be the function given by

$$x(t) = \begin{cases} a_1 + (a_1 - a_0)(t-1), & \text{if } 0 \leqslant t \leqslant 1, \\ a_{n+1} + (a_{n+1} - a_{n-1})(t-n-1), & \text{if } n < t \leqslant n+1,\ n \in \mathbb{N}. \end{cases}$$

Clearly the function x is continuous on every subinterval of type $(n, n+1)$ with $n \in \mathbb{N}_0$ and left-continuous at every $t = n \in \mathbb{N}$. Moreover, $x(n) = a_n$ and $\Delta^+ x(n) = a_{n-1} - a_n$ for all $n \in \mathbb{N}$. Consequently, $x(\cdot)$ is a regulated function with a countable set of discontinuities on $[0, \infty)$ and $x(\cdot)$ is not periodic. The boundedness of $x(\cdot)$ follows from the properties of the sequence $(a_n)_{n \in \mathbb{N}}$. In addition, for every $t \in (n, n+1]$ we obtain

$$|x(t+1) - x(t)| \leqslant 2|a_{n+2} - a_{n+1}| + |a_n - a_{n-1}|.$$

Therefore, $\lim_{t \to \infty} [x(t+1) - x(t)] = 0$, which shows that the function x is S-asymptotically 1-periodic.

We denote by $S_\omega(X)$ the space consisting of all S-asymptotically ω-periodic functions from $[0, \infty)$ to X. It is clear that $S_\omega(X)$ is a Banach space provided with the norm of uniform convergence.

Definition 5.9 Let $p\colon [0, \infty) \times \mathbb{R}^n \to \mathbb{R}^n$ be a function that satisfies the following conditions:

(i) For every $u \in \mathbb{R}^n$, the function $p(\cdot, u)$ is bounded and regulated.
(ii) For every $t \geqslant 0$, the function $p(t, \cdot)$ is continuous.

We say that p is *uniformly S-asymptotically ω-periodic on bounded sets* if there exists $\omega > 0$ such that

$$p(t+\omega, u) - p(t, u) \to 0, \ t \to \infty,$$

uniformly for u in bounded subsets of $\mathbb{R}^n$.

To establish our result on the existence of S-asymptotically ω-periodic solutions of problem MDE (5.1)–(5.2), we need to develop some previous properties.

Lemma 5.4 *Let $x \in S_\omega(\mathbb{R}^n)$. Assume that p is a uniformly S-asymptotically ω-periodic on bounded sets function that satisfies condition (A3*). Then the function $t \mapsto p(t, x(t))$ is S-asymptotically ω-periodic.*

Proof From the decomposition

$$\begin{aligned} p(t+\omega, x(t+\omega)) - p(t, x(t)) &= p(t+\omega, x(t+\omega)) - p(t, x(t+\omega)) \\ &\quad + p(t, x(t+\omega)) - p(t, x(t)), \end{aligned}$$

we infer that

$$
\begin{aligned}
\|p(t+\omega, x(t+\omega)) - p(t, x(t))\| \leqslant & \|p(t+\omega, x(t+\omega)) - p(t, x(t+\omega))\| + \\
& + \|p(t, x(t+\omega)) - p(t, x(t))\| \\
\leqslant & \|p(t+\omega, x(t+\omega)) - p(t, x(t+\omega))\| + \\
& + \varphi(\|x(t+\omega) - x(t)\|, \|x\|_\infty).
\end{aligned}
$$

Since $\|p(t+\omega, x(t+\omega)) - p(t, x(t+\omega))\| \to 0$ as $t \to \infty$, because p is uniformly S-asymptotically ω-periodic on bounded sets function and $\varphi(\|x(t+\omega) - x(t)\|, \|x\|_\infty) \to 0$ as $t \to \infty$ by the properties of $\varphi(\cdot, \|x\|_\infty)$, we conclude that $\|p(t+\omega, x(t+\omega)) - p(t, x(t))\| \to 0$ as $t \to \infty$. □

Definition 5.10 Let $y_k \colon [0, \infty) \to X$, $k \in \mathbb{N}$, be regulated functions. The sequence $(y_k)_k$ is said to be *uniformly S-asymptotically ω-periodic* if for every $\varepsilon > 0$ there exists $T_\varepsilon > 0$ such that

$$
\|y_k(t+\omega) - y_k(t)\|_X \leqslant \varepsilon
$$

for all $t \geqslant T_\varepsilon$ and $k \in \mathbb{N}$.

In what follows, until the end of this section, when we write that a sequence $(y_k)_k \subset G([0, \infty), \mathbb{R}^n)$ converges to $y \in G([0, \infty), \mathbb{R}^n)$ for the Fréchet topology on $[0, \infty)$, we mean that $(y_k)_k$ is uniformly convergent to y on bounded subsets of $[0, \infty)$. For a detailed discussion about the Fréchet topology on the space of regulated functions on unbounded intervals, we refer the reader to [15].

Lemma 5.5 *Let $(y_k)_k$ be a bounded uniformly S-asymptotically ω-periodic sequence in $S_\omega(X)$. Assume that $(y_k)_k$ converges to y for the Fréchet topology on $[0, \infty)$. Then $y \in S_\omega(X)$.*

Proof Let $\varepsilon > 0$. It follows from our hypotheses that there exists $T_\varepsilon > 0$ such that

$$
\|y_k(t+\omega) - y_k(t)\|_X \leqslant \varepsilon/3
$$

for all $t \geqslant T_\varepsilon$ and $k \in \mathbb{N}$. In addition, for each $t \geqslant T_\varepsilon$ there exists $k_1 \in \mathbb{N}$ such that $\|y(s) - y_k(s)\|_X \leqslant \varepsilon/3$ for all $k \geqslant k_1$ and for all $s \in [0, t+\omega]$. Combining these estimates, we have

$$
\begin{aligned}
\|y(t+\omega) - y(t)\|_X \leqslant & \|y(t+\omega) - y_k(t+\omega)\|_X + \\
& + \|y_k(t+\omega) - y_k(t)\|_X + \|y_k(t) - y(t)\|_X \\
\leqslant & \varepsilon
\end{aligned}
$$

for all $t \geqslant T_\varepsilon$. This implies that $y \in S_\omega(X)$. □

Theorem 5.9 *Assume that the hypotheses of Theorem 5.8 hold. Assume further that the following conditions hold:*

(a) The function h is uniformly S-asymptotically ω-periodic on bounded sets.
(b) For every $N > 0$, the function $\mu(\cdot, N)$ given by $\mu(t, N) = \int_0^t M(s, N)\mathrm{d}g(s)$, $t \geqslant 0$, is S-asymptotically ω-periodic.

Then problem MDE (5.1) *and* (5.2) *on* $[0, \infty)$ *has an S-asymptotically ω-periodic solution.*

Proof We consider the maps A and B given by (5.11) and (5.12), respectively. Using (5.19), we select $N > 0$ large enough such that

$$\|x_0 - h(0, x_0)\| + \sup_{0 \leqslant t < \infty, \|u\| \leqslant N} \|h(t, u)\| + \int_0^\infty M(s, N)\mathrm{d}g(s) \leqslant N. \tag{5.20}$$

We divide the proof in five parts.

Step (i) Let $V_N = \{x \in G([0, \infty), \mathbb{R}^n) : x(0) = x_0, \ \|x\|_\infty \leqslant N\}$. It is clear that V_N is a bounded convex closed subset of $G_b([0, \infty), \mathbb{R}^n)$. Arguing as in [11, Theorem 2.3], we will show that $(I - A)^{-1}B : V_N \to V_N$ is well defined. At first, we observe that for each $y \in V_N$ there exists at most one $x \in V_N$ such that $x = Ax + By$. In fact, if we assume that x_1, x_2 satisfy the condition with $x_1(t) \neq x_2(t)$ for some $t > 0$, then $x_1 - x_2 = Ax_1 - Ax_2$. This implies that

$$\|x_1(t) - x_2(t)\| = \|Ax_1(t) - Ax_2(t)\| \leqslant \varphi(\|x_1(t) - x_2(t)\|) < \|x_1(t) - x_2(t)\|,$$

which is a contradiction. In addition, for a fixed $y \in V_N$, we define the map $F\colon V_N \to G([0, \infty), \mathbb{R}^n)$ by $Fx = Ax + By$. It follows from (5.20) that $\|Fx(t)\| \leqslant N$ for all $t \geqslant 0$. Consequently, we can affirm that $F : V_N \to V_N$. Since F is a nonlinear contraction, applying [8, Theorem 1], we infer that F has a unique fixed point x in V_N. It is clear that $x = (I - A)^{-1}By$.

Step (ii) In what follows, and unless we specify somewhat different, we will consider the space $G_b([0, \infty), \mathbb{R}^n)$ provided with the Fréchet topology and the subsets of $G_b([0, \infty), \mathbb{R}^n)$ provided with the topology induced by $G_b([0, \infty), \mathbb{R}^n)$. We affirm that the map $(I - A)^{-1}\colon B(V_N) \to V_N$ is continuous for the Fréchet topology. In fact, assume that the sequence $(By_j)_j$ converges uniformly on bounded intervals. Let $x_j = Ax_j + By_j$ and $a > 0$. If we assume that $(x_j)_j$ is not a Cauchy sequence, we derive the existence of $\varepsilon > 0$, $t_k \in [0, a]$, $m_k, j_k \to \infty$ such that $\|x_{j_k}(t_k) - x_{m_k}(t_k)\| \geqslant \varepsilon$. In what follows, we will show that this statement is not possible. There exists $\delta > 0$ such that $\varphi(\alpha) \leqslant \alpha - \delta$ for $\varepsilon \leqslant \alpha \leqslant 2N$. Since $\|x_j(t)\| \leqslant N$ for all $t \geqslant 0$, we can apply the previous assertion with $\alpha = \|x_{j_k}(t_k) - x_{m_k}(t_k)\|$. Therefore,

$$\|x_{j_k}(t_k) - x_{m_k}(t_k)\| \leqslant \|Ax_{j_k}(t_k) - Ax_{m_k}(t_k)\| + \|By_{j_k}(t_k) - By_{m_k}(t_k)\|$$

$$\leqslant \varphi(\|x_{j_k}(t_k) - x_{m_k}(t_k)\|) + \|By_{j_k}(t_k) - By_{m_k}(t_k)\|$$
$$\leqslant \|x_{j_k}(t_k) - x_{m_k}(t_k)\| - \delta + \|By_{j_k}(t_k) - By_{m_k}(t_k)\|.$$

Hence, we have

$$\delta \leqslant \|By_{j_k}(t_k) - By_{m_k}(t_k)\| \to 0, \ k \to \infty,$$

which is a contradiction, and this proves the statement.

Step (iii) Let $x(\cdot) \in S_\omega(\mathbb{R}^n) \cap V_N$. We will show that $Ax, Bx \in S_\omega(\mathbb{R}^n)$. It follows from Lemma 5.4 that $t \mapsto h(t, x(t))$ is an S-asymptotically ω-periodic function. In addition, let $v(t) = \int_0^t f(s, x(s))\mathrm{d}g(s)$. Then

$$v(t+\omega) - v(t) = \int_t^{t+\omega} f(s, x(s))\mathrm{d}g(s),$$

which implies that

$$\|v(t+\omega) - v(t)\| = \left\| \int_t^{t+\omega} f(s, x(s))\mathrm{d}g(s) \right\| \leqslant \int_t^{t+\omega} M(s, N)\mathrm{d}g(s) \to 0, \tag{5.21}$$

as $t \to \infty$.

Step (iv) In this step, we will show that $B\colon S_\omega(\mathbb{R}^n) \cap V_N \to S_\omega(\mathbb{R}^n) \cap V_N$ is completely continuous for the Fréchet topology. First, let $(x_k)_k$ be a sequence that converges to x uniformly on bounded intervals. Applying Theorem 5.3 to the sequence $f(s, x_k(s)) - f(s, x(s))$, we conclude that $Bx_k(t) \to Bx(t)$ as $k \to \infty$ uniformly on intervals of type $[0, a]$ for all $a > 0$. This shows that B is continuous for the Fréchet topology. Let $(x_k)_k$ be a sequence in $S_\omega(\mathbb{R}^n) \cap V_N$. Proceeding as in the proof of Theorem 5.6, we can affirm that $B\colon G([0, a], \mathbb{R}^n) \to G([0, a], \mathbb{R}^n)$ is completely continuous for all $a > 0$. Consequently, there exists a subsequence $(x_k^1)_k$ of $(x_k)_k$ such that $(Bx_k^1)_k$ converges uniformly to y^1 in the interval $[0, 1]$. Repeating inductively this argument, we can affirm that there exists a subsequence $(x_k^{j+1})_k$ of $(x_k^j)_k$ such that $(Bx_k^{j+1})_k$ converges uniformly to y^{j+1} in the interval $[0, j+1]$. It is clear that $y^{j+1}(t) = y^j(t)$ for $t \in [0, j]$. This allows us to define $y(t) = y^j(t)$ for $t \in [0, j]$. In addition, using a diagonal selection process, we know that there exists a sequence $(Bx_{k_j}^j)_j$ that converges to y uniformly on compact intervals. Repeating the estimate (5.21), we deduce that the sequence $(Bx_{k_j}^j)_j$ is uniformly S-asymptotically ω-periodic. Applying Lemma 5.5, we can affirm that $y \in S_\omega(\mathbb{R}^n) \cap V_N$.

Step (v) We define $W = \{x \in S_\omega(\mathbb{R}^n) : x(0) = x_0, \ \|x\|_\infty \leqslant N\}$ endowed with the Fréchet topology. For $y \in W$, let $x \in V_N$ such that $x = Ax + By$. This implies that

$$
\begin{aligned}
x(t+\omega)-x(t) &= Ax(t+\omega)-Ax(t)+By(t+\omega)-By(t)\\
&= h(t+\omega, x(t+\omega))-h(t,x(t))+By(t+\omega)-By(t)\\
&= h(t+\omega, x(t+\omega))-h(t,x(t+\omega))+h(t,x(t+\omega))\\
&- h(t,x(t))+By(t+\omega)-By(t),
\end{aligned}
$$

which yields that

$$
\begin{aligned}
\|x(t+\omega)-x(t)\| &\leqslant \|h(t+\omega, x(t+\omega))-h(t,x(t+\omega))\|+\\
&\quad +\|h(t,x(t+\omega))-h(t,x(t))\|+\|By(t+\omega)-By(t)\|\\
&\leqslant \|h(t+\omega, x(t+\omega))-h(t,x(t+\omega))\|+\\
&\quad +\varphi(\|x(t+\omega)-x(t)\|,N)+\|By(t+\omega)-By(t)\|. \qquad (5.22)
\end{aligned}
$$

Next, we analyze each term on the right hand of (5.22). The first term converges to zero when t goes to infinity because the function h is uniformly S-asymptotically ω-periodic on bounded sets. The term $\|By(t+\omega)-By(t)\| \to 0$ as $t \to \infty$ by the assertion in Step (iii). Arguing as in Step (ii), from (5.22) we deduce that $\|x(t+\omega)-x(t)\| \to 0$ as $t \to \infty$. Hence, $x \in W$, and we can define the map $U: W \to W$ given by $Uy = (I-A)^{-1}By$. Combining steps (ii) and (iv), we derive that U is a completely continuous map. Moreover, we can show that $\overline{U(W)} \subseteq W$. In fact, for $y \in \overline{U(W)}$ we can select a sequence $(x_k)_k$ in W such that $U(x_k) \to y$ as $k \to \infty$. It follows from Step (iv) that there exists a subsequence of $(x_k)_k$, which we will continue to denote with the same index, such that $Bx_k \to z$ as $k \to \infty$ with $z \in W$. Applying again Step (ii), we obtain that $U(x_k) = (I-A)^{-1}Bx_k \to (I-A)^{-1}z \in V_N$ as $k \to \infty$. This implies that $y = (I-A)^{-1}z$ and $y = Ay+z$. Arguing as before to get the estimates (5.22), we infer that $y \in S_\omega(\mathbb{R}^n)$, which in turn implies that $y \in W$. Applying Schauder-Tychonoff's Theorem [23, Theorem II.7.1.13], we obtain that U has a fixed point x in W. Therefore, x is a fixed point of $A+B$, and $x(\cdot) \in S_\omega(\mathbb{R}^n)$ is a solution of problem MDE (5.1) and (5.2) on $[0,\infty)$.

□

5.5 Continuous Dependence on Parameters

In this section, we present a continuous dependence result on parameters for the problem MDE (5.1) and (5.2) on $J = [t_0, t_0+a]$ for functions $h, f: J \times \mathbb{R}^n \to \mathbb{R}^n$ and $g: J \to \mathbb{R}$. We assume that f satisfies conditions (A0)–(A1) and that g is a nondecreasing left-continuous function. To simplify the presentation, we will restrict ourselves to consider the type of conditions considered in Theorem 5.6. Hence, in order to present our statement, let $h_k, f_k: J \times \mathbb{R}^n \to \mathbb{R}^n$, $k \in \mathbb{N}$, be functions which satisfy conditions (A0)–(A3) with respect to functions $\varphi_k: \mathbb{R}_0^+ \to$

$\mathbb{R}_0^+$, $M_k\colon J \to \mathbb{R}^+$, for $k \in \mathbb{N}$, respectively. We denote by A_k and B_k the operators associated with functions h_k, f_k by mean of expressions (5.11) and (5.12), respectively. We assume that φ_k are $\mathfrak{D}$-functions that satisfy the condition $\varphi_k(t) < t$ uniformly on bounded intervals. To specify this requirement, we introduce the following condition:

(I) For every $0 < c < d$, there is $\delta > 0$ such that $\varphi_k(t) < t - \delta$ for all $t \in [c, d]$ and $k \in \mathbb{N}$.

As a consequence, since the hypotheses involved in the statement of Theorem 5.6 are fulfilled, we can affirm that the following problem for $t \in J$

$$x_k(t) = x_k(t_0) - h_k(t_0, x_k(t_0)) + h_k(t, x_k(t)) + \int_{t_0}^{t} f_k(s, x_k(s))\mathrm{d}g(s), \tag{5.23}$$

$$x_k(t_0) = \widehat{x}_k \tag{5.24}$$

has a solution $x_k(\cdot)$ for $k \in \mathbb{N}$.

We are in a position to present the following result of continuous dependence on parameters.

Theorem 5.10 *Under the above conditions, assume further that condition (I) holds and that the following conditions are fulfilled:*

(i) The sequence $\widehat{x}_k \to x_0$ as $k \to \infty$.
(ii) The sequence $h_k(t, u) \to h(t, u)$ as $k \to \infty$ uniformly for $t \in J$ and u in bounded subsets of $\mathbb{R}^n$.
(iii) The sequence $f_k(t, u) \to f(t, u)$ as $k \to \infty$ for $t \in J$ and uniformly for u in bounded subsets of $\mathbb{R}^n$, the sequence of functions $(M_k)_k$ converges pointwise on J to a function $M\colon J \to \mathbb{R}^n$, and there exists a constant $C > 0$ having the following property: For every subdivision $d = \{\sigma_0, \ldots, \sigma_l\}$ of the interval J and every finite subset $\{m_1, \ldots, m_l\}$ of $\mathbb{N}$, the inequality $\sum_{j=1}^{l} \int_{\sigma_{j-1}}^{\sigma_j} M_{m_j}(s)\mathrm{d}g \leqslant C$ holds.

If there exists $R > 0$ such that

$$\liminf_{k\to\infty} \sup_{r \geqslant R} \frac{\varphi_k(r)}{r} < 1, \tag{5.25}$$

then there is a subsequence of $(x_k)_k$ that converges uniformly on J to a solution $x(\cdot)$ of problem MDE (5.1) *and* (5.2).

Proof We separate the proof in four steps.

Step 1 We show that the set of functions $\{x_k : k \in \mathbb{N}\}$ is bounded on J. First, we observe that $h_k(t_0, \widehat{x}_k) \to h(t_0, x_0)$ as $k \to \infty$. In fact,

$$\|h_k(t_0, \widehat{x}_k) - h(t_0, x_0)\| \leqslant \|h_k(t_0, \widehat{x}_k) - h_k(t_0, x_0)\| + \|h_k(t_0, x_0) - h(t_0, x_0)\|$$

$$\leqslant \varphi_k(\|\widehat{x}_k - x_0\|) + \|h_k(t_0, x_0) - h(t_0, x_0)\|$$
$$< \|\widehat{x}_k - x_0\| + \|h_k(t_0, x_0) - h(t_0, x_0)\| \to 0, \ k \to \infty.$$

Moreover, from (5.23) and (5.24), we have

$$\begin{aligned}\|x_k(t)\| &\leqslant \|\widehat{x}_k - h_k(t_0, \widehat{x}_k)\| + \|h_k(t, x_k(t)) - h_k(t, 0)\| + \|h_k(t, 0)\| + \\ &\quad + \int_{t_0}^{t_0+a} M_k(s)\mathrm{d}g(s) \\ &\leqslant \|\widehat{x}_k - h(t_0, x_0)\| + \|h(t_0, x_0) - h_k(t_0, \widehat{x}_k)\| + \|h_k(t, 0)\| \\ &\quad + \varphi_k(\|x_k(t)\|) + \int_{t_0}^{t_0+a} M_k(s)\mathrm{d}g(s) \\ &\leqslant C_1 + \varphi_k(\|x_k(t)\|)\end{aligned}$$

for certain constant $C_1 \geqslant 0$ and all $t \in J$. Let $N_k = \sup_{t \in J} \|x_k(t)\|$. It follows from the above estimates that

$$1 \leqslant \frac{C_1}{N_k} + \frac{\varphi_k(\|x_k(t)\|)}{N_k} \leqslant \frac{C_1}{N_k} + \frac{\varphi_k(N_k)}{N_k} \tag{5.26}$$

for all $k \in \mathbb{N}$. Assume that the set $\{N_k : k \in \mathbb{N}\}$ is unbounded. It follows from the estimates (5.26) that $\liminf_{k\to\infty} \dfrac{\varphi_k(N_k)}{N_k} \geqslant 1$. On the other hand, it follows from (5.25) that

$$\liminf_{k\to\infty} \frac{\varphi_k(N_k)}{N_k} \leqslant \liminf_{k\to\infty} \sup_{r \geqslant R} \frac{\varphi_k(r)}{r} < 1,$$

which contradicts our previous assertion. Hence, we get that there is a constant $N > 0$ independent of k such that $\|x_k(t)\| \leqslant N$ for all $t \in J$.

Step 2 For each $v \in \mathbb{R}^n$ the function $h(\cdot, v)$ is regulated because it is the uniform limit of the sequence of regulated functions $h_k(\cdot, v)$ as $k \to \infty$. For the same reason, the set $\{h_k(\cdot, v) : k \in \mathbb{N}\}$ is equiregulated on J. In addition, for a bounded set $K \subset \mathbb{R}^n$, the set of functions $\{h_k(\cdot, v) : v \in K, \ k \in \mathbb{N}\}$ is equiregulated on J. In order to prove that the left and right limits of the functions $h_k(\cdot, v)$ exist uniformly for $v \in K$ and $k \in \mathbb{N}$, it suffices to show that we can approximate the functions $h_k(\cdot, v)$ by functions taken from an equiregulated set of functions. Specifically, for every $\varepsilon > 0$, since K is a relatively compact set in $\mathbb{R}^n$, there exist $v_i \in \mathbb{R}^n$ for $i = 1, \ldots, m$ such that for each $v \in K$ there is $i = 1, \ldots, m$ with $\|v - v_i\| \leqslant \varepsilon$. By our previous remark, the set $\{h_k(\cdot, v_i) : k \in \mathbb{N}, \ i = 1, \ldots, m\}$ is equiregulated on J. Moreover, for $v \in K$, by choosing $i = 1, \ldots, m$ such that $\|v - v_i\| \leqslant \varepsilon$, we have

$$\|h_k(t, v) - h_k(t, v_i)\| \leqslant \varphi_k(\|v - v_i\|) \leqslant \varepsilon$$

for all $k \in \mathbb{N}$ and $t \in J$. This shows that $\{h_k(\cdot, v) : v \in K,\ k \in \mathbb{N}\}$ is equiregulated on J.

Step 3 The set of functions $\{x_k(\cdot) : k \in \mathbb{N}\}$ is equiregulated on J. It follows from (5.23) that

$$x_k(t) = A_k x_k(t) + B_k x_k(t).$$

First we prove that the set $\{B_k x_k : k \in \mathbb{N}\}$ is equiregulated on J. Let

$$p_k(t) = \int_{t_0}^{t} f_k(s, x_k(s))\mathrm{d}g(s), \ \ t \in J.$$

It follows from (iii) that the set $\{p_k(t) : k \in \mathbb{N}, t \in J\}$ is bounded. Moreover, for $t_0 \leqslant t < t_0 + a$, from (5.12), and applying Theorem 5.2, we have that

$$B_k x_k(t^+) - B_k x_k(t) = f_k(t, x_k(t))\Delta^+ g(t),$$

which shows that $\{B_k x_k : k \in \mathbb{N}\}$ is right equiregulated at t. In a similar way, we can show that $\{B_k x_k : k \in \mathbb{N}\}$ is left equiregulated at $t \in (t_0, t_0 + a]$.

On the other hand, for $t_1, t_2 \in J$, we can write

$$x_k(t_2) - x_k(t_1) = A_k x_k(t_2) - A_k x_k(t_1) + B_k x_k(t_2) - B_k x_k(t_1)$$

from which we have

$$\begin{aligned} \|x_k(t_2) - x_k(t_1)\| &\leqslant \|A_k x_k(t_2) - A_k x_k(t_1)\| + \|B_k x_k(t_2) - B_k x_k(t_1)\| \\ &\leqslant \varphi_k(\|x_k(t_2) - x_k(t_1)\| + \|B_k x_k(t_2) - B_k x_k(t_1)\|. \end{aligned} \quad (5.27)$$

Assume that $\{x_k(\cdot) : k \in \mathbb{N}\}$ is not equiregulated. Specifically, to fix ideas, suppose that condition (i) from Definition 5.3 is not verified at point τ_0. Then we can affirm that there exist $\varepsilon > 0$ and two sequences $(t_1^k)_k$ and $(t_2^k)_k$ that converge to τ_0^- and $\|x_k(t_2^k) - x_k(t_1^k)\| \geqslant \varepsilon$ for all $k \in \mathbb{N}$. Using (5.27) and condition (I) with ε instead of c and $2N$ instead d, we can choose $\delta > 0$ such that

$$\|x_k(t_2^k) - x_k(t_1^k)\| \leqslant \|x_k(t_2^k) - x_k(t_1^k)\| - \delta + \|B_k x_k(t_2^k) - B_k x_k(t_1^k)\|,$$

and taking the limit above as $k \to \infty$, we derive that $\delta = 0$, which is a contradiction.

Step 4 From Step 1, Step 3, and Lemma 5.1, we infer that the set $\{x_k(\cdot) : k \in \mathbb{N}\}$ is relatively compact in the space $G(J, \mathbb{R}^n)$. Consequently, $(x_k)_{k\in\mathbb{N}}$ contains a subsequence which is uniformly convergent on J. Without loss of generality, we can denote this subsequence again by $(x_k)_{k\in\mathbb{N}}$. Let $x = \lim_{k\to\infty} x_k$. In particular,

$$x(t_0) = \lim_{k\to\infty} x_k(t_0) = \lim_{k\to\infty} \widehat{x}_k = x_0.$$

Moreover, we point out that $h_k(t, x_k(t)) \to h(t, x(t))$ as $k \to \infty$ uniformly on J. In fact,

$$\begin{aligned}\|h_k(t, x_k(t)) - h(t, x(t))\| &\leqslant \|h_k(t, x_k(t)) - h_k(t, x(t))\| \\ &\quad + \|h_k(t, x(t)) - h(t, x(t))\| \\ &< \|x_k(t) - x(t)\| + \|h_k(t, x(t)) - h(t, x(t))\|,\end{aligned}$$

and the assertion is a consequence of hypothesis (ii). In addition,

$$\int_{t_0}^{t} f_k(s, x_k(s))\mathrm{d}g \to \int_{t_0}^{t} f(s, x(s))\mathrm{d}g$$

as $k \to \infty$. Indeed, in the first place, we can see that $f_k(s, x_k(s)) \to f(s, x(s))$ as $k \to \infty$ for $s \in J$. As a matter of fact, gathering hypothesis (iii) with the continuity of the function $f(t, \cdot)$, we obtain that

$$\begin{aligned}\|f_k(t, x_k(t)) - f(t, x(t))\| &\leqslant \|f_k(t, x_k(t)) - f(t, x_k(t))\| \\ &\quad + \|f(t, x_k(t)) - f(t, x(t))\| \\ &\to 0,\end{aligned}$$

as $k \to \infty$. Furthermore, let $d = \{\sigma_0, \ldots, \sigma_l\}$ be a subdivision of the interval $[t_0, t_0 + a]$, and let $\{m_1, \ldots, m_l\}$ be a finite subset of $\mathbb{N}$. Then applying our hypothesis (iii), we have

$$\left\| \sum_{j=1}^{l} \int_{\sigma_{j-1}}^{\sigma_j} f_{m_j}(s, x_{m_j}(s))\mathrm{d}g(s) \right\| \leqslant \sum_{j=1}^{l} \int_{\sigma_{j-1}}^{\sigma_j} M_{m_j}(s)\mathrm{d}g(s) < C.$$

Thus, from Theorem 5.3, we obtain the assertion.

Finally, collecting the properties established above, taking limit in (5.23) as $k \to \infty$, we obtain

$$x(t) = x_0 - h(t_0, x_0) + h(t, x(t)) + \int_{t_0}^{t} f(s, x(s))\mathrm{d}g(s), \quad t \in J,$$

which shows that $x(\cdot)$ is a solution of problem MDE (5.1) and (5.2). □

Acknowledgments We dedicate this work to the memory of our dear coauthor Hernán R. Henríquez who passed away on Thursday, June 2, 2022.

C. A. Gallegos was partially supported by ANID/FONDECYT postdoctorado No. 3220147, and J. G. Mesquita was partially supported by CNPq grant 307582/2018-3 and FAPDF 00193-00001821/2022-21.

References

1. F.M. Atici, D.C. Biles, A. Lebedinsky, An application of time scales to economics. Math. Comput. Model. **43**(7–8), 718–726 (2006)
2. M. Benchohra, J. Henderson, S.K. Ntouyas, A. Ouahab, On first order impulsive dynamic equations on time scales. J. Difference Equ. Appl. **10**(6), 541–548 (2004)
3. M. Benchohra, J. Henderson, S.K. Ntouyas, A. Ouahab, Impulsive functional dynamic equations on time scales. Dyn. Syst. Appl. **15**(1), 43–52 (2006)
4. L. Berezansky, E. Braverman, On impulsive Beverton–Holt difference equations and their applications. J. Differ. Equ. Appl. **10**(9), 851–868 (2004)
5. L. Berezansky, E. Braverman, Impulsive equations: overview and open problems. Funct. Differ. Equ. **15**(1–2), 39–56 (2008)
6. M. Bohner, A. Peterson, *Dynamic Equations on Time Scales: An Introduction with Applications* (Birkhäuser, Boston, 2001)
7. M. Bohner, A. Peterson, *Advances in Dynamic Equations on Time Scales* (Birkhäuser, Boston, 2003)
8. B. Boyd, J.S.W. Wong, On nonlinear contractions. Proc. Amer. Math. Soc. **20**, 456–464 (1969)
9. M. Cichoń, B. Satco, Existence theory for semilinear evolution inclusions involving measures. Math. Nachr. **290**(7), 1004–1016 (2017)
10. P.C. Das, R.R. Sharma, On optimal controls for measure delay-differential equations. SIAM J. Control **9**, 43–61 (1971)
11. B.C. Dhage, Some variants of two basic hybrid fixed point theorems of Krasnoselskii and Dhage with applications. Nonlinear Stud. **25**(3), 559–573 (2018)
12. B.C. Dhage, N.S. Jadhav, Basic results in the theory of hybrid differential equations with linear perturbations of second type. Tamkang J. Math. **44**(2), 171–186 (2013)
13. L. Di Piazza, V. Marraffa, B. Satco, Closure properties for integral problems driven by regulated functions via convergence results. J. Math. Anal. Appl. **466**(1), 690–710 (2018)
14. L. Di Piazza, V. Marraffa, B. Satco, Approximating the solutions of differential inclusions driven by measures. Ann. Mat. Pura Appl. (4) **198**(6), 2123–2140 (2019)
15. S. Dudek, L. Olszowy, Measures of noncompactness in the space of regulated functions on an unbounded interval. Ann. Funct. Anal. **13**(4), Paper No. 63, 13 pp. (2022)
16. M. Federson, J.G. Mesquita, A. Slavík, Measure functional differential equations and functional dynamic equations on time scales. J. Differ. Equ. **252**(6), 3816–3847 (2012)
17. M. Federson, J.G. Mesquita, A. Slavík, Basic results for functional differential and dynamic equations involving impulses. Math. Nachr. **286**(2–3), 181–204 (2013)
18. M. Federson, R. Grau, J.G. Mesquita, E. Toon, Lyapunov stability for measure differential equations and dynamic equations on time scales. J. Differ. Equ. **267**(7), 4192–4223 (2019)
19. M. Fleury, J.G. Mesquita, A. Slavík, Massera's theorems for various types of equations with discontinuous solutions. J. Differ. Equ. **269**(12), 11667–11693 (2020)
20. D. Fraňková, Regulated functions. Math. Bohem. **116**(1), 20–59 (1991)
21. C.A. Gallegos, H.R. Henríquez, J.G. Mesquita, Growth of solutions for measure differential equations and dynamic equations on time scales. J. Math. Anal. Appl. **479**(1), 941–962 (2019)
22. C.A. Gallegos, R. Grau, J.G. Mesquita, Stability, asymptotic and exponential stability for various types of equations with discontinuous solutions via Lyapunov functionals. J. Differ. Equ. **299**, 256–283 (2021)
23. A. Granas, J. Dugundji, *Fixed Point Theory* (Springer, New York, 2003)
24. H.R. Henríquez, J.G. Mesquita, Self-accessible states for linear systems on time scales. Proc. Amer. Math. Soc. **146**(3), 1257–1269 (2018)
25. H.R. Henríquez, M. Pierri, P. Taboas, On S-asymptotically ω–periodic functions on Banach spaces and applications. J. Math. Anal. Appl. **343**(2), 1119–1130 (2008)
26. H.R. Henríquez, M. Pierri, P. Taboas, Existence of S-asymptotically ω-periodic solutions for abstract neutral functional-differential equations. Bull. Aust. Math. Soc. **78**(3), 365–382 (2008)

27. R. Henstock, Definitions of Riemann-type of the variational integrals. Proc. Lond. Math. Soc. **11**(3), 402–418 (1961)
28. D.S. Kurtz, C.W. Swartz, *Theories of Integration: The Integrals of Riemann, Lebesgue, Henstock–Kurzweil, and McShane*, Ser. Real Anal., vol. 13 (World Scientific, Singapore, 2012)
29. X. Li, M. Bohner, C-K. Wang, Impulsive equations: periodic solutions and applications. Autom. J. IFAC **52**, 173–178 (2015)
30. R. López Pouso, I. Márquez Albés, General existence principles for Stieltjes differential equations with applications to mathematical biology. J. Differ. Equ. **264**(8), 5388–5407 (2018)
31. H. Lu, S. Sun, D. Yang, H. Teng, Theory of fractional hybrid differential equations with linear perturbations of second type. Bound. Value Probl. **2013**, Article ID 23 (2013)
32. G.A. Monteiro, A. Slavík, Extremal solutions of measure differential equations. J. Math. Anal. Appl. **444**(1), 568–597 (2016)
33. G.A. Monteiro, A. Slavík, M. Tvrdý, *Kurzweil–Stieltjes Integral Theory and Applications*, Ser. Real Anal., vol. 15 (World Scientific, Singapore, 2018)
34. M.Z. Nashed, J.S.W. Wong, Some variants of a fixed point theorem of Krasnoselskii and applications to nonlinear integral equations. J. Math. Mech. **18**, 767–777 (1969)
35. R. Precup, *Methods in Nonlinear Integral Equations* (Springer Science+Business Media, Dordrecht, 2002)
36. B. Satco, Ulam-type stability for differential equations driven by measures. Math. Nachr. **293**(1), 147–157 (2020)
37. W.W. Schmaedeke, Optimal control theory for nonlinear vector differential equations containing measures. SIAM J. Control **3**, 231–280 (1965)
38. Š. Schwabik, *Generalized Ordinary Differential Equations*, Ser. Real Anal., vol. 5 (World Scientific, Singapore, 1992)
39. R.R. Sharma, An abstract measure differential equation. Proc. Amer. Math. Soc. **32**, 503–510 (1972)
40. A. Slavík, Dynamic equations on time scales and generalized ordinary differential equations. J. Math. Anal. Appl. **385**(1), 534–550 (2012)
41. Y. Zhao, Y. Sun, Z. Liu, Z. Bai, Basic theory of differential equations with linear perturbations of second type on time scales. Bound. Value Probl. **2019**, Article ID 183 (2019)

Chapter 6
LaSalle-Type Steady-State Oscillation for Generalized Ordinary Differential Equations and Applications

Rogelio Grau Acuña

Abstract In this chapter, we are interested in investigating steady-state oscillation (in the sense of LaSalle) in the framework of generalized ordinary differential equations (generalized ODEs in short) and their applications to measure differential equations. First, we establish steady-state oscillation for generalized ODEs. Second, we use the well-known correspondence between solutions of generalized ODEs and solutions of measure differential equations, obtaining analog results for the last equations.

Keywords Steady-state oscillation · Measure differential equations · Generalized ordinary differential equations · Henstock-Kurzweil-Stieltjes integral · Periodic solution · Barotropic fluid

MSC 2020 34D20, 34C25, 26A39, 34A06, 34D45

6.1 Introduction

It is a known fact that generalized ordinary differential equations can encompass stochastic differential equation, Volterra integral equations, functional differential equations, impulsive functional differential equations at variable time and fixed time, measure functional differential equations, functional dynamic equations on time scales, and measure neutral functional dynamic equations on time scales, among others. See [1–7, 13, 14].

Furthermore, the advantage when we are dealing with all these types of equations via generalized ODEs is that these last ones represent a tool that allows us to obtain more general conditions than the usual ones found in the literature to the proved

R. Grau Acuña (✉)
Departamento de Matemáticas y Estadística, Universidad del Norte, Barranquilla, Atlántico, Colombia
e-mail: graur@uninorte.edu.co

P. Benevieri, J. G. Mesquita (eds.), *Functional Differential Equations and Dynamic Equations on Time Scales*, Advances in Continuum Mechanics 52,
https://doi.org/10.1007/978-3-031-83327-4_6

results. It happens because when we translate the obtained results for generalized ODEs to these other equations, we get results concerning the integral of the function that appears on the right-hand side of the equation, not on the proper function. These types of conditions allow that the function in the right-hand side of the equation may have bad behavior, with several discontinuities and high oscillation, see, e.g., [1, 10, 11, 14], and the references therein.

Hence, our motivation and the main goal of this chapter are to investigate the steady-state oscillation (in the sense of LaSalle) in the framework of generalized ODEs. Specifically, we show that a class of generalized ODEs has a unique periodic solution, and all solutions approach this one as t approaches infinity. Furthermore, to illustrate our main result we present an example applied to a model of the quasi-static pressure distribution of a perfect and barotropic fluid (see Example 6.1 in Sect. 6.3). Moreover, we present its applicability to measure differential equations of type

$$x(t) = x(0) + \int_0^t \mathcal{G}(x(s), s)\mathrm{d}g(s), \qquad t \geqslant 0, \tag{6.1}$$

where $\mathcal{G} : \mathbb{R}^n \times [0, +\infty) \to \mathbb{R}^n$ and $g : [0, +\infty) \to \mathbb{R}$ are functions and the integral in the right-hand side is understood in the sense of Henstock-Kurzweil-Stieltjes.

It is important to highlight that we can apply the previous steady-state oscillation results to dynamic equations on a time scale of the type

$$x^{\Delta}(t) = f(x^*(t), t), \qquad t \in \mathbb{T}, \tag{6.2}$$

using the well-known correspondence (see [1]) between the solutions of (6.1) and solutions of (6.2).

We organize the next sections of this chapter as follows. The second section is dedicated to recalling the basic background and main results concerning generalized ODEs. In the third section, we are going to present a result about LaSalle-type steady-state oscillation for generalized ODEs and an example to illustrate our result. Finally, in the fourth section, we apply the previous steady-state oscillation results to measure differential equations.

6.2 Preliminaries

In this section, we present the basic concept for generalized ODEs. For more details, see [1, 14].

Throughout this section, let us assume that X is a Banach space with norm $\|\cdot\|$.

We recall that a *tagged division* of a compact interval $[a, b] \subset \mathbb{R}$ is a finite collection given by

$$D = \{(\tau_i, [\alpha_{i-1}, \alpha_i]) : i = 1, 2, \ldots, n\},$$

where $a = \alpha_0 < \alpha_1 < \ldots < \alpha_n = b$ is a division of $[a, b]$ and $\tau_i \in [s_{i-1}, s_i]$, for $i = 1, 2, \ldots, n$.

Let $\delta : [a, b] \to \mathbb{R}^+$ be a function (called a *gauge* on $[a, b]$). Any *tagged division* D of the interval $[a, b]$ is called δ*-fine*, if

$$[\alpha_{i-1}, \alpha_i] \subset (\tau_i - \delta(\tau_i), \tau_i + \delta(\tau_i)), \quad i = 1, \ldots, n.$$

We start by presenting the concept of the Kurzweil integral (see [14]).

Definition 6.1 Let $U : [a, b]\times[a, b] \to X$ be a function. We say that U is *Kurzweil integrable* on $[a, b]$, if there exists $I \in X$ such that given $\varepsilon > 0$, there exists a gauge δ on $[a, b]$ such that

$$\left\| \sum_{i=1}^{k} [U(\tau_i, s_i) - U(\tau_i, s_{i-1})] - I \right\| < \varepsilon$$

for all δ-fine tagged partition of $[a, b]$. In this case, I is called the Kurzweil integral of U on $[a, b]$ and is symbolized by $\int_a^b DU(\tau, t)$.

To the next result, let $\Omega := X \times [0, +\infty)$.

Lemma 6.1 *Let $G : \Omega \to X$, $H : X \to X$ and $x : [a, b] \to X$, $[a, b] \subset [0, +\infty)$ be given. If one of the integrals*

$$\int_a^b D[G(x(\tau), t) + H(x(\tau))], \qquad \int_a^b DG(x(\tau), t)$$

exists, then the other also exists, and we have

$$\int_a^b D[G(x(\tau), t) + H(x(\tau))] = \int_a^b DG(x(\tau), t).$$

Proof Define $U(\tau, t) := G(x(\tau), t) + H(x(\tau))$ and $V(\tau, t) := G(x(\tau), t)$ for all $(\tau, t) \in [a, b] \times [a, b]$. Let $D = \{(\tau_i, [s_{i-1}, s_i]) : i = 1, 2, \ldots, k\}$ be an arbitrary tagged partition of the interval $[a, b]$. Then,

$$\begin{aligned} S(U, D) &= \sum_{i=1}^{k} [G(x(\tau_i), s_i) + H(x(\tau_i)) - G(x(\tau_i), s_{i-1}) - H(x(\tau_i))] \\ &= \sum_{i=1}^{k} [G(x(\tau_i), s_i) - G(x(\tau_i), s_{i-1})] \\ &= S(V, D). \end{aligned}$$

Hence, the integral $\int_a^b D[G(x(\tau), t) + H(x(\tau))]$ exists if and only if $\int_a^b DG(x(\tau), t)$ exists, and they have the same value. □

Next, we present the concept of a solution for generalized ordinary differential equations. For more details, the reader can refer to [1, 14].

Definition 6.2 Let $I \subset [0, +\infty)$ be an interval and $G : X \times [0, +\infty) \to X$ be a function. We say that a function $x : I \to X$ is *a solution of the generalized ordinary differential equation*

$$\frac{\mathrm{d}x}{\mathrm{d}\tau} = DG(x, t)$$

on the interval I, if

$$x(v) = x(u) + \int_u^v DG(x(\tau), s), \quad u, \ v \in I.$$

In order to establish results for generalized ODE, we define a specific class of functions, introduced by J. Kurzweil (see, e.g., [10, 11]).

Definition 6.3 Let $\Omega = X \times [0, +\infty)$ and $h : [0, +\infty) \to \mathbb{R}$ be a nondecreasing function. We say that a function $F : \Omega \to X$ belongs to the class $\mathcal{F}(\Omega, h)$, if the following inequalities hold:

$$\|F(x, s_2) - F(x, s_1)\| \leqslant |h(s_2) - h(s_1)| \tag{6.3}$$

and

$$\|F(x, s_2) - F(x, s_1) - F(y, s_2) + F(y, s_1)\| \leqslant \|x - y\||h(s_2) - h(s_1)| \tag{6.4}$$

for $s_1, \ s_2 \in [0, +\infty)$ and all $x, y \in X$.

Remark 6.1 Note that if $F(x, t) = tH(x)$ and H is a continuous function, then the equation

$$\frac{\mathrm{d}x}{\mathrm{d}\tau} = DF(x, t) = D\,[tH(x)] \tag{6.5}$$

is the generalized ODE associated with the ODE

$$\frac{\mathrm{d}x}{\mathrm{d}t} = H(x). \tag{6.6}$$

Also, x is a solution of the generalized ODE (6.5) if, and only if, x is a solution of the ODE (6.6). For more details, the reader can see [7].

The next result is a Substitution Theorem for the Kurzweil integral [14, Theorem 1.18].

Theorem 6.1 *Assume that* $\phi : [c, d] \to \mathbb{R}$ *is a continuous strictly monotone function on* $[c, d]$. *Let* $U : [\phi(c), \phi(d)] \times [\phi(c), \phi(d)] \to X$ *be given. If one of the integrals*

$$\int_{\phi(c)}^{\phi(d)} DU(\tau, t), \qquad \int_{c}^{d} DU(\phi(\tau), \phi(t))$$

exists, then the other also exists, and we have

$$\int_{\phi(c)}^{\phi(d)} DU(\tau, t) = \int_{c}^{d} DU(\phi(\tau), \phi(t)).$$

The next result can be found in [5, Corollary 3.14].

Theorem 6.2 *Let* $\Omega = X \times [t_0, +\infty)$ *and* $F \in \mathcal{F}(\Omega, h)$, *where the function* h *is nondecreasing and left-continuous. Then for every* $(x_0, s_0) \in \Omega$, *there exists a unique solution of the IVP*

$$\frac{\mathrm{d}x}{\mathrm{d}\tau} = DF(x, t), \qquad x(s_0) = x_0$$

defined in $[s_0, +\infty)$.

The next result gives us another useful estimate. See [1, Lemma 4.6].

Lemma 6.2 *Suppose that* $F \in \mathcal{F}(\Omega, h)$, *where* h *is a nondecreasing function. If* $x, y : [a, b] \to X$ *are regulated functions, then*

$$\left\| \int_{s_1}^{s_2} D[F(x(\tau), t) - F(y(\tau), t)] \right\| \leqslant \int_{s_1}^{s_2} \|x(s) - y(s)\| dh(s),$$

whenever $a \leqslant s_1 \leqslant s_2 \leqslant b$.

In the sequel, we present a Gronwall-type inequality. See [14, Corollary, 1.43].

Lemma 6.3 *Let* $g : [a, b] \to [0, \infty)$ *be a nondecreasing and left-continuous function,* $k > 0$, *and* $l \geqslant 0$. *Assume that* $\psi : [a, b] \to [0, \infty)$ *is bounded and satisfies*

$$\psi(\xi) \leqslant k + l \int_{a}^{\xi} \psi(s) \mathrm{d}g(s), \quad \xi \in [a, b].$$

Then $\psi(\xi) \leqslant k e^{l(g(\xi) - g(a))}$ *for all* $\xi \in [a, b]$.

6.3 Main Result

Throughout this section, let us assume that X is a Banach space with a norm $\|\cdot\|$.

Let $\Omega = X \times [0, +\infty)$, and consider the generalized ODE given by

$$\frac{\mathrm{d}x}{\mathrm{d}\tau} = DF(x, t), \tag{6.7}$$

where we assume that:

(B1) $F \in \mathcal{F}(\Omega, h)$.

(B2) There exist $\omega > 0$ and a function $C : X \to X$ such that

$$F(x, t + \omega) - F(x, t) = C(x),$$

for all $t \geqslant 0$ and all $x \in X$.

For each $x_0 \in X$, let us denote by $x(s, x_0)$ the unique solution of the IVP

$$\begin{cases} \frac{\mathrm{d}x}{\mathrm{d}\tau} = DF(x, t) \\ x(0) = x_0 \end{cases} \tag{6.8}$$

defined for all $s \in [0, +\infty)$. The existence of such a solution is ensured by Theorem 6.2.

The following lemma is crucial in the proof of our main result. See [8, Lemma 2.7].

Lemma 6.4 *If* (B1) *and* (B2) *hold, then the function* h *in* (B1) *can be chosen in such a way that* $t \mapsto h(t + \omega) - h(t)$ *is a constant function on* $[0, +\infty)$.

In the sequel, we introduce the concept of steady-state oscillation (in the sense of LaSalle) in the framework of generalized ODEs. This concept is inspired by [12, Section 26].

Definition 6.4 We say that the generalized ODE (6.7) has a *steady-state oscillation* (in the sense of LaSalle) if (6.7) has a unique periodic solution, and all solutions approach this one as t approaches infinity.

The following results will be crucial to prove that the generalized ODE (6.7) has a steady-state oscillation.

Lemma 6.5 *Consider the generalized ODE* (6.7)*, and assume that* $(B1)$ *and* $(B2)$ *are satisfied. Moreover, suppose that* $x_0 \in X$*. Then,*

$$x(s + \omega, x_0) = x(s, x(\omega, x_0)), \tag{6.9}$$

for all $s \geqslant 0$.

Proof Define a function $y : [0, +\infty) \to X$ given by $y(s) = x(s + \omega, x_0)$, for all $s \geqslant 0$.

We will prove that y is a solution of the IVP

$$\frac{\mathrm{d}x}{\mathrm{d}\tau} = DF(x, t), \qquad x(0) = x(\omega, x_0). \tag{6.10}$$

In fact, let $s \geqslant 0$. Then, by condition $(B2)$, Theorem 6.1, and Lemma 6.1, we have

$$\begin{aligned}
y(s) &= x(s + \omega, x_0) \\
&= x_0 + \int_0^{s+\omega} DF(x(\tau, x_0), t) \\
&= x_0 + \int_0^{\omega} DF(x(\tau, x_0), t) + \int_{\omega}^{s+\omega} DF(x(\tau, x_0), t) \\
&= x(\omega, x_0) + \int_{\phi(0)}^{\phi(s)} DF(x(\tau, x_0), t), \quad \phi(\xi) := \xi + \omega, \ \xi \in [0, s] \\
&= y(0) + \int_0^{s} DF(x(\phi(\tau), x_0), \phi(t)) \\
&= y(0) + \int_0^{s} DF(x(\tau + \omega, x_0), t + \omega) \\
&= y(0) + \int_0^{s} DF(y(\tau), t + \omega) \\
&= y(0) + \int_0^{s} DF(y(\tau), t).
\end{aligned}$$

Hence, $y(s)$ and $x(s, x(\omega, x_0))$ are the solution of the IVP (6.10). By uniqueness, we have

$$y(s) = x(s, x(\omega, x_0)),$$

and, therefore,

$$x(s + \omega, x_0) = x(s, x(\omega, x_0)),$$

for all $s \geqslant 0$. □

Corollary 6.1 *Consider the generalized ODE* (6.7)*, and assume that* $(B1)$ *and* $(B2)$ *are satisfied. Moreover, suppose that* $x_0 \in X$*. Then,*

$$x(s + n\omega, x_0) = x(s, x(n\omega, x_0)), \tag{6.11}$$

for all $s \geqslant 0$ *and all* $n \in \mathbb{N}$.

Proof The proof is immediate using induction and Lemma 6.5. □

Corollary 6.2 *Consider the generalized ODE* (6.7)*, and assume that* $(B1)$ *and* $(B2)$ *are satisfied. Moreover, suppose that* $x_0 \in X$*. Then,*

$$x(n\omega, x_0) = x(\omega, x((n-1)\omega, x_0)), \tag{6.12}$$

for all $n \in \mathbb{N}$.

Proof We proceed by induction on n.

Firstly, for $n = 1$, we have

$$x(\omega, x_0) = x(\omega, x(0, x_0)) = x(\omega, x((1-1)\omega, x_0)).$$

For $n = 2$, by Lemma 6.5, we obtain

$$x(2\omega, x_0) = x(\omega + \omega, x_0) = x(\omega, x(\omega, x_0)).$$

Assume that

$$x(i\omega, x_0) = x(\omega, x((i-1)\omega, x_0)), \text{ for all integers } 1 \leqslant i \leqslant n.$$

Let us prove the case $i = n+1$. Indeed, by Lemma 6.5 and the induction hypothesis, we have

$$\begin{aligned}
x((n+1)\omega, x_0) &= x(n\omega + \omega, x_0) \\
&= x(n\omega, x(\omega, x_0)) \\
&= x((n-1)\omega + \omega, x(\omega, x_0)) \\
&= x((n-1)\omega, x(\omega, x(\omega, x_0))) \\
&= x((n-1)\omega, x(2\omega, x_0)) \\
&= x((n-2)\omega + \omega, x(2\omega, x_0)) \\
&= x((n-2)\omega, x(\omega, x(2\omega, x_0))) \\
&= x((n-2)\omega, x(3\omega, x_0)) \\
&\ \ \vdots \\
&= x((n-(n-1))\omega, x(((n-1)+1)\omega, x_0)) \\
&= x(\omega, x(n\omega, x_0)).
\end{aligned}$$

Therefore, the identity also holds when $i = n+1$. This completes the induction. □

The next result ensures that the generalized ODE (6.7) has a steady-state oscillation.

Theorem 6.3 *Consider the generalized ODE* (6.7), *and assume that* $(B1)$ *and* $(B2)$ *are satisfied. Let* $\omega > 0$ *and* $a : [0, +\infty) \to (0, +\infty)$ *be a continuous function such that* $r = \lim_{n\to\infty} a(n\omega) < 1$. *Moreover, assume the following condition:*

(H1) *Any two solutions* z *and* y *of the generalized ODE* (6.7) *satisfy*

$$\|z(t) - y(t)\| \leqslant a(t)\|z(0) - y(0)\|. \tag{6.13}$$

Then generalized ODE (6.7) *has a steady-state oscillation.*

Proof Firstly we prove that the generalized ODE (6.7) has a unique ω-periodic solution. Indeed, we define the operator (called Poincaré map) $\mathcal{P}\colon X \to X$ given by $\mathcal{P}(z_0) := x(\omega, z_0)$ for all $z_0 \in X$.

Since $r = \lim_{k\to\infty} a(k\omega) < 1$, there exist $m_0 \in \mathbb{N}$ and $r < \lambda < 1$ such that

$$a(k\omega) < \lambda, \quad \text{for all } k \geqslant m_0. \tag{6.14}$$

We are going to prove by induction that

$$\mathcal{P}^n(z_0) = x(n\omega, z_0), \quad \text{for all } n \in \mathbb{N} \text{ and all } z_0 \in X, \tag{6.15}$$

where $\mathcal{P}^n := \underbrace{\mathcal{P} \circ \mathcal{P} \circ \ldots \circ \mathcal{P}}_{n-\text{times}}$. Indeed, when $n = 1$, we have

$$\mathcal{P}^1(z_0) = \mathcal{P}(z_0) = x(\omega, z_0).$$

Suppose that the equality $\mathcal{P}^i(z_0) = x(i\omega, z_0)$ holds for all integers $1 \leqslant i \leqslant n$. We want to show that it also holds when $i = n + 1$. In fact, by Corollary 6.2 and the induction hypothesis, we have

$$\mathcal{P}^{n+1}(z_0) = \mathcal{P}(\mathcal{P}^n(z_0)) = \mathcal{P}(x(n\omega, z_0)) = x(\omega, x(n\omega, z_0)) = x((n+1)\omega, z_0),$$

getting (6.15).

Now, for $z_0, y_0 \in X$, by (6.13), (6.14), and (6.15), we get

$$\begin{aligned} \|\mathcal{P}^{m_0}(z_0) - \mathcal{P}^{m_0}(y_0)\| &= \|x(m_0\omega, z_0) - x(m_0\omega, y_0)\| \\ &\leqslant a(m_0\omega)\|z_0 - y_0\| \\ &< \lambda\|z_0 - y_0\|. \end{aligned}$$

Thus, $\mathcal{P}^{m_0}$ is a contraction. Using this fact together with [9, Lemma 5.4-3], we conclude that $\mathcal{P}$ has a unique fixed point $x_0 \in X$, that is, $x(\omega, x_0) = x_0 = x(0, x_0)$. By the uniqueness of solutions, we have

$$x(t + \omega, x_0) = x(t, x_0), \quad \text{for all } t \geqslant 0,$$

which proves that $x(t, x_0)$ is the unique ω-periodic solution of the generalized ODE (6.7).

Now, we want to show that every solution of the generalized ODE (6.7) approaches $x(s, x_0)$ as $s \to \infty$.

Let $y_0 \in X$ arbitrary. By induction, one can show that

$$\|x(s + nm_0\omega, x_0) - x(s + nm_0\omega, y_0)\| \leqslant a(s)\|x_0 - y_0\|\lambda^n, \tag{6.16}$$

for all $n \in \mathbb{N}$ and all $s \geqslant 0$. In fact, when $n = 1$, by (6.13), (6.14), and Corollary 6.1, we get

$$\begin{aligned}
\|x(s + m_0\omega, x_0) - x(s + m_0\omega, y_0)\| &= \|x(s, x(m_0\omega, x_0)) - x(s, x(m_0\omega, y_0))\| \\
&\leqslant a(s)\|x(m_0\omega, x_0) - x(m_0\omega, y_0)\| \\
&\leqslant a(s)a(m_0\omega)\|x_0 - y_0\| \\
&\leqslant a(s)\|x_0 - y_0\|\lambda.
\end{aligned}$$

Assume that for all integers $1 \leqslant i \leqslant n$, the inequality holds. Let us prove the case $i = n + 1$. In fact, by (6.13), (6.14), Corollary 6.1, and the induction hypothesis, we have

$$\begin{aligned}
&\|x(s + (n + 1)m_0\omega, x_0) - x(s + (n + 1)m_0\omega, y_0)\| \\
&= \|x(s, x((n + 1)m_0\omega, x_0)) - x(s, x((n + 1)m_0\omega, y_0))\| \\
&\leqslant a(s)\|x((n + 1)m_0\omega, x_0) - x((n + 1)m_0\omega, y_0)\| \\
&= a(s)\|x(m_0\omega + nm_0\omega, x_0) - x(m_0\omega + nm_0\omega, y_0)\| \\
&\leqslant a(s)a(m_0\omega)\|x_0 - y_0\|\lambda^n \\
&\leqslant a(s)\|x_0 - y_0\|\lambda^{n+1},
\end{aligned}$$

which proves the assertion.

Let $\varepsilon > 0$ arbitrary. Since

$$\lim_{n\to\infty} a(0)\|x_0 - y_0\|\lambda^n e^{h(\omega)-h(0)} = 0,$$

there exists $n_0 \in \mathbb{N}$ such that

$$a(0)\|x_0 - y_0\|\lambda^n e^{h(\omega)-h(0)} < \varepsilon,$$

for all $n \geqslant n_0$.

Now, let $s \geqslant n_0m_0\omega$ be given. Since $[n_0m_0\omega, +\infty) = \bigcup_{k=n_0}^{\infty} [km_0\omega, (k+1)m_0\omega)$, there is a unique natural number $k \geqslant n_0$ such that

$$km_0\omega \leqslant s < (k+1)m_0\omega.$$

According to the definition of solution and Lemma 6.2, we have

$$\begin{aligned}
&\|x(s, x_0) - x(s, y_0)\| \\
&= \|x(km_0\omega, x_0) + \int_{km_0\omega}^{s} DF(x(\tau, x_0), t) - x(km_0\omega, y_0) - \int_{km_0\omega}^{s} DF(x(\tau, y_0), t)\| \\
&\leqslant \|x(km_0\omega, x_0) - x(km_0\omega, x_0)\| + \|\int_{km_0\omega}^{s} D[F(x(\tau, x_0), t) - F(x(\tau, y_0), t)]\| \\
&\leqslant a(0)\|x_0 - y_0\|\lambda^k + \|\int_{km_0\omega}^{s} D[F(x(\tau, x_0), t) - F(x(\tau, y_0), t)]\| \\
&\leqslant a(0)\|x_0 - y_0\|\lambda^k + \int_{km_0\omega}^{s} \|x(\xi, x_0) - x(\xi, y_0)\|\, dh(\xi).
\end{aligned}$$

Hence, by Lemmas 6.3 and 6.4 and the fact that h is a nondecreasing function, we have

$$\begin{aligned}
\|x(s, x_0) - x(s, y_0)\| &\leqslant a(0)\|x_0 - y_0\|\lambda^k e^{h(s) - h(km_0\omega)} \\
&\leqslant a(0)\|x_0 - y_0\|\lambda^k e^{h((k+1)m_0\omega) - h(km_0\omega)} \\
&= a(0)\|x_0 - y_0\|\lambda^k e^{h(\omega) - h(0)} \\
&< \varepsilon,
\end{aligned}$$

which holds for each $s \geqslant n_0 m_0\omega$. Since $\varepsilon > 0$ is arbitrary, it follows that

$$\lim_{s\to\infty} \|x(s, x_0) - x(s, y_0)\| = 0,$$

and the proof is complete. □

Example 6.1 Let $\omega > 0$, $U := \{a \in \mathbb{R} : |a| < 1\}$, and $\Omega = U \times [0, +\infty)$. Consider the function $F : \Omega \to \mathbb{R}$ given by $F(p, z) = -g_0 z p$ with $g_0 > 0$.

Consider the generalized ODE

$$\frac{\mathrm{d}p}{\mathrm{d}\tau} = DF(p, z) = D[-g_0 z p]. \tag{6.17}$$

By Remark 6.1, Eq. (6.17) is equivalent to the ODE

$$\frac{\mathrm{d}p}{\mathrm{d}z} = -g_0 p. \tag{6.18}$$

Equation (6.18) represents a model (see Remark 6.2) of the quasi-static pressure distribution of a perfect and barotropic fluid that has a kinetic equation of state defined by $\rho = p$ with, under the action of the gravitational field given by $b = (0, 0, -g_0)$.

Next, we will show that (B1), (B2), and (H1) hold.

In fact, for all $(p, z_2), (p, z_1), (q, z_2), (q, z_1) \in \Omega$, we have

$$\begin{aligned}|F(p, z_2) - F(p, z_1)| &= |-g_0 z_2 p + g_0 z_1 p| = g_0 |(z_2 - z_1)p| \leqslant g_0 |(z_2 - z_1)| \\ &= |h(z_2) - h(z_1)|\end{aligned}$$

and

$$\begin{aligned}|F(p, z_2) - F(p, z_1) - F(q, z_2) + F(q, z_1)| &= |-g_0 z_2 p + g_0 z_1 p + g_0 z_2 q - g_0 z_1 q| \\ &= g_0 \,|(z_2 - z_1)|\, |p - q| \\ &= |h(z_2) - h(z_1)||p - q|,\end{aligned}$$

getting (B1), with $h(z) := g_0 z$.

Now, we will prove (B2). Indeed, defining $C(p) := -g_0 \omega p$, we get $F(p, z + \omega) - F(p, z) = C(p)$, obtaining (B2).

On the other hand, it is easy to verify that if u and v are solutions of (6.17), then

$$u(z) = u(0)e^{-g_0 z} \quad \text{and} \quad v(z) = v(0)e^{-g_0 z}.$$

Put $a(z) := 2e^{-g_0 z}$, and we have that a is a continuous function, $\lim_{n\to\infty} a(n\omega) = 0 < 1$ and

$$|u(z) - v(z)| \leqslant a(z)|u(0) - v(0)|,$$

proving the condition (H1). Since all hypothesis of Theorem 6.3 are satisfied, the generalized ODE (6.7) has a steady-state oscillation.

Remark 6.2 The Momentum Conservation Equation for the quasi-static case is

$$-\nabla p + \rho b = 0. \tag{6.19}$$

Using this fact together with $b = (0, 0, -g_0)$, we find $\frac{\partial p}{\partial x} = \frac{\partial p}{\partial y} = 0$. Hence, $p = p(z)$ and (6.19) reduces to

$$\frac{\mathrm{d}p}{\mathrm{d}z} = -g_0 p,$$

since $\rho = p$.

6.4 Applications to Measure Differential Equations

In this section, we apply the results on steady-state oscillation for a class of measure differential equations (MDEs, for short). We will use the correspondence between generalized ODEs and MDEs to obtain the results.

Consider the integral form of an MDE of type:

$$x(t) = x(0) + \int_0^t \mathcal{G}(x(s), s)\mathrm{d}g(s), \qquad t \geqslant 0, \tag{6.20}$$

where $\mathcal{G} : \mathbb{R}^n \times [0, +\infty) \to \mathbb{R}^n$ and $g\colon [0, +\infty) \to \mathbb{R}$ are functions, and the integral in the right-hand side is understood in the sense of Kurzweil-Henstock-Stieltjes.

We recall the reader that $G([0, +\infty), \mathbb{R}^n)$ denotes the vector space of functions $x\colon [0, +\infty) \to \mathbb{R}^n$ such that $x|_{[a,b]}$ belongs to the space $G([a, b], \mathbb{R}^n)$, for all $[a, b] \subset [0, +\infty)$. On the other hand, the symbol $G_0([0, +\infty), \mathbb{R}^n)$ denotes the vector space of all functions $x \in G([0, +\infty), \mathbb{R}^n)$ such that $\sup\limits_{s\in[0,+\infty)} e^{-s} |x(s)| < +\infty$. This space is endowed with the norm

$$\|x\|_{[0,+\infty)} = \sup_{s\in[0,+\infty)} e^{-s} |x(s)| , \qquad x \in G_0([0, +\infty), \mathbb{R}^n),$$

and it becomes a Banach space (see [1, Proposition 1.9]).

We will suppose the following conditions:

(D1) The function $g : [0, +\infty) \to \mathbb{R}$ is nondecreasing and left-continuous on $(0, +\infty)$.

(D2) The Kurzweil-Henstock-Stieltjes integral

$$\int_{s_1}^{s_2} \mathcal{G}(x(t), t)\mathrm{d}g(t)$$

exists, for all $x \in G([0, +\infty), \mathbb{R}^n)$ and all $0 \leqslant s_1 \leqslant s_2 < +\infty$.

(D3) There exists a locally Kurzweil-Henstock-Stieltjes integrable function $M : [0, +\infty) \to \mathbb{R}^+$ with respect to g such that

$$\left| \int_{s_1}^{s_2} \mathcal{G}(x(s), s)\mathrm{d}g(s) \right| \leqslant \int_{s_1}^{s_2} M(s)\mathrm{d}g(s),$$

for all $x \in G([0, +\infty), \mathbb{R}^n)$ and all $0 \leqslant s_1 \leqslant s_2 < +\infty$.

(D4) There exists a locally Kurzweil-Henstock-Stieltjes integrable function $L : [0, +\infty) \to \mathbb{R}^+$ with respect to g such that

$$\left| \int_{s_1}^{s_2} [\mathcal{G}(x(s), s) - \mathcal{G}(z(s), s)]\mathrm{d}g(s) \right| \leqslant \|x - z\|_{[0,+\infty)} \int_{s_1}^{s_2} L(s)\mathrm{d}g(s),$$

for all $x,\ z \in G_0([0, +\infty), \mathbb{R}^n)$ and all $0 \leqslant s_1 \leqslant s_2 < +\infty$.

(D5) There exist constants $\omega, \beta > 0$ such that $g(t+\omega)-g(t) = \beta$ and $\mathcal{G}(x, t+\omega) = \mathcal{G}(x, t)$ for all $t \geqslant 0$.

In the next two results, we describe how we can translate Theorem 6.3 to measure differential equations. For proof of these results, the reader may consult [5, Theorems 4.2 and 4.8].

Theorem 6.4 *Assume that* $\mathcal{G} : \mathbb{R}^n \times [0, +\infty) \to \mathbb{R}^n$ *satisfies the conditions* (D2), (D3), *and* (D4), *and* $g : [0, +\infty) \to \mathbb{R}$ *satisfies the condition* (D1). *Define* $F : \mathbb{R}^n \times [0, +\infty) \to \mathbb{R}^n$ *by*

$$F(x, t) = \int_0^t \mathcal{G}(x, s)\mathrm{d}g(s), \qquad (x, t) \in \mathbb{R}^n \times [0, +\infty). \tag{6.21}$$

Then $F \in \mathcal{F}(\Omega, h)$, *where* $\Omega = \mathbb{R}^n \times [0, +\infty)$, *and* $h : [0, +\infty) \to \mathbb{R}$ *given by*

$$h(t) = \int_0^t (M(s) + L(s))\mathrm{d}g(s), \qquad t \in [0, +\infty), \tag{6.22}$$

is a nondecreasing and left-continuous function.

Theorem 6.5 *Suppose* $\mathcal{G} : \mathbb{R}^n \times [0, +\infty) \to \mathbb{R}^n$ *satisfies conditions (D2), (D3) and (D4), and* $g : [0, +\infty) \to \mathbb{R}$ *satisfies the condition (D1). Then, the function* $x : I \to \mathbb{R}^n$ *is a solution of the MDE* (6.20) *on* $I \subset [0, +\infty)$ *if, and only if,* x *is a solution of the generalized ODE*

$$\frac{\mathrm{d}x}{\mathrm{d}\tau} = DF(x, t)$$

on I, *where* F *is given by* (6.21).

In the sequel, we present the concept of steady-state oscillation for measure differential equations.

Definition 6.5 We say that the MDE (6.20) has a *steady-state oscillation* (in the sense of LaSalle) if (6.20) has a unique periodic solution and all solutions approach this one as t approaches infinity.

Theorem 6.6 (Steady-State Oscillation for MDEs) *Suppose that* $\mathcal{G} : \mathbb{R}^n \times [0, +\infty) \to \mathbb{R}^n$ *satisfies the conditions (D2),* (D3), (D4), *and* (D5), *and* $g : [0, +\infty) \to \mathbb{R}$ *satisfies the condition* (D1) *and* (D5). *Let* $a : [0, +\infty) \to (0, +\infty)$

be a continuous function such that $r = \lim_{n\to\infty} a(n\omega) < 1$. *Moreover, assume the following condition:*

($\mathcal{J}$) *Any two solutions u and v of the MDE* (6.20) *satisfy*

$$\|u(t) - v(t)\| \leqslant a(t)\|u(0) - v(0)\|. \tag{6.23}$$

Then the MDE (6.20) *has a steady-state oscillation.*

Proof Define the function $F: \mathbb{R}^n \times [0, +\infty) \to \mathbb{R}^n$ by

$$F(x,t) = \int_0^t \mathcal{G}(x,s)\mathrm{d}g(s), \qquad (x,t) \in \mathbb{R}^n \times [0, +\infty). \tag{6.24}$$

According to Theorem 6.4, there is a nondecreasing and left-continuous function $h : [0, +\infty) \to \mathbb{R}$ such that $F \in \mathcal{F}(\Omega, h)$, where $\Omega = \mathbb{R}^n \times [0, +\infty)$ and

$$h(t) = \int_{t_0}^t (L(s) + M(s))\mathrm{d}g(s).$$

Thus, F satisfies condition (B1) from Theorem 6.3. On the other hand, by (6.24) and [1, Lemma 10.3], we have

$$F(x, t+\omega) - F(x,t) = \int_t^{t+\omega} \mathcal{G}(x,s)\,dg(s) = \int_0^{\omega} \mathcal{G}(x,s)\,dg(s).$$

Hence, F satisfies condition (B2) (from Theorem 6.3) with $C(x) := \int_0^{\omega} \mathcal{G}(x,s)\,dg(s)$. Also, by Theorem 6.5 and by hypothesis ($\mathcal{J}$), the condition (H1) from Theorem 6.3 holds. Since all the hypotheses from Theorem 6.3 are fulfilled, the generalized ODE $\frac{\mathrm{d}x}{\mathrm{d}\tau} = DF(x,t)$ has a steady-state oscillation. Using this fact together with Theorem 6.5, we have that the MDE (6.20) has a steady-state oscillation. □

References

1. E.M. Bonotto, M. Federson, J.G. Mesquita, *Generalized Ordinary Differential Equations in Abstract Spaces and Applications* (John Wiley & Sons, Hoboken, 2021)
2. E.M. Bonotto, R. Collegari, M. Federson, T.Q. Gill, Operator-valued stochastic differential equations in the context of Kurzweil-like equations. J. Math. Anal. Appl. **527**, 127464 (2023)
3. M. Federson, S. Šchwabik, Generalized ODE approach to impulsive retarded functional differential equations. Differ. Integral Equ. **19**, 1201–1234 (2006)
4. M. Federson, J.G. Mesquita, A. Slavík, Measure functional differential equations and functional dynamic equations on time scales. J. Differ. Equ. **252**, 3816–3847 (2012)
5. M. Federson, R. Grau, J.G. Mesquita, Prolongation of solutions of measure differential equations and dynamical equations on time scales. Math. Nachr. **291**, 22–25 (2019)

6. M. Federson, R. Grau, J.G. Mesquita, E. Toon, Lyapunov stability for measure differential equations and dynamic equations on time scales. J. Differ. Equ. **267**, 4192–4223 (2019)
7. M. Federson, R. Grau, J. G. Mesquita, E. Toon, Permanence of equilibrium points in the basin of attraction and existence of periodic solutions for autonomous measure differential equations and dynamic equations on time scales via generalized ODEs. Nonlinearity **35**, 3118 (2022)
8. M. Fleury, J.G. Mesquita, A. Slavík, Massera's theorems for various types of equations with discontinuous solutions. J. Differ. Equ. **269**, 11667–11693 (2020)
9. E. Kreyszig, *Introductory Functional Analysis with Applications* (Wiley, New York, 1978)
10. J. Kurzweil, Generalized ordinary differential equations and continuous dependence on a parameter. Czechoslovak Math. J. **7**, 418–449 (1957)
11. J. Kurzweil, Generalized ODEs. Czechoslovak Math. J. **8**, 360–388 (1958)
12. J. LaSalle, S. Lefschetz, *Stabillity by Lyapunov's Direct Method, with Application*, Mathematics in Science and Engineering (Academic Press, New York, 1961)
13. S. Šchwabik, Generalized Volterra integral equations. Czechoslovak Math. J. **32**, 245–270 (1982)
14. S. Šchwabik, *Generalized Ordinary Differential Equations* (World Scientific, Singapore, 1992)

Chapter 7
Controllability for Linear Singular Systems on Time Scales

Hernán Henríquez and Jaqueline Godoy Mesquita

Abstract In this chapter, we introduce singular control systems on time scales. We discuss the existence of solutions for homogeneous and nonhomogeneous singular systems on time scales. We apply the Variation of Constants Formula to characterize the controllability of singular control systems on time scales in terms of its parameters.

Keywords Singular control systems · Time scales · Controllability · Reachability

MSC 2020 93B05, 93B07, 34N05

7.1 Introduction

The singular differential systems are usually described by

$$\begin{cases} M\dot{x}(t) = Ax(t) + Bu(t), \\ y(t) = Cx(t), \end{cases} \tag{7.1}$$

where $x(t) \in \mathbb{R}^n$, $u(t) \in \mathbb{R}^m$, $y(t) \in \mathbb{R}^r$, $M, A \in \mathbb{R}^{n\times n}$, $B \in \mathbb{R}^{n\times m}$, $C \in \mathbb{R}^{r\times n}$ are constant matrices, and M is a singular matrix. If we assume that M is a nonsingular matrix, then the system (7.1) becomes

---

H. Henríquez
Departamento de Matemática, Universidad de Santiago de Chile, Santiago, Chile
e-mail: hernan.henriquez@usach.cl

J. G. Mesquita (✉)
Department of Mathematics, University of Brasília, Brasilia, Brazil
e-mail: jgmesquita@unb.br

P. Benevieri, J. G. Mesquita (eds.), *Functional Differential Equations and Dynamic Equations on Time Scales*, Advances in Continuum Mechanics 52,
https://doi.org/10.1007/978-3-031-83327-4_7

$$\begin{cases} \dot{x}(t) = M^{-1}Ax(t) + M^{-1}Bu(t), \\ y(t) = Cx(t), \end{cases} \tag{7.2}$$

which is the well-known classical linear system. However, there are several important models in the fields of economy, science, and technology, among others which cannot be described in a realistic way by the system (7.2). In such a case, they are better established by equations with the form (7.1), where M is a singular matrix. For instance, consider a dynamic Leontief model of a multisector economy [19] given by

$$x(k) = Ax(k) + B[x(k+1) - x(k)] + d(k). \tag{7.3}$$

We claim that if B is a nonsingular matrix in (7.3), then the dynamic Leontief model cannot describe the reality. Indeed, in (7.3), the components (b_{ij}) of the matrix B represent the amount of the stock of commodity i, as a capital good, that sector j must achieve for each unit of production. However, not every sector produces significant capital goods which implies that some rows of the matrix B usually contain only null elements.

Thus, the singularity of the matrix B is often required by the structure of capital in a multisector of the economy, and the system (7.3) cannot be viewed as a classical linear system. For more details, see [19].

As a matter of fact, singular systems and theory of control represent a powerful tool for applications in several fields of knowledge, e.g., engineering systems (such as power system, electrical networks, aerospace engineering, and chemical processes), economic models, network analysis, biological systems, system modeling, and others. See, for instance, [1, 2, 5, 7, 8, 10–15, 20, 22], and the references therein.

On the other hand, the theory of time scales plays an important role to model interesting real-world problems in fields such as economy, biology, and technology (see [9, 17], for instance). This theory was introduced by Stefan Hilger, on his doctoral thesis (see [16]), and since then, it has been intensively investigated by several authors (see [1, 3–5, 20, 21], for instance, and the references therein), and the interest in the subject remains increasing.

Since time scale is any closed nonempty subset of $\mathbb{R}$, the theory of dynamic equations on time scales allows unifying several theories, depending on the chosen time scale. For instance, if $\mathbb{T} = \mathbb{Z}$, we have a result for difference equations. On the other hand, choosing $\mathbb{T} = \mathbb{R}$, we obtain a result for differential equations. Further, this theory allows describing continuous-discrete hybrid processes, which have many applications. It is a known fact that several important phenomena do not possess either only continuous property or only discrete aspects, but they also contain processes that feature elements of both the continuous and the discrete. See, for instance, [6, 9, 17, 18, 21].

Therefore, to investigate the singular systems in the framework of time scales enables us to obtain a powerful tool to address various applications, since it allows to describe more realistically the practical phenomena, e.g., economic models, population models, and biological problems, among others. However, to the best of our knowledge, the singular system has been treated only in the framework of continuous and discrete theories, not in the framework of time scales theory.

Motivated by this fact, the main goal of this chapter is to investigate the *singular dynamic systems on time scales* and study their properties such as controllability and reachability. More precisely, we focus our attention in systems given by

$$\begin{cases} Mx^{\Delta}(t) = Ax(t) + Bu(t), \\ y(t) = Cx(t), \end{cases}$$

where $x(t) \in \mathbb{R}^n$, $u(t) \in \mathbb{R}^m$, $y(t) \in \mathbb{R}^r$, $M, A \in \mathbb{R}^{n\times n}$, $B \in \mathbb{R}^{n\times m}$, and $C \in \mathbb{R}^{r\times n}$ are constant matrices, and M is a singular matrix.

This chapter is organized as follows. In Sect. 7.2, we recall some basic aspects of dynamic systems on time scales, in order to ensure that it is self-contained. In Sect. 7.3, we investigate singular linear systems on time scales. Finally, in Sect. 7.4, we characterize the controllability of singular control systems on times scales.

7.2 Preliminaries

In this section, we present some basic concepts and results concerning time scales which will be essential to prove our main results. For more details, see [3, 4].

Let $\mathbb{T}$ be a time scale, that is, a closed and nonempty subset of $\mathbb{R}$.

Definition 7.1 For every $t \in \mathbb{T}$, we define the *forward* and *backward jump operators* $\sigma, \rho\colon \mathbb{T} \to \mathbb{T}$, respectively, as follows:

$$\sigma(t) = \inf\{s \in \mathbb{T},\ s > t\} \quad \text{and} \quad \rho(t) = \sup\{s \in \mathbb{T},\ s < t\}.$$

In this definition, we consider $\inf\emptyset = \sup\mathbb{T}$ and $\sup\emptyset = \inf\mathbb{T}$.

If $\sigma(t) > t$, t is called *right-scattered*. If $\sigma(t) = t$ and $t < \sup\mathbb{T}$, then t is called *right-dense*. Similarly, if $\rho(t) < t$, then we say that t is *left-scattered*, whereas if $\rho(t) = t$ and $t > \inf\mathbb{T}$, then t is *left-dense*.

The *graininess function* $\mu : \mathbb{T} \to \mathbb{R}^+$ is defined by $\mu(t) = \sigma(t) - t$.

Definition 7.2 A function $f\colon \mathbb{T} \to \mathbb{R}$ is called *rd-continuous* if it is regulated on $\mathbb{T}$ and continuous at right-dense points of $\mathbb{T}$. If the function $f : \mathbb{T} \to \mathbb{R}$ is continuous at each right-dense point and each left-dense point, then the function f is said to be *continuous* on $\mathbb{T}$.

Throughout this chapter, given a pair of numbers $a, b \in \mathbb{T}$, the symbol $[a, b]_{\mathbb{T}}$ will be used to denote a closed interval in $\mathbb{T}$, that is, $[a, b]_{\mathbb{T}} = \{t \in \mathbb{T};\ a \le t \le b\}$. On the other hand, $[a, b]$ is the usual closed interval on the real line.

We define the set $\mathbb{T}^{\kappa}$ which is derived from $\mathbb{T}$ as follows: If $\mathbb{T}$ has a left-scattered maximum m, then $\mathbb{T}^{\kappa} = \mathbb{T} - \{m\}$; otherwise, $\mathbb{T}^{\kappa} - \mathbb{T}$.

For $y\colon \mathbb{T} \to \mathbb{R}$ and $t \in \mathbb{T}^\kappa$, we define y^Δ to be the number (provided it exists) with the following property: Given $\varepsilon > 0$, there exists a neighborhood U of t such that

$$|y(\sigma(t)) - y(s) - y^\Delta(t)[\sigma(t) - s]| < \varepsilon|\sigma(t) - s|,$$

for all $s \in U$. In this case, $y^\Delta(t)$ denotes the delta derivative of y at t.

Moreover, we say that f is delta-differentiable on $\mathbb{T}^\kappa$ provided $f^\Delta(t)$ exists for all $t \in \mathbb{T}^\kappa$. In this case, $f^\Delta : \mathbb{T}^\kappa \to \mathbb{R}$ is called *Δ-derivative* of f on $\mathbb{T}^\kappa$. For the properties of the delta derivatives, see [3, Theorem 1.20].

A function $f\colon \mathbb{T} \to \mathbb{R}$ has a second delta derivative $f^{\Delta\Delta}$ provided f^Δ is delta-differentiable on $\mathbb{T}^{\kappa^2} = (\mathbb{T}^\kappa)^\kappa$ with delta derivative $f^{\Delta\Delta} = (f^\Delta)^\Delta\colon \mathbb{T}^{\kappa^2} \to \mathbb{R}$. Similarly, we define higher order delta derivatives $f^{\Delta^n} : \mathbb{T}^{\kappa^n} \to \mathbb{R}$. Moreover, we denote $\sigma^2(t) = \sigma(\sigma(t))$ and $\rho^2(t) = \rho(\rho(t))$, and $\sigma^n(t)$ and $\rho^n(t)$ for $n \in \mathbb{N}$ are defined accordingly. See [3, 4].

In the sequel, we present the definition of Riemann Δ-integrals. In order to do it, we give some preliminary concepts.

Definition 7.3 A *partition* of $[a, b]_\mathbb{T}$ is a finite sequence of points $\{t_0, t_1, \ldots, t_m\} \subset [a, b]_\mathbb{T}, \quad a = t_0 < t_1 < \ldots < t_m = b$.

Given such a partition, we put $\Delta t_i = t_i - t_{i-1}$. A *tagged partition* consists of a partition and a sequence of tags $\{\xi_1, \ldots, \xi_m\}$ such that $\xi_i \in [t_{i-1}, t_i)$ for every $i \in \{1, \ldots, m\}$. The set of all tagged partitions of $[a, b]_\mathbb{T}$ will be denoted by the symbol $D(a, b)$.

If $\delta > 0$, then $D_\delta(a, b)$ denotes the set of all tagged partitions of $[a, b]_\mathbb{T}$ such that for every $i \in \{1, \ldots, m\}$, either $\Delta t_i \le \delta$ or $\Delta t_i > \delta$ and $\sigma(t_{i-1}) = t_i$. It is clear that, in the last case, the only way to choose a tag in $[t_{i-1}, t_i)$ is to take $\xi_i = t_{i-1}$.

Definition 7.4 We say that f is *Riemann Δ-integrable* on $[a, b]_\mathbb{T}$, if there exists a number I with the following property: For every $\varepsilon > 0$, there exists $\delta > 0$ such that

$$\left|\sum_i f(\xi_i)(t_i - t_{i-1}) - I\right| < \varepsilon,$$

for every $P \in D_\delta(a, b)$ independently of $\xi_i \in [t_{i-1}, t_i)_\mathbb{T}$ for $1 \le i \le n$. It is clear that such a number I is unique and is the Riemann Δ-integral of f from a to b.

The next result contains an important property of Riemann Δ-integrable functions. A proof of it can be found in [4, Theorem 5.34].

Theorem 7.1 (Fundamental Theorem of Calculus) *Let $[a, b]_\mathbb{T}$ be a time scale interval and g be a continuous function on $[a, b]_\mathbb{T}$ such that g is delta-differentiable on $[a, b)_\mathbb{T}$. If g^Δ is Riemann Δ-integrable from a to b, then*

$$\int_a^b g^\Delta(t)\Delta t = g(b) - g(a).$$

In what follows, we present Taylor's Formula. A proof of it can be found in [3, Theorem 1.111].

Theorem 7.2 (Taylor's Formula) *Let $h \in \mathbb{N}$, and suppose f is h times differentiable on $\mathbb{T}^{\kappa^h}$. Let $\alpha \in \mathbb{T}^{\kappa^{h-1}}$, $t \in \mathbb{T}$, and define the functions g_k by*

$$g_0(r,s) \equiv 1 \quad \text{and} \quad g_{k+1}(r,s) = \int_s^r g_k(\sigma(\tau), s)\Delta\tau \quad \text{for } k \in \mathbb{N}_0. \tag{7.4}$$

Then, we have

$$f(t) = \sum_{k=0}^{h-1} (-1)^k g_k(\alpha, t) f^{\Delta^k}(\alpha) + \int_\alpha^{\rho^{h-1}(t)} (-1)^{h-1} g_{h-1}(\sigma(\tau), t) f^{\Delta^h}(\tau)\Delta\tau.$$

As usual, the concepts of delta derivative and Riemann Δ-integral can be generalized to vector functions applying them to each component.

Definition 7.5 We say that a function $p\colon \mathbb{T} \to \mathbb{R}$ is *regressive* provided $1 + \mu(t)p(t) \neq 0$, for all $t \in \mathbb{T}^\kappa$. The set of all regressive and rd-continuous functions will be denoted by $\mathcal{R} = \mathcal{R}(\mathbb{T}) = \mathcal{R}(\mathbb{T}, \mathbb{R})$.

Suppose that $p, q \in \mathcal{R}$; then we define $p \oplus q$ and $\ominus p$ as follows: $(p \oplus q)(t) := p(t) + q(t) + \mu(t)p(t)q(t)$, for all $t \in \mathbb{T}^\kappa$ and

$$(\ominus p)(t) := \frac{-p(t)}{1 + \mu(t)p(t)}, \quad \text{for all } t \in \mathbb{T}^\kappa.$$

It is clear that $(\mathcal{R}, \oplus)$ is an Abelian group (see [3]).

In what follows, we present some definitions about matrix-valued functions.

Definition 7.6 Let A be an $m \times n$ matrix-valued function on $\mathbb{T}$. A is called *rd-continuous* on $\mathbb{T}$ if each entry of A is rd-continuous on $\mathbb{T}$. We denote the class of all rd-continuous $m \times n$ matrix-valued function on $\mathbb{T}$ by $C_{\text{rd}} = C_{\text{rd}}(\mathbb{T}) = C_{\text{rd}}(\mathbb{T}, \mathbb{R}^{m\times n})$.

We say that A is *delta-differentiable* at $\mathbb{T}^\kappa$ if each entry of A is delta-differentiable on $\mathbb{T}^\kappa$. And, in this case, we have $A^\sigma(t) = A(t) + \mu(t)A^\Delta(t)$.

Definition 7.7 An $n \times n$ matrix-valued function A on a time scale $\mathbb{T}$ is called *regressive* (with respect to $\mathbb{T}$) provided $I + \mu(t)A(t)$ is invertible for all $t \in \mathbb{T}^\kappa$, and the class of all such regressive rd-continuous matrix-valued functions is denoted by $\mathcal{R} = \mathcal{R}(\mathbb{T}) = \mathcal{R}(\mathbb{T}, \mathbb{R}^{n\times n})$. Here, I denotes as usual the $n \times n$ identity matrix.

If A and B are regressive $n \times n$ matrix-valued functions on $\mathbb{T}$, then we define the following operations: $(A \oplus B)(t) = A(t) + B(t) + \mu(t)A(t)B(t)$ for all $t \in \mathbb{T}^\kappa$, and we define $\ominus A$ by $(\ominus A)(t) = -[I + \mu(t)A(t)]^{-1}A(t)$, for all $t \in \mathbb{T}^\kappa$. It is clear that $(\mathcal{R}(\mathbb{T}, \mathbb{R}^{n\times n}), \oplus)$ is a group (see [3]).

Definition 7.8 (Fundamental Matrix) Let $t_0 \in \mathbb{T}$, and assume that $A \in \mathcal{R}$ is an $n \times n$ matrix-valued function. The unique matrix-valued solution of the IVP

$$\begin{cases} Y^{\Delta}(t) = A(t)Y(t) \\ Y(t_0) = I \end{cases}$$

is called the *fundamental matrix* at t_0, and it is denoted by $e_A(\cdot, t_0)$. For the properties of the fundamental matrix, see [3, Theorem 5.21].

Using these notions, one can obtain the following result, which is a Variation of Constants Formula that can be found in [3, Theorem 5.24].

Theorem 7.3 (Variation of Constants Formula) *Let $A \in \mathcal{R}$ be an $n \times n$ matrix-valued function on $\mathbb{T}$, and suppose that $f : \mathbb{T} \to \mathbb{R}^n$ is rd-continuous. Let $t_0 \in \mathbb{T}$ and $y_0 \in \mathbb{R}^n$. Then the initial value problem*

$$\begin{cases} y^{\Delta} = A(t)y + f(t), \\ y(t_0) = y_0 \end{cases} \tag{7.5}$$

has a unique solution $y : \mathbb{T} \to \mathbb{R}^n$. Moreover, this solution is given by

$$y(t) = e_A(t, t_0)y_0 + \int_{t_0}^{t} e_A(t, \sigma(\tau))f(\tau)\Delta\tau.$$

7.3 Existence of Solutions of Linear Singular Control Systems

Throughout this chapter, we denote by $\mathbb{R}^n$ the space of n-dimensional column vectors $x = (x_1, \ldots, x_n)$ endowed with a norm $\|\cdot\|$. The set consisting of rd-continuous functions $f : \mathbb{T} \to \mathbb{R}^n$ will be denoted by $C_{\mathrm{rd}}(\mathbb{T}, \mathbb{R}^n)$, and we denote by $C^k_{\mathrm{rd}}(\mathbb{T}, \mathbb{R}^n)$ the set consisting of functions $f : \mathbb{T} \to \mathbb{R}^n$ that are k-times differentiable, and their delta derivatives $f, f^{\Delta}, f^{\Delta^2}, \ldots, f^{\Delta^k}$ are rd-continuous.

Let $M, A \in \mathbb{R}^{n \times n}$. In this section, we are concerned with the singular system

$$\begin{cases} Mx^{\Delta}(t) = Ax(t) + f(t), \; t \in \mathbb{T}^{\kappa}, \\ x(t_0) = z, \end{cases} \tag{7.6}$$

where $t_0 \in \mathbb{T}$ and $f \in C_{\mathrm{rd}}(\mathbb{T}, \mathbb{R}^n)$. In order to ensure the existence of solutions of system (7.6), we assume that (M, A) is a regular pencil of matrices, that is, there exists $\lambda \in \mathbb{R}$ such that $\det(\lambda M - A) \neq 0$, and moreover, the matrices A and M are regressive. In this case ([10]), there exists a pair of invertible matrices P and Q such that

$$QMP = \begin{bmatrix} I_{n_1} & 0 \\ 0 & N \end{bmatrix}; \quad QAP = \begin{bmatrix} A_1 & 0 \\ 0 & I_{n_2} \end{bmatrix},$$

where A_1 is a matrix of order $n_1 \times n_1$, N is a nilpotent matrix of order $n_2 \times n_2$ and nilpotent index h, that is, $N^{h-1} \neq 0$ and $N^h = 0$, and I_{n_i}, $i = 1, 2$, denotes the identity matrix of order n_i, and $n = n_1 + n_2$. Therefore, applying the transformation of coordinates defined by the matrices P and Q, with

$$x = P \begin{bmatrix} x_1 \\ x_2 \end{bmatrix}; \quad Qf = \begin{bmatrix} f_1 \\ f_2 \end{bmatrix},$$

we obtain that the system (7.6) is equivalent to the system defined by

$$x_1^{\Delta}(t) = A_1 x_1(t) + f_1(t), \; t \in \mathbb{T}^{\kappa}, \tag{7.7}$$

$$N x_2^{\Delta}(t) = x_2(t) + f_2(t), \; t \in \mathbb{T}^{\kappa}, \tag{7.8}$$

satisfying the following initial conditions:

$$x_1(t_0) = z_1, \tag{7.9}$$

$$x_2(t_0) = z_2, \tag{7.10}$$

where

$$z = P \begin{bmatrix} z_1 \\ z_2 \end{bmatrix}.$$

As usual for normal systems, we say that (7.7) is the *slow subsystem*, while (7.8) is the *fast subsystem*. Moreover, Eqs. (7.7) and (7.8) are said to be the *standard form* of system (7.6). Since (7.7) is a normal system, by Theorem 7.3, the solution of (7.7)–(7.9) is given by

$$x_1(t) = e_{A_1}(t, t_0) z_1 + \int_{t_0}^{t} e_{A_1}(t, \sigma(\tau)) f_1(\tau) \Delta\tau. \tag{7.11}$$

Furthermore, proceeding as in [10], we can establish the following result.

Theorem 7.4 *Assume $f \in C_{\mathrm{rd}}^{h-1}(\mathbb{T}, \mathbb{R}^n)$, (M, A) is a regular pencil of matrices, and the matrices A and M are regressive. If the consistent initial condition*

$$z_2 = -\sum_{i=0}^{h-1} N^i f_2^{\Delta^i}(t_0)$$

holds, then the unique solution of Eq. (7.8) with initial condition (7.10) is given by

$$x_2(t) = -\sum_{i=0}^{h-1} N^i f_2^{\Delta^i}(t). \tag{7.12}$$

Proof Using the properties of Δ derivative, it follows from (7.8) that

$$N^2 x^{\Delta^2}(t) = N(x_2^{\Delta}(t) + f_2^{\Delta}(t)) = x_2(t) + f_2(t) + N f_2^{\Delta}(t), \ t \in \mathbb{T}^{\kappa},$$

and proceeding inductively, we get

$$N^h x^{\Delta^h}(t) = 0 = x_2(t) + f_2(t) + N f_2^{\Delta}(t) + \cdots + N^{h-1} f_2^{\Delta^{h-1}}(t), \ t \in \mathbb{T}^{\kappa},$$

using the fact that N is nilpotent matrix of index h. □

Consequently, under the hypotheses of Theorem 7.4, the unique solution of system (7.6) is given by

$$x(t) = P \begin{bmatrix} e_{A_1}(t, t_0) z_1 + \displaystyle\int_{t_0}^{t} e_{A_1}(t, \sigma(\tau)) f_1(\tau) \Delta\tau \\ -\displaystyle\sum_{i=0}^{h-1} N^i f_2^{\Delta^i}(t) \end{bmatrix}. \tag{7.13}$$

7.4 Controllability of Systems

In this section, we study the controllability of a singular system on time scales described by

$$\begin{cases} M x^{\Delta}(t) = A x(t) + B u(t), \ t \in \mathbb{T}, \\ x(t_0) = z, \end{cases} \tag{7.14}$$

where $t_0 \in \mathbb{T}$, with states $x(t) \in \mathbb{R}^n$ and controls $u(t) \in \mathbb{R}^m$. Throughout this section, we keep the notations and assumptions introduced in Sect. 7.3 to ensure the existence of solutions of the system (7.14). Moreover, B is an $n \times m$ constant real matrix, and we write $QB = \begin{bmatrix} B_1 \\ B_2 \end{bmatrix}$. We study the system (7.14) on the interval $[t_0, t_1]_{\mathbb{T}}$, where $t_0, t_1 \in \mathbb{T}$, $t_0 < t_1$. In order to do that, we restrict us to consider as admissible control to functions $u \in C_{\mathrm{rd}}^{h-1}([t_0, t_1]_{\mathbb{T}}, \mathbb{R}^m)$. We denote by $x(\cdot; z, u)$ the solution of (7.14) defined by (7.13) with $Bu(t)$ instead of $f(t)$.

Definition 7.9 The set $\mathcal{R}(t_0, t_1, z) = \{x(t_1; z, u) : u \in C_{\mathrm{rd}}^{h-1}([t_0, t_1]_{\mathbb{T}}, \mathbb{R}^m)\}$ is called the *reachable set* of system (7.14) from the state z on the interval $[t_0, t_1]_{\mathbb{T}}$.

As a consequence of Taylor's Formula (Theorem 7.2), we have the following property.

Lemma 7.1 *Let* $x_0, x_1, \ldots, x_{h-1} \in \mathbb{R}^n$ *and* $t_1 \in \mathbb{T}^{\kappa^{h-1}}$. *Then, there exists* $f \in C_{\rm rd}^{h-1}(\mathbb{T}, \mathbb{R}^n)$ *such that* $f^{\Delta^k}(t_1) = x_k$ *for* $k = 0, 1, \ldots, h-1$.

Proof Indeed, we can define a function $f\colon \mathbb{T} \to \mathbb{R}^n$ as follows:

$$f(t) = \sum_{k=0}^{h-1} (-1)^k g_k(t_1, t)x_k + \int_{t_1}^{\rho^{h-1}(t)} (-1)^{h-1} g_{h-1}(\sigma(\tau), t) f^{\Delta^h}(\tau)\Delta\tau,$$

where g_k is given by (7.4). Then, by Taylor's Formula (Theorem 7.2), we obtain

$$x_k = f^{\Delta^k}(t_1), \quad \text{for} \quad k = 0, 1, 2, \ldots, h-1,$$

and the desired result follows as well. □

For a pair of regressive matrices $F \in \mathbb{R}^{n\times n}$ and $G \in \mathbb{R}^{n\times m}$, we use the notation

$$\langle F|G\rangle = \operatorname{Im}[G, FG, \ldots, F^{n-1}G].$$

In our next result, we assume that the system (7.14) is in the standard form.

Theorem 7.5 *Under the above conditions,*

$$\mathcal{R}(t_0, t_1, 0) = \langle A_1|B_1\rangle \oplus \langle N|B_2\rangle.$$

Proof Let $col[q_1, q_2] \in \mathcal{R}(t_0, t_1, 0)$. This implies that there exists $u \in C_{rd}^{h-1}([t_0, t_1]_{\mathbb{T}}, \mathbb{R}^m)$ such that

$$q_1 = \int_{t_0}^{t_1} e_{A_1}(t_1, \sigma(\tau))B_1 u(\tau)\Delta\tau,$$

and

$$q_2 = -\sum_{i=0}^{h-1} N^i B_2 u^{i\Delta}(t_1).$$

It is immediate that $q_2 \in \langle N|B_2\rangle$. In addition, if we assume that $q_1 \notin \langle A_1|B_1\rangle$, applying the Hahn-Banach Theorem, we infer the existence of $w_1 \in \mathbb{R}^{n_1}$, $w_1 \neq 0$, such that $w_1^* A_1^i B_1 = 0$ for $i = 0, 1, \ldots, n_1 - 1$ and $w_1^* q_1 \neq 0$. Using the first assertion and applying the construction in [20, Theorem 2], we obtain that $w_1^* q_1 = 0$. From this contradiction, we deduce that $q_1 \in \langle A_1|B_1\rangle$. Consequently, $col[q_1, q_2] \in \langle A_1|B_1\rangle \oplus \langle N|B_2\rangle$.

Conversely, let $col[q_1, q_2] \in \langle A_1|B_1\rangle \oplus \langle N|B_2\rangle$. If we assume that $q_1 \notin \mathcal{R}(t_0, t_1, 0)$, then there is $w_1 \in \mathbb{R}^{n_1}$, $w_1 \neq 0$, $w_1^* q_1 \neq 0$, such that

$$\int_{t_0}^{t_1} w_1^* e_{A_1}(t_1, \sigma(\tau)) B_1 B_1^* e_{A_1}(t_1, \sigma(\tau))^* w_1 \Delta\tau = 0.$$

Arguing now as in the proof of [20, Theorems 1 and 2], we can assert that $w_1^* A_1^i B_1 = 0$ for $i = 0, 1, \ldots, n_1 - 1$, which in turn implies that $w_1^* q_1 = 0$. Hence, we obtain that $q_1 \in \mathcal{R}(t_0, t_1, 0)$. Furthermore, we can write

$$q_2 = \sum_{i=0}^{h-1} N^i B_2 v_i$$

for some $v_i \in \mathbb{R}^m$ for $i = 0, 1, \ldots, h - 1$. Applying Lemma 7.1, there exists $u \in C_{rd}^{h-1}([t_0, t_1]_{\mathbb{T}}, \mathbb{R}^m)$ such that $u^{i\Delta}(t_1) = -v_i$ $i = 0, 1, \ldots, h - 1$. This implies that

$$q_2 = -\sum_{i=0}^{h-1} N^i B_2 u^{i\Delta}(t_1) \in \mathcal{R}(t_0, t_1, 0),$$

which completes the proof. □

In what follows, let us consider the following concept of controllability.

Definition 7.10 System (7.14) is called *controllable* on $[t_0, t_1]_{\mathbb{T}}$ if for all $z, w \in \mathbb{R}^n$, there exists $u \in C_{rd}^{h-1}([t_0, t_1]_{\mathbb{T}}, \mathbb{R}^m)$ such that the solution $x(\cdot)$ of the system (7.14) satisfies $x(t_1) = w$.

The following result characterizes the controllable systems and generalizes [10, Theorem 2-2.1], [1, Corollary 3], and [11, Theorem 2.7].

Theorem 7.6 *Under the above conditions, the following conditions are equivalent.*

(a) *System (7.14) is controllable on* $[t_0, t_1]_{\mathbb{T}}$.
(b) $rank[B_1, A_1 B_1, \ldots, A_1^{n_1-1} B_1] = n_1$ *and* $rank[B_2, N B_2, \ldots, N^{h-1} B_2] = n_2$.
(c) $rank[\lambda M - A, B] = n$ *for every* $\lambda \in \mathbb{C}$, *and* $rank[M, B] = n$.

Proof We can assume that system (7.14) is in the standard form. Suppose the condition (a) is fulfilled, then by choosing $z = 0$, we have

$$\mathcal{R}(t_0, t_1, 0) = \mathbb{R}^n = \mathbb{R}^{n_1} \oplus \mathbb{R}^{n_2}.$$

On the other hand, applying Theorem 7.5, we get

$$\mathcal{R}(t_0, t_1, 0) = \langle A_1|B_1\rangle \oplus \langle N|B_2\rangle,$$

which implies that $\mathbb{R}^{n_1} = \langle A_1|B_1\rangle$ and $\mathbb{R}^{n_2} = \langle N|B_2\rangle$. Hence, we infer that the condition (b) holds.

The equivalence between the conditions (b) and (c) is established by [10, Theorem 2-2.1]. Assume finally that condition (b) holds. Using again Theorem 7.5, we infer that $\mathcal{R}(t_0, t_1, 0) = \mathbb{R}^n$. Let $z = col(z_1, z_2)$, $w = col(w_1, w_2) \in \mathbb{R}^n$. We define $\tilde{w} = col(w_1 - e_{A_1}(t_1, t_0)z_1, w_2)$. Then there is $u \in C_{rd}^{h-1}([t_0, t_1]_{\mathbb{T}}, \mathbb{R}^m)$ such that the solution $x(\cdot; 0, u)$ verifies

$$x_1(t_1) = \tilde{w}_1 = w_1 - e_{A_1}(t_1, t_0)z_1,$$

$$x_2(t_1) = w_2.$$

This implies that $x(t_1; z, u) = w$, where x is a solution of (7.14). Hence, we conclude that system (7.14) is controllable on $[t_0, t_1]_{\mathbb{T}}$. It completes the proof. □

Remark 7.1 It is clear from Theorem 7.6 that the controllability of system (7.14) does not depend on t_0, t_1 when our assumptions are fulfilled.

Acknowledgments I dedicate this work to the memory of my dear coauthor Hernán R. Henríquez who passed away in 2022. J. G. Mesquita was partially supported by CNPq grant 310583/2021-7 and FAPDF 0193-00001821/2022-21.

References

1. Z. Bartosiewicz, E. Pawluszewicz, Realizations of linear control systems on time scales. Control Cybernet. **35**, 769–786 (2006)
2. D.J. Bell, D.H. Jacobson, *Singular Optimal Control Problems* (Academic Press, 1975)
3. M. Bohner, A. Peterson, *Dynamic Equations on Time Scales: An Introduction with Applications* (Birkhäuser, Boston, 2001)
4. M. Bohner, A. Peterson, *Advances in Dynamic Equations on Time Scales* (Birkhäuser, Boston, 2003)
5. M. Bohner, N. Wintz, Controllability and observability of time-invariant linear dynamic systems. Math. Bohem. **137**(2), 149–163 (2012)
6. D. Brigo, F. Mercurio, Discrete time vs continuous time stock-price dynamics and implications for option pricing. Finance Stoch. **4**, 147–159 (2000)
7. S.L. Campbell, *Singular Systems of Differential Equations* (Pitman Publishing, New York, 1980)
8. S.L. Campbell, *Singular Systems of Differential Equations II* (Pitman Publishing, New York, 1982)
9. F.B. Christiansen, T.M. Fenchel, *Theories of Populations in Biological Communities*, Lect. Notes in Ecological Studies 20 (Springer, Berlin, 1977)
10. L. Dai, *Singular Control Systems*, Lect. Notes in Control and Information Sci. 118 (Springer, Berlin, 1989)
11. J.M. Davis, I.A. Gravagne, B.J. Jackson, R.J. Marks II, Controllability, observability, realizability, and stability of dynamical linear systems. Electron. J. Differ. Equ. **37**, 1–32 (2009)
12. J. Diblík, D. Ya Khusainov, M. Ružičková, Controllability of linear discrete systems with constant coefficients and pure delay. SIAM J. Control Optim. **47**(3), 1140–1149 (2008)

13. J. Diblík, D. Ya Khusainov, J. Lukáčová, M. Ružičková, Control of oscillating systems with a single delay. Adv. Differ. Equ. **2010**, Art ID 108218, 15 pp. (2010)
14. J. Diblík, M. Fečkan, M. Pospíšil, On the new control functions for linear discrete delay systems. SIAM J. Control Optim. **52**(3), 1745–1760 (2014)
15. L.V. Fausett, N.K. Murty, Controllability, observability and realizability criteria on time scale dynamic systems. Nonlinear Stud. **11**(4), 627–638 (2004)
16. S. Hilger, *Ein Maßkettenkalkül mit Anwendung auf Zentrumsmanningfaltigkeiten*, PhD thesis, Universität Würzburg, 1988
17. S. Keller, *Asymptotisches Verhalten invarianter Faserbündel bei Diskretisierung und Mittelwertbildung in Rahmen der Analysis auf Zeitskalin*, PhD thesis, Universität Augsburg, 1999
18. I. Klapper, H. Qian, Remarks on discrete and continuous large-scale models of DNA dynamics. Biophys. J. **74**, 2504–2514 (1998)
19. D.G. Luenberger, A. Arbel, Singular dynamic Leontief systems. Econometrica **45**(4), 991–995 (1977)
20. V. Lupulescu, A. Younus, On controllability and observability for a class of linear impulsive dynamic systems on time scales. Math. Comput. Model. **54**, 1300–1310 (2011)
21. C.C. Tisdell, A. Zaidi, Basic qualitative and quantitative results for solutions to nonlinear, dynamic equations on time scales with an application to economic modelling. Nonlinear Anal. **68**(11), 3504–3524 (2008)
22. M. Tvrdý, Generalized differential equations in the space of regulated functions (Boundary value problems and controllability). Math. Bohem. **116**, 225–244 (1991)

Chapter 8
Massera's Theorem for Nonlinear Dynamic Equations on Time Scales

Hernán Henríquez, Jaqueline Godoy Mesquita, and Eduard Toon

Abstract In this chapter, we present a new result concerning the existence of solutions of a dynamic equation on time scales and also a result on uniqueness of solutions for this type of equations. Moreover, we prove Massera's type results in order to ensure the existence of periodic solutions for a dynamic equation on time scales defined by a periodic forcing function and taking values in an infinite-dimensional Banach space. Finally, we investigate the real almost periodic solutions of dynamic equations on time scales defined by a periodic forcing function. All these results are completely new in the literature.

Keywords Dynamic equations on time scales · Almost periodic · Periodic · Existence · Uniqueness · Massera's theorem

8.1 Introduction

In the last years, the theory of dynamic equations on time scales has been attracting the attention of several researchers (see [1, 2, 6, 17, 21, 26], and the references therein). The main reason behind this comes from the fact that these equations encompass differential equations, difference equations, quantum difference equations, and hybrid equations, among others, depending on the chosen time scale,

H. Henríquez
Departamento de Matemática, Universidad de Santiago de Chile, Santiago, Chile
e-mail: hernan.henriquez@usach.cl

J. G. Mesquita (✉)
Department of Mathematics, University of Brasília, Brasilia, Brazil
e-mail: jgmesquita@unb.br

E. Toon
Departamento de Matemática, Universidade Federal de Juiz de Fora, Minas Gerais, Brazil
e-mail: eduard.toon@ufjf.edu.br

P. Benevieri, J. G. Mesquita (eds.), *Functional Differential Equations and Dynamic Equations on Time Scales*, Advances in Continuum Mechanics 52,
https://doi.org/10.1007/978-3-031-83327-4_8

which is a powerful tool for modeling various applications. Nowadays, most of the physical phenomena are described using differential equations. However, it is a known fact that, depending on the phenomena, these equations may not be the best choice to describe them. These equations can offer only an approximation of the real behavior of the phenomena, since most of the phenomena in the environment do not possess only continuous behavior as we have to assume when we are dealing with the theory of differential equations.

Similarly, most systems do not have only discrete behavior, but they share both behaviors, that is, they behave as both discrete and continuous systems, and this must be taken into account in the analysis for a more precise description of the problem in study. Considering this fact, dynamic equations on time scales are the best choice to describe several real-world models, since they can model this type of behavior. Moreover, these equations can be used in several different fields of knowledge, such as in the description of the population models, physical models, biological models, economic problems, and engineers problems, among others. See [1, 7, 9, 17, 26], and the references therein.

On the other hand, the subject of the existence of periodic solutions for differential equations is one of the most interesting and studied aspects of the qualitative theory of differential equations, so from the theoretical point of view as from their numerous applications to concrete systems. In this context, the Massera theorem (see [22]) is one of the most important results in this direction found in the literature. For this reason, many authors have extended this theorem to several types of equations such as functional differential equations and neutral functional differential equations, among others (see [3, 12, 13, 20, 23, 24, 27]). However, to the best of our knowledge, the extension of this theorem in the setting of the theory of dynamic equations on time scales has not been fully developed yet. There are very few references concerning this topic (see [19]), they do not present a version of this theorem for functions taking values in a Banach space of infinite dimension, and also, they do not investigate this theorem in the setting of almost periodic functions on time scales, which is a new concept introduced in the literature (see [18]). Also, it is important to notice that in [19] the authors only presented a version of Massera's theorem for functions taking values in $\mathbb{R}$, and the results presented are obtained by the means of fixed point theorems. On the other hand, in [3], the authors presented a version of Massera's theorem only for quantum difference equations by considering functions taking value in $\mathbb{R}$. Recently, in [15], the authors investigated the version of Massera's theorem for generalized ODEs, and as a consequence, they obtained results for measure differential equations, impulsive differential equations, and dynamic equations on time scales as special cases. However, all the results presented there were for functions taking value in finite-dimensional spaces, as well as the authors have not considered any version of Massera's theorem involving almost periodic functions. Therefore, this lack in the literature motivates us to investigate all these open problems mentioned above. The aim of this chapter is to address these issues. Specifically, in this chapter, we will study the problem of the existence and uniqueness of periodic and almost periodic solutions of the dynamic equation on time scales given by

$$x^{\Delta}(t) = f(t, x(t)), \quad t \in \mathbb{T}, \; x(t) \in X, \tag{8.1}$$

where $\mathbb{T} \subseteq \mathbb{R}$ is a time scale, X is a Banach space, and $f : \mathbb{T} \times X \to X$ satisfies appropriate conditions which will be specified later. We point out that our conditions to establish the existence and uniqueness of solutions of (8.1) are more general than the usual ones found in the literature. For instance, instead of the usual Lipschitz condition, we employ the following condition to ensure the uniqueness of solutions:

- There exists a nondecreasing continuous function $M : \mathbb{R}^+ \to [0, \infty)$ such that $M(0) = 0$ and

$$\|f(t, x) - f(t, y)\| \leq M(\|x - y\|), \quad \forall t \in \mathbb{T}, \; \forall\, x, y \in X.$$

This chapter is organized as follows. In the second section, we present the basic concepts concerning the theory of dynamic equations on time scales. The third section is devoted to prove our results concerning existence and uniqueness of solutions of Eq. (8.1). In the fourth section, we prove a version of Massera's type result on time scales for functions with values in infinite-dimensional Banach spaces. Finally, we prove a result that relates the real almost periodic solution of Eq. (8.1) with the p-periodic solutions of this equation. Some notations used throughout the text are the following: For a function u, the notation $Im(u)$ stands for the image of u; for a Banach space X, we denote by $B_r[x, X]$ the closed ball with center at $x \in X$ and radius $r \geq 0$.

8.2 Preliminaries

In this section, we present some basic concepts and results concerning the theory of time scales which will be essential to prove our results. For more details, see [4, 5].

Let $\mathbb{T}$ be a time scale, that is, a closed and nonempty subset of $\mathbb{R}$. For every $t \in \mathbb{T}$, we define the *forward jump operator* $\sigma : \mathbb{T} \to \mathbb{T}$ as follows:

$$\sigma(t) = \inf\{s \in \mathbb{T} : \; s > t\} \quad \text{and} \quad \rho(t) = \sup\{s \in \mathbb{T} : \; s < t\}.$$

In this definition, we consider $\inf \emptyset = \sup \mathbb{T}$. If $\sigma(t) > t$, we say that t is *right-scattered*. If $t < \sup \mathbb{T}$ and $\sigma(t) = t$, then t is called *right-dense*. On the other hand, if $\rho(t) < t$, we say that t is *left-scattered*, while if $\rho(t) = t$, then t is called *left-dense*. We also define the *graininess function* $\mu : \mathbb{T} \to \mathbb{R}^+$ by $\mu(t) = \sigma(t) - t$.

We define the set $\mathbb{T}^{\kappa}$ which is derived from $\mathbb{T}$ as follows: If $\mathbb{T}$ has a left-scattered maximum m, then $\mathbb{T}^{\kappa} = \mathbb{T} \setminus \{m\}$; otherwise, $\mathbb{T}^{\kappa} = \mathbb{T}$.

In the rest of this chapter, X stands for a Banach space endowed with the norm $\|\cdot\|$.

Definition 8.1 A function $f: \mathbb{T} \to X$ is called *rd-continuous* if it is regulated on $\mathbb{T}$ and continuous at right-dense points of $\mathbb{T}$. We denote the class of all rd-continuous functions from $\mathbb{T}$ into X by $C_{\mathrm{rd}}(\mathbb{T}, X)$. Consider also the space

$$C^b_{\mathrm{rd}}(\mathbb{T}, X) = \{f \in C_{\mathrm{rd}}(\mathbb{T}, X) : f \text{ is bounded}\}$$

endowed with the norm of uniform convergence on $\mathbb{T}$, that is, for each $f \in C^b_{\mathrm{rd}}(\mathbb{T}, X)$, we define

$$\|f\|_\infty = \sup\{\|f(t)\| : t \in \mathbb{T}\}.$$

It is well known that $C^b_{\mathrm{rd}}(\mathbb{T}, X)$ is a Banach space. In particular, if $\mathbb{T}$ is compact, then $C_{rd}(\mathbb{T}, X)$ endowed with the norm of uniform convergence is a Banach space.

Given a pair of numbers $a, b \in \mathbb{T}$, the symbol $[a, b]_{\mathbb{T}}$ will be used to denote the closed interval $[a, b]$ in $\mathbb{T}$, that is, $[a, b]_{\mathbb{T}} = \{t \in \mathbb{T} : a \le t \le b\}$. On the other hand, we keep the notation $[a, b]$ for the usual closed interval on the real line, that is, $[a, b] = \{t \in \mathbb{R} : a \le t \le b\}$.

Definition 8.2 For $y: \mathbb{T} \to X$ and $t \in \mathbb{T}^\kappa$, we define the *Δ-derivative* of y at t as the element $y^\Delta(t) \in X$ (if it exists) with the following property: Given $\varepsilon > 0$, there exists a neighborhood U of t such that

$$\|y(\sigma(t)) - y(t) - y^\Delta(t)[\sigma(t) - s]\| \le \varepsilon|\sigma(t) - s|,$$

for all $s \in U$. We say that a function $y(\cdot)$ is *Δ-differentiable* at t provided its Δ-derivative $y^\Delta(t)$ exists.

In the sequel, we remember the definition of a partition of the interval $[a, b]_{\mathbb{T}}$ (see [4, 5]).

Definition 8.3 A *partition* of $[a, b]_{\mathbb{T}}$, for $a,\ b \in \mathbb{T}$, is a finite sequence of points

$$\{t_0, t_1, \ldots, t_m\} \subset [a, b]_{\mathbb{T}}, \quad a = t_0 < t_1 < \ldots < t_m = b.$$

Given such a partition, we put $\Delta t_i = t_i - t_{i-1}$. A tagged partition consists of a partition and a sequence of tags $\{\xi_1, \ldots, \xi_m\}$ such that $\xi_i \in [t_{i-1}, t_i)_{\mathbb{T}}$ for every $i \in \{1, \ldots, m\}$. The set of all tagged partitions of $[a, b]_{\mathbb{T}}$ will be denoted by the symbol $D(a, b)$.

If $\delta > 0$, then $D_\delta(a, b)$ denotes the set of all tagged partitions of $[a, b]_{\mathbb{T}}$ such that for every $i \in \{1, \ldots, m\}$, either $\Delta t_i \le \delta$ or $\Delta t_i > \delta$ and $\sigma(t_{i-1}) = t_i$.

Next, we present the definition of Riemann Δ-integral. It is a generalization from the definition from [4, 5] for a Banach space X. Also, in the sequel, we present some results from this integral for a Banach space X, which are a generalization from the results found in [4, 5]. We do not present a proof here, since it is very similar to the ones found in [4, 5] with obvious adaptations.

Definition 8.4 We say that $f : [a, b]_{\mathbb{T}} \to X$ is *Riemann* Δ*-integrable* on $[a, b]_{\mathbb{T}}$ if there exists an element $I \in X$ with the following property: For every $\varepsilon > 0$, there exists $\delta > 0$ such that

$$\left\| \sum_{i=1}^{m} f(\xi_i)(t_i - t_{i-1}) - I \right\| < \varepsilon,$$

for every $P \in D_\delta(a, b)$ independently of $\xi_i \in [t_{i-1}, t_i)_{\mathbb{T}}$ for $1 \leq i \leq n$. It is clear that such element I is unique, and it is called the Riemann Δ-integral of f from a to b.

Lemma 8.1 ([5, Theorem 5.32]) *Let* $g : [a, b]_{\mathbb{T}} \to X$ *be a function. Assume there exists a sequence* $g_n : [a, b]_{\mathbb{T}} \to X$, $n \in \mathbb{N}$, *of* Δ*-integrable functions on* $[a, b]_{\mathbb{T}}$ *such that* $g_n \to g$ *uniformly on* $[a, b)_{\mathbb{T}}$. *Then* g *is* Δ*-integrable on* $[a, b]_{\mathbb{T}}$ *and*

$$\int_a^b g(\tau)\Delta\tau = \lim_{n\to\infty} \int_a^b g_n(\tau)\Delta\tau.$$

For vector functions, we can establish the following mean value theorem. We denote by $c(S)$ the convex hull of the set S.

Lemma 8.2 ([16, Theorem 3.5]) *Let* $g : [a, b]_{\mathbb{T}} \to X$ *be a* Δ*-integrable function. Then*

$$\int_a^b g(t)\Delta t \in (b - a)\overline{c(Im(g))}.$$

Lemma 8.3 (Fundamental Theorem of Calculus, First Version [5]) *Let* $g : [a, b]_{\mathbb{T}} \to X$ *be a continuous function on* $[a, b]_{\mathbb{T}}$ *such that* g *is* Δ*-differentiable on* $[a, b)_{\mathbb{T}}$. *If* g^Δ *is* Δ*-integrable on* $[a, b]_{\mathbb{T}}$, *then*

$$\int_a^b g^\Delta(t)\Delta t = g(b) - g(a).$$

Now, we present the definition of invariant under translations time scales (see [21]).

Definition 8.5 A time scale $\mathbb{T}$ is called *invariant under translations* if

$$\Pi := \{\tau \in \mathbb{R} : t \pm \tau \in \mathbb{T},\ \forall t \in \mathbb{T}\} \neq \{0\}. \tag{8.2}$$

Note that $\Pi \neq \emptyset$ since $0 \in \Pi$.

The next result is an immediate consequence of [4, Theorem 1.98].

Lemma 8.4 *Let $\mathbb{T}$ be an invariant under translations time scale, and assume $g\colon \mathbb{T} \to X$ is an rd-continuous function. Let $a, b \in \mathbb{T}$, $a < b$, and let $h \in \Pi$. Then*

$$\int_{a+h}^{b+h} g(\tau)\Delta\tau = \int_{a}^{b} g(\tau + h)\Delta\tau.$$

In what follows, we present a definition of almost periodic function on time scales, which can be found in [18].

Definition 8.6 Let $\mathbb{T}$ be an invariant under translation time scale. A function $f \in C(\mathbb{T}, X)$ is called *almost periodic* if the ε-translation set of f given by

$$E\{\varepsilon, f\} = \{\tau \in \Pi; \; \|f(t+\tau) - f(t)\| < \varepsilon, \; \forall t \in \mathbb{T}\}$$

is a relatively dense set in $\mathbb{T}$ for all $\varepsilon > 0$.

Before to present the definition of an rd-equicontinuous set of functions and a version of Arzelà-Ascoli's Theorem for functions defined on time scales, we present the definition of an equiregulated set of functions.

Definition 8.7 A subset $\mathcal{F} \subset C_{rd}([a, b]_{\mathbb{T}}, X)$ is said to be *equiregulated* if for every $\varepsilon > 0$ and every $t_0 \in [a, b]_{\mathbb{T}}$ there exists $\delta > 0$ such that for all $x \in \mathcal{F}$:

(i) For any $t_0 - \delta < s < t_0$, $\|x(s) - x(t_0^-)\| < \varepsilon$.
(ii) For any $t_0 < \tau < t_0 + \delta$, $\|x(\tau) - x(t_0^+)\| < \varepsilon$.

Now, we introduce the definition of an rd-equicontinuous set of functions.

Definition 8.8 A subset $\mathcal{F} \subset C_{rd}([a, b]_{\mathbb{T}}, X)$ is said to be *rd-equicontinuous* if $\mathcal{F}$ is equiregulated on $[a, b]$, and the following condition is satisfied: For every $\varepsilon > 0$ and for all right-dense point $t' \in [a, b]_{\mathbb{T}}$, there exists a $\delta > 0$ such that

$$\|f(t) - f(t')\| < \varepsilon$$

for all $f \in \mathcal{F}$ and $t \in [t', t' + \delta) \cap [a, b]_{\mathbb{T}}$.

Next, we present a version of Arzelà-Ascoli's Theorem for functions defined on time scales. The proof of this result is very similar to the proof found in [25, Theorem 3.8] with obvious adaptations. For this reason, we omit it here.

Theorem 8.1 (Arzelà-Ascoli's Theorem on Time Scales) *A set $H \subset C_{\mathrm{rd}}([a, b]_{\mathbb{T}}, X)$ is relatively compact, if and only if H is rd-equicontinuous and for each $t \in [a, b]_{\mathbb{T}}$, $H(t)$ is a relatively compact set in X.*

8.3 Existence and Uniqueness of Solutions

In this section, our goal is to prove some results concerning the existence and uniqueness of solutions of the dynamic equation on time scales

$$x^{\Delta}(t) = f(t, x(t)), \ t \in \mathbb{T}, \tag{8.3}$$

where $f : \mathbb{T} \times X \to X$ is an appropriate function.

We begin by establishing the definition of a solution of (8.3).

Definition 8.9 A function $x : \mathbb{T} \to X$ is said to be a *solution of the dynamic equation* (8.3) if $x : \mathbb{T} \to X$ is a Δ-differentiable function and Eq. (8.3) is satisfied.

Throughout this section, we consider the following conditions for a function $f : \mathbb{T} \times X \to X$.

(F1) For each $x \in X$, the map $t \mapsto f(t, x)$ is rd-continuous on $\mathbb{T}$.
(F2) There exists a nondecreasing continuous function $M : \mathbb{R}^+ \to [0, \infty)$ such that $M(0) = 0$ and

$$\|f(t, x) - f(t, y)\| \leq M(\|x - y\|), \ \ \forall t \in \mathbb{T}, \ \forall x, y \in X.$$

(F3) For every bounded subset $I \times B \subset \mathbb{T} \times X$, the set $f(I \times B)$ is relatively compact in X.

Now, we present a result that ensures the existence of a global solution of Eq. (8.3) on $[a, b]_{\mathbb{T}}$ such that $x(t_0) = x_0$.

Theorem 8.2 *Let $a, \ b \in \mathbb{T}$, $a < b$, $t_0 \in [a, b]_{\mathbb{T}}$, and $x_0 \in X$. Assume that $f : \mathbb{T} \times X \to X$ satisfies conditions (F1)–(F3), and suppose the following inequality is fulfilled:*

$$\liminf_{\xi \to \infty} \frac{M(\xi)}{\xi} < \frac{1}{b - a}. \tag{8.4}$$

Then there exists a solution of (8.3) *defined on $[a, b]_{\mathbb{T}}$ such that $x(t_0) = x_0$.*

Proof We define $\Gamma : C_{\mathrm{rd}}([a, b]_{\mathbb{T}}, X) \to C_{\mathrm{rd}}([a, b]_{\mathbb{T}}, X)$ by

$$(\Gamma x)(t) = x_0 + \int_{t_0}^{t} f(s, x(s))\Delta s, \ t \in [a, b]_{\mathbb{T}}. \tag{8.5}$$

Clearly, Γ is well defined. At first, we affirm that Γ is continuous. Indeed, let $x, y \in C_{\mathrm{rd}}([a, b]_{\mathbb{T}}, X)$; then

$$\begin{aligned} \|\Gamma x(t) - \Gamma y(t)\| &\leq \left| \int_{t_0}^{t} \|f(s, x(s)) - f(s, y(s))\|\Delta s \right| \\ &\leq M(\|x - y\|_{\infty})|t - t_0|, \end{aligned}$$

where $\|x-y\|_\infty = \sup_{t\in[a,b]_{\mathbb{T}}} \|x(t)-y(t)\|$. Taking the supremum with respect to t when $t \in [a,b]_{\mathbb{T}}$, we have

$$\|\Gamma x - \Gamma y\|_\infty \le M(\|x-y\|_\infty)\max\{t_0-a, b-t_0\} \le M(\|x-y\|_\infty)(b-a). \tag{8.6}$$

It follows from (8.6) that Γ is continuous. Let now $\beta > 0$, and define $A_\beta \subseteq C_{\mathrm{rd}}([a,b]_{\mathbb{T}}, X)$ by the set consisting of functions $x(\cdot)$ such that $x(t_0) = x_0$ and

$$\|x(t) - x_0\| \le \beta \quad \text{for every} \quad t \in [a,b]_{\mathbb{T}}.$$

It is clear that A_β is a closed and convex subset of $C_{\mathrm{rd}}([a,b]_{\mathbb{T}}, X)$. We next show that there exists $\beta > 0$ such that $\Gamma(A_\beta) \subseteq A_\beta$. Assume the contrary, i.e., for every $r > 0$, there exists $x^r(\cdot) \in A_r$ such that $\Gamma(x^r) \notin A_r$. This implies that there exists $t^r \in [a,b]_{\mathbb{T}}$ such that

$$\begin{aligned}
r &< \|\Gamma x^r(t^r) - x_0\| \\
&= \left\| \int_{t_0}^{t^r} f(s, x^r(s))\Delta s \right\| \\
&\le \left| \int_{t_0}^{t^r} M(\|x^r(s) - x_0\|)\Delta s \right| + \left| \int_{t_0}^{t^r} \|f(s, x_0)\|\Delta s \right| \\
&\le |t^r - t_0| M(r) + \left| \int_{t_0}^{t^r} \|f(s, x_0)\|\Delta s \right| \\
&\le (b-a)M(r) + \int_a^b \|f(s, x_0)\|\Delta s.
\end{aligned}$$

This implies that

$$1 < (b-a)\frac{M(r)}{r} + \frac{1}{r}\int_a^b \|f(s, x_0)\|\Delta s, \ \forall r > 0,$$

which contradicts (8.4), proving the assertion.

Let $\beta > 0$ such that $\Gamma(A_\beta) \subseteq A_\beta$. We next show that $\Gamma(A_\beta)$ is a relatively compact set in $C_{\mathrm{rd}}([a,b]_{\mathbb{T}}, X)$. In fact, it follows from (F3) and Lemma 8.2 that

$$\Gamma(A_\beta)(t) \subseteq x_0 + (t - t_0)\overline{c(f([a,b]_{\mathbb{T}} \times B_\beta[x_0; X]))},$$

which is a relatively compact set in X for each $t \in [a,b]_{\mathbb{T}}$. Moreover, if $x \in A_\beta$, $t_1, t_2 \in [a,b]_{\mathbb{T}}$, $t_1 < t_2$, then using again condition (F2), we get

$$\|\Gamma x(t_2)-\Gamma x(t_1)\| = \left\|\int_{t_1}^{t_2} f(s,x(s))\Delta s\right\| \le (t_2-t_1)(M(\beta)+\sup_{s\in[a,b]_{\mathbb{T}}}\|f(s,x_0)\|),$$

which shows that $\Gamma(A_\beta)$ is an rd-equicontinuous set. From Theorem 8.1, we conclude that $\Gamma(A_\beta)$ is a relatively compact set. Finally, applying Schauder's Fixed Point Theorem, we conclude that Γ has a fixed point $x(\cdot) \in A_\beta$ and $x(\cdot)$ is a solution of (8.3) with the initial condition $x(t_0) = x_0$. □

The previous result does not allow us to conclude the uniqueness of the solution. To guarantee this property, we need to impose an additional condition on the function M involved in the condition (F2).

(F4) Let $a,\ b,\ t_0 \in \mathbb{T}$ such that $a \le t_0 \le b$. If $\alpha\colon [a,b]_{\mathbb{T}} \to \mathbb{R}_+$ is a positive rd-continuous function such that

$$\alpha(t) \le \int_{t_0}^{t} M(\alpha(s))\Delta s,$$

then $\alpha \equiv 0$ in $[a,b]_{\mathbb{T}}$.

Theorem 8.3 *Let $a, t_0, b \in \mathbb{T}$ with $a \le t_0 < b$ and $x_0 \in X$. Assume the hypotheses of Theorem 8.2 and condition (F4) are fulfilled. Then there exists a unique solution of* (8.3) *with initial condition $x(t_0) = x_0$ defined on $[a,b]_{\mathbb{T}}$.*

Proof The existence of solutions of Eq. (8.3) with the initial condition $x(t_0) = x_0$ is guaranteed by Theorem 8.2. Suppose $x, y\colon [a,b]_{\mathbb{T}} \to X$ are solutions of (8.3) satisfying $x(t_0) = y(t_0) = x_0$; then

$$x(t) = x_0 + \int_{t_0}^{t} f(s,x(s))\Delta s, \quad t \in [a,b]_{\mathbb{T}}$$

$$y(t) = x_0 + \int_{t_0}^{t} f(s,y(s))\Delta s, \quad t \in [a,b]_{\mathbb{T}}.$$

Thus, for every $t \in [a,b]_{\mathbb{T}}$, we have

$$\begin{aligned}\|x(t)-y(t)\| &= \left\|\int_{t_0}^{t} f(s,x(s))\Delta s - \int_{t_0}^{t} f(s,y(s))\Delta s\right\| \\ &\le \left|\int_{t_0}^{t} M(\|x(s)-y(s)\|)\Delta s\right|.\end{aligned}$$

Applying (F4), we obtain that $x(t) = y(t)$ for every $t \in [a,b]_{\mathbb{T}}$. This implies that the solution of Eq. (8.3) with initial condition $x(t_0) = x_0$ on $[a,b]_{\mathbb{T}}$ is unique. □

Remark 8.1 A careful examination about the conditions (F1)–(F3) reveals that they are more general than the classical ones considered extensively in the literature, which justifies to present this type of existence result in the present chapter. Notice that condition (F3) encompasses the classical Lipschitz condition, but it can be read in a much more general setting, depending on the definition of function M. On the other hand, condition (F4), which appears in the uniqueness theorem, plays an important role to ensure this property for the solution of equation, having no need to employ Gronwall's Inequality and presenting a more elegant proof of it. The condition (F4) was already considered previously in the literature by other authors; see, for instance, [10, 11].

8.4 Periodic Solutions

In this section, we are interested in the study of the existence of periodic solutions of Eq. (8.3). Throughout this section, let us consider that $\mathbb{T}$ is an invariant under translations time scale, and let us fix a positive $p \in \Pi$.

As usual, we say that a function $f\colon \mathbb{T} \times X \to X$ is *p-periodic with respect to the first variable* if $f(t+p,x) = f(t,x)$ for all $t \in \mathbb{T}$ and $x \in X$.

In what follows, we present our first result about the existence of periodic solutions. It is a version of Massera's Theorem on time scales for functions taking values in Banach spaces.

Theorem 8.4 *Let $\mathbb{T}$ be an invariant under translations time scale such that $0 < p \in \Pi$, $t_0 \in \mathbb{T}$ with $a \leq t_0 < b$, $a, b \in \mathbb{T}$. Assume that $f : \mathbb{T} \times X \to X$ is continuous, $f(\cdot, x)$ is p-periodic for all $x \in X$, and conditions (F1)–(F4) hold. Assume further that there exists a solution $u(\cdot)$ of (8.3) on $\mathbb{T}$ such that $Im(u)$ is a bounded subset of X and $u(t_0 + np) \to y$ as $n \to \infty$. Then there exists a p-periodic solution $v\colon [a,b]_{\mathbb{T}} \to X$ of Eq. (8.3) with $v(t_0) = y$.*

Proof We abbreviate the notation by writing $B = \mathrm{Im}(u)$. Let $t_0 \in [a,b]_{\mathbb{T}}$. We define the sequence of functions $v_n\colon [a,b]_{\mathbb{T}} \to X$, $n \in \mathbb{N}$, by

$$v_n(t) := u(t+np) \qquad \text{for every } \ t \in [a,b]_{\mathbb{T}}. \tag{8.7}$$

Since u is a solution of (8.3), u is rd-continuous which implies that v_n is also rd-continuous for each $n \in \mathbb{N}$. Hence, $v_n \in C_{\mathrm{rd}}([a,b]_{\mathbb{T}}, X)$. Moreover, since $\mathrm{Im}(v_n) \subset B$, we obtain that $(v_n)_n$ is bounded.

We next establish some additional properties of the sequence $(v_n)_n$.

(i) Assertion 1 $(v_n)_n$ is an rd-equicontinuous sequence of functions.

Indeed, for every $s_1, s_2 \in [a,b]_{\mathbb{T}}$, $s_1 < s_2$, we have

$$\begin{aligned} v_n(s_2) - v_n(s_1) &= u(s_2+np) - u(s_1+np) \\ &= \int_{s_1+np}^{s_2+np} f(\tau, u(\tau))\Delta\tau \end{aligned} \tag{8.8}$$

$$= \int_{s_1}^{s_2} f(\tau + np, u(\tau + np))\Delta\tau$$

$$= \int_{s_1}^{s_2} f(\tau, u(\tau + np))\Delta\tau \tag{8.9}$$

$$= \int_{s_1}^{s_2} f(\tau, v_n(\tau))\Delta\tau, \tag{8.10}$$

where to obtain (8.8), we use the fact that u is a solution of (8.3), and (8.9) follows from the p-periodicity of f with respect to the first variable. Since $[s_1, s_2]_{\mathbb{T}} \times B$ is a bounded set of $\mathbb{T} \times X$, it follows from (F3) that the set $f([s_1, s_2]_{\mathbb{T}} \times B)$ is relatively compact in X. Applying Lemma 8.2 to the equality (8.10), we obtain

$$v_n(s_2) - v_n(s_1) \in (s_2 - s_1)\overline{c(f([s_1, s_2]_{\mathbb{T}} \times B))},$$

which shows that $(v_n)_n$ is an rd-equicontinuous sequence of functions.
(ii) Assertion 2 For $t \in \mathbb{T}$, the set $V(t) = \{v_n(t) : n \in \mathbb{N}\}$ is relatively compact in X.

In fact, assuming that $t_0 \leq t$ and proceeding as above, we can write

$$v_n(t) - v_n(t_0) = \int_{t_0}^{t} f(\tau, u(\tau + np))\Delta\tau \in (t - t_0)\overline{c(f([t_0, t]_{\mathbb{T}} \times B))},$$

which implies that

$$v_n(t) \in u(t_0 + np) + (t - t_0)\overline{c(f([t_0, t]_{\mathbb{T}} \times B))}.$$

Since $u(t_0 + np) \to y$ as $n \to \infty$, it follows that $V(t)$ is a relatively compact set. A similar calculation holds for $t \leq t_0$.

Now, we define the limit of $(v_n)_n$. Combining (i) and (ii), it follows by Theorem 8.1 that the set $\{v_n : n \in \mathbb{N}\}$ is relatively compact in $C_{\text{rd}}([a, b]_{\mathbb{T}}, X)$ for all $a,\ b \in \mathbb{T}$. Therefore, there exists a subsequence $(v_{n_k})_k$ that converges uniformly to a function $v(\cdot)$ on $[a, b]_{\mathbb{T}}$. Using again expression (8.10), we infer

$$v_{n_k}(t) - v_{n_k}(t_0) = \int_{t_0}^{t} f(\tau, u(\tau + n_k p))\Delta\tau$$

$$= \int_{t_0}^{t} f(\tau, v_{n_k}(\tau))\Delta\tau.$$

By taking the limit as $k \to \infty$ above and by applying Lemma 8.1, we obtain

$$v(t) - y = \int_{t_0}^{t} f(\tau, v(\tau))\Delta\tau,$$

which implies that $v(\cdot)$ is a solution of Eq. (8.3) with $v(t_0) = y$ on $[a, b]_{\mathbb{T}}$.

A similar argument allows us to conclude that every subsequence of $(v_n)_n$ contains a subsequence which converges to a solution $w(\cdot)$ of Eq. (8.3) with initial condition $w(t_0) = y$. Using conditions (F1)–(F4), the solution of problem (8.3) with initial condition y at t_0 is unique. Consequently, the original sequence $(v_n)_n$ converges to $v(\cdot)$, and $v(\cdot)$ is a solution of (8.3) with $v(t_0) = y$.

(iii) Assertion 3 The function $v(\cdot)$ is p-periodic on $[a, b]_{\mathbb{T}}$. In fact, for every $t \in [a, b]_{\mathbb{T}}$, we have

$$\begin{aligned} v(t+p) &= \lim_{n\to\infty} v_n(t+p) \\ &= \lim_{n\to\infty} u(t+p+np) \\ &= \lim_{n\to\infty} u(t+(n+1)p) \\ &= \lim_{n\to\infty} v_{n+1}(t) = v(t), \end{aligned}$$

which implies that $v(\cdot)$ is a p-periodic function on $[a, b]_{\mathbb{T}}$. □

For spaces X of finite dimension, we can weaken the hypotheses of Theorem 8.4. In particular, we can avoid assumption (F3).

Corollary 8.1 *Assume* $\dim(X) < \infty$. *Let* $\mathbb{T}$ *be an invariant under translations time scale such that* $0 < p \in \Pi$ *and* $t_0 \in [a, b]_{\mathbb{T}}$, $a, b \in \mathbb{T}$. *Assume that* $f : \mathbb{T} \times X \to X$ *is continuous,* $f(\cdot, x)$ *is p-periodic for all* $x \in X$, *and conditions (F1), (F2), and (F4) hold. Assume further that there exists a solution* $u(\cdot)$ *of* (8.3) *on* $\mathbb{T}$ *such that* $Im(u)$ *is a bounded subset of* X *and* $u(t_0 + np) \to y$ *as* $n \to \infty$. *Then there exists a p-periodic solution* $v : [a, b]_{\mathbb{T}} \to X$ *of Eq.* (8.3) *with* $v(t_0) = y$.

Proof We only need to observe that condition (F3) also holds, since every bounded set in X is relatively compact. □

Finally, we study the relation between almost periodic solutions and periodic solutions of (8.3) in the case $X = \mathbb{R}$. Our last result establishes conditions under which any almost periodic solution of Eq. (8.3) is a p-periodic solution of (8.3).

We next use similar notations than the ones found in the book of Fink [14] and some ideas from [8]. Specifically, if $u : \mathbb{T} \to \mathbb{R}$ is an almost periodic function, we call *hull of* u to the set consisting of functions $v(\cdot)$ for which there exists a sequence $(\alpha_n)_n$ in Π such that $v(t) = \lim_{n\to\infty} u(t + \alpha_n)$ uniformly for $t \in \mathbb{T}$. We consider the following condition:

(F5) For every $t \in \mathbb{T}$, $f(t, x) \geq 0$ for all $x \in \mathbb{R}$, and for every right-scattered $t \in \mathbb{T}$, the function $f(t, \cdot)$ is increasing with respect to the second variable.

Theorem 8.5 *Let* $\mathbb{T}$ *be an invariant under translations time scale such that* $0 < p \in \Pi$. *Assume that* $f : \mathbb{T} \times \mathbb{R} \to \mathbb{R}$ *is continuous,* $f(\cdot, x)$ *is p-periodic for all* $x \in \mathbb{R}$, *and conditions (F1), (F2), (F4), and (F5) hold. Then any almost periodic solution of Eq.* (8.3) *on* $\mathbb{T}$ *is a p-periodic function.*

Proof Let $v: \mathbb{T} \to \mathbb{R}$ be an almost periodic solution of (8.3). Using that $np \in \Pi$ for all $n \in \mathbb{N}$, we can define a sequence of functions $u_n: \mathbb{T} \to \mathbb{R}$ as follows:

$$u_n(t) := v(t + np) \quad \text{for every } \; t \in \mathbb{T}. \tag{8.11}$$

Using (8.11), it is not difficult to show that u_n is also a solution of (8.3) for each $n \in \mathbb{N}$.

Assertion 1 If there exists $t^* \in \mathbb{T}$ such that $v(t^*) = v(t^* + p)$, then $v(t) = v(t + p)$ for all $t \in \mathbb{T}$.

In fact, suppose there exists a $t^* \in \mathbb{T}$ such that $v(t^*) = v(t^* + p)$. This implies that $v(t^*) = u_1(t^*)$. Since v and u_1 are solutions of (8.3), the uniqueness of solutions implies that $v(t) = u_1(t) = v(t + p)$ for all $t \in \mathbb{T}$. Hence, it follows that v is a p-periodic function on $[a, b]_{\mathbb{T}}$, proving the assertion.

Assertion 2 Assume that $v(\cdot)$ is not p-periodic. Initially we will show that $v(t) < v(t + p)$ or $v(t) > v(t + p)$ for all $t \in \mathbb{T}$.

In fact, since v is not p-periodic, we can assume, without loss of generality, that there exists $t^* \in \mathbb{T}$ such that $v(t^*) < v(t^* + p)$. Clearly, $v(t) \neq v(t + p)$ for all $t \in \mathbb{T}$; otherwise by Assertion 1, t^* should not exist. Now, consider t^* is right-dense, we can affirm that there exists a sequence $t_n \to t^*$ such that $t_n \in \mathbb{T}$ and $t_n > t^*$ such that by the continuity of function v, and it follows that $v(t_n) < v(t_n + p)$ for each $n \in \mathbb{N}$. On the other hand, if t^* is right-scattered, then

$$\begin{aligned} v(\sigma(t^*)) &= v(t^*) + \mu(t^*) f(t^*, v(t^*)) \\ &< v(t^* + p) + \mu(t^*) f(t^*, v(t^* + p)) \\ &= v(t^* + p) + \mu(t^* + p) f(t^* + p, v(t^* + p)) \\ &= v(\sigma(t^*) + p), \end{aligned}$$

where the second inequality follows from hypothesis (F5) and the third equality follows from the invariance of the time scale and by the p-periodicity of the function f with respect to the first variable.

Therefore, combining these two cases, we obtain that $v(t) < v(t + p)$ for all $t \in \mathbb{T}$. Proceeding similarly, if we assume that $v(t^*) > v(t^* + p)$, we obtain that $v(t) > v(t + p)$ for all $t \in \mathbb{T}$.

Since v is almost periodic, it follows that $(u_n)_n$ possesses a convergent subsequence which converges to some function x. Moreover, by the almost periodicity of v, it follows that the sequence $(u_n)_n$ is bounded, and as remarked before, $(u_n)_n$ is monotone; then we conclude that the entire sequence $(u_n)_n$ itself converges uniformly to the function x.

Next, we will show that the function $x(\cdot)$ is p-periodic. By the construction of x, for $t \in \mathbb{T}$, we have

$$\begin{aligned} x(t + p) &= \lim_{n\to\infty} u_n(t + p) = \lim_{n\to\infty} v(t + p + np) = \lim_{n\to\infty} v(t + (n + 1)p) \\ &= \lim_{n\to\infty} u_{n+1}(t) = x(t), \end{aligned}$$

which implies that x is a p-periodic function. Since x belongs to the hull of $v(\cdot)$, it follows from [14, Theorem 1.1.8] that $v(\cdot)$ belongs to the hull of $x(\cdot)$. Consequently, $v(\cdot)$ is a p-periodic function, which contradicts our initial assumption and completes the proof of the theorem. □

Acknowledgments Part of this work was done when the coauthor Eduard Toon visited the Universidad de Santiago de Chile, whose hospitality he gratefully acknowledges.

H. Henríquez was supported by Vicerrectoría de Investigación, Desarrollo e Innovación of the Universidad de Santiago, under grant DICYT-USACH 041733HM, and J. G. Mesquita was partially supported by CNPq grant 310583/2021-7 and FAPDF 0193-00001821/2022-21. E. Toon gratefully acknowledges the financial support from Universidad de Santiago de Chile, USACH, and Proyecto POSTDOC_DICYT, Código 041733UL_POSTDOC, Vicerrectoría de Investigación, Desarrollo e Innovación.

We dedicate this work to the memory of our dear coauthor Hernán R. Henríquez who passed away on Thursday, June 2, 2022.

References

1. F.M. Atici, D.C. Biles, A. Lebedinsky, An application of time scales to economics. Math. Comput. Model. **43**(7–8), 718–726 (2006)
2. M. Bohner, G. Guseinov, Line integrals and Green's formula on time scales. J. Math. Anal. Appl. **326**(2), 1124–1141 (2007)
3. M. Bohner, J.G. Mesquita, Massera's theorem in quantum calculus. Proc. Am. Math. Soc. **146**(11), 4755–4766 (2018)
4. M. Bohner, A. Peterson, *Dynamic Equations on Time Scales: An Introduction with Applications* (Birkhäuser Boston, Inc., Boston, 2001), 358p.
5. M. Bohner, A. Peterson, *Advances in Dynamic Equations on Time Scales* (Birkhäuser Boston, Inc., Boston, 2003), 348p.
6. M. Bohner, N. Wintz, The Kalman filter for linear systems on time scales. J. Math. Anal. Appl. **406**(2), 419–436 (2013)
7. D. Brigo, F. Mercurio, Discrete time vs continuous time stock-price dynamics and implications for option pricing. Finance Stochastics **4**, 147–159 (2000)
8. T.A. Burton, *Stability and Periodic Solutions of Ordinary and Functional-Differential Equations*. Mathematics in Science and Engineering, vol. 178 (Academic Press, Inc., Orlando, 1985), 337p.
9. F.B. Christiansen, T.M. Fenchel, *Theories of Populations in Biological Communities* (Springer, Berlin, 1977)
10. B.C. Dhage, Quadratic perturbations of periodic boundary value problems of second order ordinary differential equations. Differ. Equ. Appl. **2**(4), 465–486 (2010)
11. B.C. Dhage, Some variants of two basic hybrid fixed point theorems of Krasnoselskii and Dhage with applications. Nonlinear Stud. **25**(3), 559–573 (2018)
12. K. Ezzinbi, G.M. N'Guérékata, Massera type theorem for almost automorphic solutions of functional differential equations of neutral type. J. Math. Anal. Appl. **316**(2), 707–721 (2006)
13. K. Ezzinbi, S. Fatajou, G. N'Guérékata, Massera-type theorem for the existence of $C^{(n)}$-almost-periodic solutions for partial functional differential equations with infinite delay. Nonlinear Anal. **69**(4), 1413–1424 (2008)
14. A.M. Fink, *Almost Periodic Differential Equations*. Lecture Notes in Mathematics, vol. 377. (Springer, Berlin, 1974), 336p.

15. M. Fleury, J.G. Mesquita, A. Slavík, Massera's theorems for various types of equations with discontinuous solutions. J. Differ. Equations **269**(12), 11667–11693 (2020)
16. H.R. Henríquez, J.G. Mesquita, Self-accessible states for linear systems on time scales. Proc. Am. Math. Soc. **146**(3), 1257–1269 (2018)
17. I. Klapper, H. Qian, Remarks on discrete and continuous large-scale models of DNA dynamics. Biophys. J. **74** 2504–2514 (1998)
18. Y. Li, C. Wang, Uniformly almost periodic functions and almost periodic solutions to dynamic equations on time scales. Abstr. Appl. Anal., 1–22 (2011). Art. ID 341520
19. X-L. Liu, W-T. Li, Periodic solutions for dynamic equations on time scales. Nonlinear Anal. **67**(5), 1457–1463 (2007)
20. J. Liu, G. N'Guérékata, N.V. Minh, A Massera type theorem for almost automorphic solutions of differential equations. J. Math. Anal. Appl. **299**(2), 587–599 (2004)
21. C. Lizama, J.G. Mesquita, Almost automorphic solutions of dynamic equations on time scales. J. Funct. Anal. **265**(10), 2267–2311 (2013)
22. J.L. Massera, The existence of periodic solutions of systems of differential equations. Duke Math. J. **17**, 457–475 (1950)
23. S. Murakami, T. Naito, N.V. Minh, Massera's theorem for almost periodic solutions of functional differential equations. J. Math. Soc. Japan **56**(1), 247–268 (2004)
24. T. Naito, N.V. Minh, J.S. Shin, A Massera type theorem for functional differential equations with infinite delay. Japan. J. Math. (N.S.) **28**(1), 31–49 (2002)
25. Y. Su, Z. Feng, Positive solutions to the singular p-Laplacian BVPs with sign-changing nonlinearities and higher-order derivatives in Banach spaces on time scales. Dyn. Partial Differ. Equ. **8**(2), 149–171 (2011)
26. C. Tisdell, A. Zaidi, Basic qualitative and quantitative results for solutions to nonlinear, dynamic equations on time scales with an application to economic modelling. Nonlinear Anal. **68**(11), 3504–3524 (2008)
27. L. Yong, L. Zhenghua, L. Zhaoxing, A Massera type criterion for linear functional-differential equations with advance and delay. J. Math. Anal. Appl. **200**(3), 717–725 (1996)

Chapter 9
Attractors for Locally Damped Bresse Systems and a Unique Continuation Property

To Fu Ma, Rodrigo N. Monteiro, and Paulo N. Seminario-Huertas

Abstract This chapter is devoted to Bresse systems, a robust model for circular beams, given by a set of three coupled wave equations. The main objective is to establish the existence of global attractors for the dynamics of semilinear problems with localized damping. In order to deal with localized damping, a unique continuation property (UCP) is needed. Therefore we also provide a suitable UCP for Bresse systems. Our strategy is to set the problem in a Riemannian geometry framework and see the system as a single equation with different Riemann metrics. Then we perform Carleman-type estimates to get our result.

Keywords Bresse system · Unique continuation · Localized damping · Riemannian manifold · Global attractor

MSC 2010 35L05, 35B40, 35B41, 35B60, 35R01

9.1 Introduction

The Bresse system is a model for circular beams given by three coupled wave equations, namely,

T. F. Ma (✉)
Departamento de Matemática, Universidade de Brasília, Brasília, Brazil
e-mail: matofu@mat.unb.br

R. N. Monteiro
Department of Mathematics, State University of Londrina, Londrina, PR, Brazil

P. N. Seminario-Huertas
Departamento de Matemática Aplicada a las Tecnologías de la Información y las Comunicaciones, Universidad Politécnica de Madrid, Madrid, Spain

P. Benevieri, J. G. Mesquita (eds.), *Functional Differential Equations and Dynamic Equations on Time Scales*, Advances in Continuum Mechanics 52,
https://doi.org/10.1007/978-3-031-83327-4_9

$$\begin{cases} \rho_1\varphi_{tt} - k(\varphi_x + \psi + \ell w)_x - k_0\ell(w_x - \ell\varphi) = 0 \text{ in } (0, L) \times (0, \infty), \\ \rho_2\psi_{tt} - b\psi_{xx} + k(\varphi_x + \psi + \ell w) = 0 \text{ in } (0, L) \times (0, \infty), \\ \rho_1 w_{tt} - k_0(w_x - \ell\varphi)_x + k\ell(\varphi_x + \psi + \ell w) = 0 \text{ in } (0, L) \times (0, \infty). \end{cases}$$

The functions $\varphi = \varphi(x, t)$, $\psi = \psi(x, t)$, $w = w(x, t)$ correspond to the vertical displacement, rotation angle, and longitudinal displacement at a point $x \in (0, L)$ and time $t \geqslant 0$, respectively. The coefficients are all positive constants defined by $\rho_1 = \rho A$, $\rho_2 = \rho I$, $k = KAG$, $b = EI$, and $k_0 = AE$, where the quantities ρ, A, I, K, G, and E denote, respectively, material density, cross-sectional area, second moment of the cross-section area, a shear factor, shear modulus, and modulus of elasticity. In addition, $\ell > 0$ denotes the beam's curvature. Its mathematical modeling can be found in [4, 17], and it is worth observing that when $\ell = 0$ the arched beam reduces to the Timoshenko beam [30].

The Bresse system and its viscoelastic and thermoelastic extensions were studied by many authors. Roughly speaking, most of the results are concerned with a certain asymptotic stability dichotomy. Indeed, by analogy to the Timoshenko system, the Bresse system having damping terms in one or two of its equations is exponentially stable if and only if it satisfies the equal wave speeds condition

$$\frac{\rho_1}{k} = \frac{\rho_2}{b} \quad \text{and} \quad k = k_0. \tag{9.1}$$

Such a condition was firstly observed for Timoshenko systems in [28]. Otherwise only polynomial stability can be obtained. See, e.g., [1, 3, 11–13, 19, 26, 29, 32].

In a different direction, Charles et al. [8] proved the exponential stability of Bresse systems by adding a localized damping in each one of its three equations, without assuming the speed condition (9.1). This is quite interesting since the equal speed assumption cannot be realized physically, cf. [22].

The main objective of this chapter is to establish existence of global attractors for dynamics of semilinear Bresse systems with locally defined damping (see Problem 9.32), without assuming condition (9.1). Our approach is very different from [8]. Indeed, one of the ingredients for obtaining exponential stability of wave equations with localized damping is a unique continuation property (UCP). To our purpose, the UCP says whether a wave equation that vanishes in a subdomain must be identically null. In [8] they have used a UCP derived from Holmgren uniqueness theorem, which is only valid for equations with analytic coefficients. Because of nonlinear terms, our problem (9.32) has no longer analytic coefficients. To overcome this difficult, we propose a new UCP for Bresse systems.

More precisely, we discuss the unique continuation property for coupled wave equations of the form

$$
\begin{cases}
\partial_t^2 u_1 - \Delta_1 u_1 = f_1(u_1, u_2, \cdots, u_n) \text{ in } (0, L) \times (0, \infty), \\
\partial_t^2 u_2 - \Delta_2 u_2 = f_2(u_1, u_2, \cdots, u_n) \text{ in } (0, L) \times (0, \infty), \\
\qquad \vdots \\
\partial_t^2 u_n - \Delta_n u_n = f_n(u_1, u_2, \cdots, u_n) \text{ in } (0, L) \times (0, \infty),
\end{cases}
\tag{9.2}
$$

where for $i = 1, \cdots, n$ the following is assumed:

We consider on the system (9.2) the Dirichlet boundary conditions

$$
u_i(0, t) = u_i(L, t) = 0 \quad \forall t \in (0, \infty) \tag{9.3}
$$

and initial data

$$
(u_1, \partial_t u_1, \cdots, u_n, \partial_t u_n)\big|_{t=0} = (u_1^0, u_1^1, \cdots, u_n^0, u_n^1) \ \text{ in } \ (0, L). \tag{9.4}
$$

Given $\gamma_i > 0$, the operator Δ_i represents the one-dimensional Laplacian operator with wave propagation velocity $\sqrt{\gamma_i}$, defined by

$$
\Delta_i = \gamma_i \partial_x^2. \tag{9.5}
$$

The functions f_i are coupling terms at energy level such that $f_i \in L^2(0, T; L^2(0, L))$ and

$$
f_i(u_1, u_2, \cdots, u_n) = \sum_{j=1}^{n} p_j^i \partial_x u_j + \sum_{j=1}^{n} q_j^i u_j, \tag{9.6}
$$

where $p_j^i, q_j^i \in L^2(0, T; L^2(0, L))$. Additionally, given $T > 0$ there exists a constant $C_T > 0$ such that

$$
\int_0^T \int_0^L |f_i(u_1, u_2, \cdots, u_n)|^2 \mathrm{d}x\mathrm{d}t \le C_T \int_0^T F_u(t)\mathrm{d}t, \tag{9.7}
$$

where $F_u(t)$ represents the energy of the system (9.2) defined by

$$
F_u(t) = \sum_{i=1}^{n} F_{u,i}(t), \tag{9.8}
$$

and $F_{u,i}(t)$ is the energy of the i-th equation of the system given by

$$
F_{u,i}(t) = \int_0^L \Big[|u_i|^2 + \gamma_i |\partial_x u_i|^2 + |\partial_t u_i|^2 \Big] \mathrm{d}x. \tag{9.9}
$$

Then, if a weak solution of problem (9.2)–(9.4) vanishes in an open subset of $(0, L)$, for $0 < t < T$, then it vanishes over all $[0, L] \times [0, T]$. See Theorem 9.3.2.

The UCP has been extensively used in the analysis of exact controllability, exponential stability, and in the theory of attractors for locally damped scalar wave equations. See for instance [2, 5, 6, 23, 24].

The study of global attractors for wave equations with critical nonlinearities and locally distributed damping has been established in [7, 10, 14, 21]. Despite these results are concerned with wave equations, the significant difference between them is the damping regions imposed on the damping coefficient. Therefore different types of UCP are needed. Reference [14] uses the UCP given in [25]. In [10, 21], the authors apply the UCP corresponding to Carleman-type estimates for wave equations with linear potential, and in [7] the authors introduced a new UCP using the techniques of [31].

Here, in order to deal with our system (9.2), the main idea is to set the problem in a Riemannian geometry framework and see the system as a single equation with different Riemann metrics. Then we show how Carleman-type estimates obtained in [31] can be used to obtain a UCP for our system (9.2) under above assumptions on the functions $f^i(\cdot)$, $(i = 1, 2, \cdots, n)$. The result is formalized in Theorem 9.3.2. To the best of our knowledge, this is the first UCP for systems of coupled wave equations.

In the second part of the chapter, we establish the existence of global attractors for a semilinear Bresse system with nonlinear localized dissipation (Problem (9.32)). The UCP is strategically used to prove that corresponding dynamical system is gradient. The main result is Theorem 9.4.3 that improves substantially the ones in [20] where the analysis was given under the presence of full damping. In the final section, we discuss the well posedness of wave equations with overdetermined conditions.

9.2 A Riemannian Geometry Framework

9.2.1 Some Notations

Let (M, g) be an n-dimensional, compact Riemannian manifold, with smooth boundary and smooth metric. The tangent space on M at p is denoted by T_pM and fixes a coordinate system $(x_1, \cdots, x_n)$ and then $(\partial_{x_1}, \cdots, \partial_{x_n})$ represents the associated coordinate vector fields. In this case

$$g(X, Y) = \langle X, Y\rangle = \sum_{i,j=1}^{n} g_{ij}\alpha_i\beta_j, \ |X|^2 = \langle X, X\rangle,$$

where

$$X = \sum_{i=1}^{n} \alpha_i \partial_{x_i}, \ Y = \sum_{i=1}^{n} \beta_i \partial_{x_i} \text{ in } T_pM \text{ for some } p \in M, \tag{9.10}$$

and

$$g_{ij} = \left\langle \partial_{x_i}, \partial_{x_j} \right\rangle.$$

Note that $|\cdot|$ represents the norm with respect to the metric $g(\cdot,\cdot)$. In particular, we denote the inner product $g(\cdot,\cdot)$ by the matrix $(g_{ij})_{n\times n}$ and its inverse by $(g^{ij})_{n\times n}$. The tangent and cotangent bundle of M are respectively denoted by TM and T^*M. The symbol D denotes the Levi-Civita connection of M such that for two vector fields X and Y on M, given by (9.10), the following equality holds true

$$D_X Y = \sum_{i,k=1}^{n} \left[\alpha_i \partial_{x_i} \beta_k \partial_{x_k} + \sum_{j=1}^{n} \beta_j \Gamma_{ij}^k \partial_{x_k} \right],$$

where Γ_{ij}^k represent the Christoffel symbols. Let $f : M \to \mathbb{R}$ and $H \in T_pM$ for all $p \in M$.

If $f \in C^1(M)$, then the differential $Df : TM \to \mathbb{R}$ represents the gradient of the connection D on f and

$$Df(H) = D_H f = H(f) = \langle \nabla f, H \rangle,$$

where ∇ is the usual gradient defined in a coordinate system by

$$\nabla f = \sum_{i,j=1}^{n} g^{ij} \partial_{x_i} f \partial_{x_j}. \tag{9.11}$$

Thanks to the musical isomorphism we will identify Df with ∇f. Here, we often denote Df by ∇f. Let $\{E_1, \cdots, E_n\}$ be an orthonormal basis of T_pM and $H = \sum_{i=1}^n h_i E_i$. Then, in particular,

$$Df(H) = H(f) = \sum_{i=1}^{n} h_i E_i(f).$$

If $f \in C^2(M)$, then $D^2 f$ represents the Hessian of f such that for all $Y \in TM$

$$D^2 f(\cdot, Y) = D(Df)(\cdot, Y) = D_Y(\nabla f(\cdot)) : TM \to \mathbb{R},$$

with

$$D_Y(\nabla f(X)) = \langle D_X(\nabla f), Y \rangle, \ \forall X \in TM.$$

In particular

$$D^2 f(X, X) = \langle D_X(\nabla f), X \rangle, \ \forall X \in TM.$$

For $f \in C^3(M)$, then it is strictly convex in the metric g if and only if $D^2 f(X, X) > 0$ for all $X \in TM$. As before, for an orthonormal basis $\{E_1, \cdots, E_n\}$ of T_pM, the divergent of H is defined by

$$\operatorname{div}(H) = \sum_{i=1}^{n} \langle D_{E_i} H, E_i \rangle .$$

If $f \in C^1(M)$, then

$$\operatorname{div}(fH) = f\operatorname{div}(H) + H(f).$$

For a function $f \in C^2(M)$, we define the Laplace-Beltrami operator Δ by

$$\Delta f = \operatorname{div}(\nabla f) = \sum_{i=1}^{n} \langle D_{E_i}(\nabla f), E_i \rangle = \sum_{i=1}^{n} D^2 f(E_i, E_i).$$

The covariant derivative DH is the bilinear form given by

$$DH(X, Y) = \langle D_X H, Y \rangle, \ \forall X, Y \in TM.$$

In particular if $f \in C^2(M)$,

$$D(\nabla f)(X, Y) = D^2 f(X, Y), \ \forall X, Y \in TM.$$

If $f \in C^1(M)$, then

$$\langle \nabla f, \nabla (H(f)) \rangle = DH(\nabla f, \nabla f) + \frac{1}{2}\operatorname{div}(|\nabla f|^2 H) - \frac{1}{2}|\nabla f|^2 \operatorname{div} H.$$

Remark 9.2.1 Let Ω be an open bounded, connected, compact subset of M with smooth boundary $\partial\Omega$ and $f \in C^3(M)$ a strictly convex function in the metric g. Then by translating and rescaling [31, Remark 1.2], the function f satisfies the following conditions

$$D^2 f(X, X) \geq |X|^2, \ \forall p \in \Omega, \forall X \in T_pM \text{ and } \min_{\overline{\Omega}} f(x) \equiv m > 0.$$

9.2.2 *Geometry of the Wave System*

Let us consider $\mathbb{R}$ with usual topology and x the natural coordinate system. In particular, for each $x \in \mathbb{R}$ the tangent space is $T_x\mathbb{R} = \mathbb{R}_x = \mathbb{R}$.

For a fixed $i \in \{1, \cdots, n\}$, we begin defining the metrics g_i associated with Δ_i by

$$g_i(X, Y) = \langle X, Y \rangle_{g_i} = \gamma_i^{-1} \alpha\beta, \tag{9.12}$$

with corresponding norm

$$|X|_{g_i}^2 = \langle X, X \rangle_{g_i},$$

where $X = \alpha\partial_x$, $Y = \beta\partial_x \in \mathbb{R}_x$, for each $x \in \mathbb{R}$.

Recalling assumption $\gamma_i > 0$, one can see that the pairs $(\mathbb{R}, g_i)$ are Riemannian manifolds. In this case, the Levi-Civita connection of $(\mathbb{R}, g_i)$ will be denoted by D_{g_i}. Here, the symbol $(\mathbb{R}, g_0)$ denotes the $\mathbb{R}$ space with Euclidean metric, and we use the following notations for the metric and norm

$$g_0(X, Y) = X \cdot Y \text{ and } |X|_{g_0} = |X|, \quad \forall X, Y \in \mathbb{R}_x, \quad \forall x \in \mathbb{R}. \tag{9.13}$$

Important properties of the above metrics are stated in the following lemma. Although most of these results are followed straightforwardly from the known results, they are crucial for what follows. So for the convenience of the reader, we give their proofs here.

Lemma 9.2.1 *Let* $(0, L) \subset \mathbb{R}$ *for some* $L > 0$. *Let* x *be the natural coordinate system in* $\mathbb{R}$, $f, h \in C^1([0, L])$ *and* $X = \alpha\partial_x$, $Y = \beta\partial_x$ *vector fields. Then:*

1. $\langle X, \gamma_i Y \rangle_{g_i} = X \cdot Y$.
2. $\nabla_{g_i} f = \gamma_i \partial_x f \partial_x$.
3. $X(f) = \langle \nabla_{g_i} f, X \rangle_{g_i} = \alpha\partial_x f$.
4. $\langle \nabla_{g_i} f, \nabla_{g_i} h \rangle_{g_i} = \gamma_i \partial_x f \partial_x h$.
5. *If* ν *represents the unit outward normal vector for* $([0, L], g_i)$, *then*

$$\langle \nabla_{g_i} f(0), \nu(0) \rangle_{g_i} = -\sqrt{\gamma_i}\partial_x f(0), \quad \langle \nabla_{g_i} f(L), \nu(L) \rangle_{g_i} = \sqrt{\gamma_i}\partial_x f(L).$$

Proof In light of (9.12) and (9.13), one can easily see that Item 1 holds true. In fact, by definition

$$\langle X, \gamma_i Y \rangle_{g_i} = \gamma_i^{-1} \gamma_i \alpha\beta = X \cdot Y. \tag{9.14}$$

To prove Item 2, we recall the definition (9.11) to find

$$\nabla_{g_i} f = \gamma_i \partial_x f \partial_x.$$

Reasoning analogously to (9.14) and assuming $h \in C^1([0, L])$, we have the validity of Items 2–4. To conclude, let us show Item 5. Firstly, note that if $p \in \{0, L\}$ and $\{\partial_x\}$ represents the associated coordinate vector field for $T_p[0, L]$, then

$$\left\langle \pm\sqrt{\gamma_i}\partial_x, \pm\sqrt{\gamma_i}\partial_x \right\rangle_{g_i} = 1.$$

In particular $E = \pm\sqrt{\gamma_i}\partial_x$ represents an orthonormal basis for $p \in T_pM$. Let $\nu = E$, and therefore

$$\nabla_{g_i} f = \gamma_i \partial_x f \partial_x = \pm\sqrt{\gamma_i}\partial_x f \nu$$

and

$$\left\langle \nabla_{g_i} f, \nu \right\rangle_{g_i} = \pm\sqrt{\gamma_i}\partial_x f.$$

Note that $\nu(0) = -\sqrt{\gamma_i}\partial_x$ and $\nu(L) = \sqrt{\gamma_i}\partial_x$. We complete the proof of Lemma. □

Remark 9.2.2 The Hessian of $f \in C^2(\mathbb{R})$ with respect to the metric g_i is given by

$$D^2_{g_i} f(X, X) = \left\langle D_{g_i,X}(\nabla_{g_i} f), X \right\rangle_{g_i} = \alpha^2 \gamma_i^{-1} \gamma_i \partial_x^2 f = \alpha^2 \partial_x^2 f,$$

where $X = \alpha\partial_x$. We observe that the Hessian of f is positive if and only if $\partial_x^2 f$ is positive.

9.3 Unique Continuation Property (UCP)

As already observed, Carleman estimates can be an important tool for proving unique continuation property for solutions to partial differential equations [7, 18, 31]. In this section, we prove the Carleman estimates for Problem (9.2). The following notations will be used.

Let L_0 a real number in $[0, L]$. We define Ω_1 and Ω_2 the subsets of $[0, L]$ as follows:

$$\Omega_1 = (0, L_0) \text{ and } \Omega_2 = (L_0, L). \tag{9.15}$$

Note that $\overline{\Omega_1} \cup \overline{\Omega_2} = [0, L]$. For $\varepsilon > 0$ small, we will also consider subsets of Ω_1 and Ω_2

$$V_1 = \left(0, L_0 - \frac{\varepsilon}{4}\right), \; V_2 = \left(L_0 + \frac{\varepsilon}{4}, L\right),$$

and

$$\omega = \left(L_0 - \frac{\varepsilon}{2}, L_0 + \frac{\varepsilon}{2}\right).$$

There exist strictly convex functions $d_1 : [0, L] \to \mathbb{R}$ and $d_2 : [0, L] \to \mathbb{R}$ such that

$$d_1(x) = \frac{1}{2}(x + L)^2, \text{ and } d_2(x) = \frac{1}{2}(x - 2L)^2.$$

In particular, if x denotes the natural coordinate system, Lemma 9.2.1 implies that

$$\nabla_{g_i} d_1 = \gamma_i(x + L)\partial_x, \; \nabla_{g_i} d_2 = \gamma_i(x - 2L)\partial_x \;\; (i = 1, \cdots, n),$$

and

$$D^2_{g_i} d_j(X, X) = \alpha^2, \text{ for all } x \in [0, L] \text{ and } X = \alpha\partial_x \in T_x[0, L] \;\; (j = 1, 2).$$

The functions $d_j(\cdot)$ $(j = 1, 2)$ have the following properties:

Lemma 9.3.1 *Under the above definitions, the functions $d_j(\cdot)$ $(j = 1, 2)$ satisfy:*

1. $d_j \in C^\infty([0, L])$.
2. $D^2_{g_i} d_j(X, X) = |X|_{g_i}$, *for all* $x \in \Omega_j$ *and* $X \in T_x[0, L]$.
3. $\inf_{\Omega_j} |\nabla_{g_i} d_j|_{g_i} > 0$.
4. $\min_{\overline{\Omega}_j} d_j = \dfrac{L^2}{2} > 0$.

Moreover, if ν represents the unit outward normal vector, then:

5. $\left\langle \nabla_{g_i} d_j(x), \nu(x) \right\rangle_{g_i} < 0$ *on* $\{0, L\} \cap \overline{V_j}$.

The next section is devoted to the proof the Carleman estimate compatible with the system (9.2). To this aim, we allocated the above notations in the same context as Section 1 in [31]. First, without loss of generality (by rescaling), we can assume that

$$k_{ij} \equiv \inf_{x \in \Omega_j} \frac{|\nabla_{g_i} d_j|^2_{g_i}}{d_j} > 4, \;\; \forall i = 1, \cdots, n, \;\; \forall j = 1, 2.$$

In what follows, for fixed $i \in \{1, \cdots, n\}$ and $j = 1, 2$, we define

$$T^2_{0,j} \equiv 4 \max_{x \in \overline{\Omega}_j} d_j(x). \tag{9.16}$$

Let $T > \max\{T_{0,1}, T_{0,2}\}$, and by (9.16) there exist $\delta > 0$ and $c = c_\delta \in (0, 1)$ satisfying

$$cT^2 > 4 \max_{x \in \overline{\Omega}_j} d_j(x) + 4\delta. \tag{9.17}$$

In the above context, we define functions $\phi_j : \Omega_j \times \mathbb{R} \to \mathbb{R} \in C^3(\Omega_j)$ by

$$\phi_j(x,t) \equiv d_j(x) - c\left(t - \frac{T}{2}\right)^2, \quad (x,t) \in \Omega_j \times [0,T].$$

The following properties are valid for $\phi_j(\cdot)$:

(ϕ.1) For the constant $\delta > 0$

$$\phi_j(x,0) = \phi_j(x,T) = d_j(x) - c\frac{T^2}{4} \leq \max_{j=\{1,2\}}\left(\max_{x\in\overline{\Omega}_j} d_j(x)\right) - c\frac{T^2}{4} \leq -\delta,$$

uniformly in $\overline{\Omega}_j$.

(ϕ.2) There are t_0 and t_1 with $0 < t_0 < T/2 < t_1 < T$, such that

$$\min_{j=1,2}\left(\min_{(x,t)\in\overline{\Omega}_j\times[t_0,t_1]} \phi_j(x,t)\right) \geq \sigma, \tag{9.18}$$

for $\sigma \in \left(0, \min_{\overline{\Omega}_j} d_j\right)$, where $\min_{\overline{\Omega}_j} d_j = \frac{L^2}{2}$ (Lemma 9.3.1).

As a consequence of Corollary 4.2 in [31],

$$\max_{i=1,\cdots,n}\max_{j=1,2}\left(\max_{(x,t)\in\overline{\Omega}_j\times[0,T]}\left(|\partial_t\phi_j|^2 + |\nabla_{g_i}\phi_j|^2_{g_i}\right)\right) \leq \frac{4(1+7c)\sigma^*}{\varepsilon(1-c)},$$

for any $\varepsilon \in (0, \min\{2-2c, 1\})$ with $\sigma^* \in (0,\sigma)$ and $c \in (0,1)$.

We end this section defining

$$Q(\sigma) = \left\{(x,t) \in [0,L]\times[0,T] \mid \min_{j=1,2}\phi_j(x,t) \geq \sigma\right\}. \tag{9.19}$$

This set will play an important role in Carleman estimates by being able to separate the set $[0,L]\times[0,T]$ from the level surface generated by the pseudo-convex function ϕ_j at height of σ. Additionally, note that

$$\Omega_j \times [t_0,t_1] \subset Q(\sigma) \subset [0,L]\times[0,T], \quad \forall j = 1,2.$$

9.3.1 *Analysis of the Coupled System—Problem (9.2)*

In this section, we will study Problem (9.2) under the new decomposition Ω_1 and Ω_2 (9.15). This decomposition will allow us to define the boundary terms for the solutions of this system. We begin with the following definition.

Definition 9.3.1 Let $T > 0$. The vector function $(u_1, \cdots, u_n)$ is a weak solution to Problem (9.2)–(9.4) if the function u_i solves the variational form of Eq. $(9.2)_i$ and possesses regularity

$$u_i \in C(0, T; (H_0^1(M_i))) \cap C^1(0, T; (L^2(M_i))),$$

where M_i denotes the Riemannian manifold $([0, L], g_i)$ $(i = 1, \cdots, n)$.

If $u_i \in C^2(M_i \times \mathbb{R}^+)$, the vector function $(u_1, \cdots, u_n)$ is a regular solution to Problem (9.2)–(9.4).

Now, let $(u_1, \cdots, u_n)$ be a regular solution to Problem (9.2)–(9.4), and we define

$$u_{i,j}(x,t) = \chi_j(x,t)u_i(x,t), \; i = 1, \cdots, n \text{ and } j = 1, 2,$$

where χ_j is a smooth cutoff function such that

$$\chi_j = \begin{cases} 1 \text{ in } \overline{V_j} \times [0, T], \\ 0 \text{ in } \big([0, L] \setminus \Omega_j\big) \times [T + 1, \infty), \end{cases}$$

where T is a positive constant satisfying (9.17).

By definition, we can observe that $u_{i,j} \in C^2(M_{i,j} \times \mathbb{R}^+)$, where $M_{i,j} \equiv (\overline{\Omega}_j, g_i)$ represents a 1-dimensional compact connected smooth Riemannian manifold with boundary $\partial\Omega_j$ and with metric g_i.

Recall that the objective of the present section is to show a unique continuation property for Problem (9.2). For this purpose we shall assume

$$u_i(x,t) = 0, \quad \forall (x,t) \in \omega \times [0, \infty), \quad i = 1, \cdots, n. \tag{9.20}$$

Under the above notations, the function $u_{i,j} \in C^2(M_{i,j} \times \mathbb{R}^+)$ solves the problem

$$\begin{cases} \partial_t^2 u_{i,j} - \Delta_i u_{i,j} = f_i(u_{1,j}, u_{2,j}, \cdots, u_{n,j}) \text{ in } M_{i,j} \times (0, T], \\ u_{i,j}(x,t) \qquad = 0 \text{ on } \partial M_{i,j} \times (0, T], \\ u_{i,j}(x,0) \qquad = \chi_j(x,0)u_i^0(x) \text{ in } M_{i,j}, \\ \partial_t u_{i,j}(x,0) \qquad = \chi_j(x,0)u_i^1(x) \text{ in } M_{i,j}. \end{cases} \tag{9.21}$$

In particular, the following decomposition is valid

$$u_i = u_{i,1} + u_{i,2}, \quad (x,t) \in [0, L] \times [0, T] \;\; (i = 1, \cdots, n). \tag{9.22}$$

On the other hand, note that the system

$$\begin{cases} \partial_t^2 v_{i,j} - \Delta_i v_{i,j} = f_i(v_{1,j}, v_{2,j}, \cdots, v_{n,j}) \text{ in } M_{i,j} \times (0, T], \\ v_{i,j}(x,t) \qquad = 0 \text{ on } \partial M_{i,j} \times (0, T], \\ v_{i,j}(x,0) \qquad = \chi_j(x,0)u_i^0(x) \text{ in } M_{i,j}, \\ \partial_t v_{i,j}(x,0) \qquad = \chi_j(x,0)u_i^1(x) \text{ in } M_{i,j} \end{cases} \tag{9.23}$$

is well posed, for all $i = 1, \cdots, n$ and $j = 1, 2$.

Due to the uniqueness of solutions in the previous system (by density and continuity), we will focus on studying the problem (9.23), in such a way that when assuming (9.20) we have that $v_{i,j} = u_{i,j}$.

9.3.2 Carleman Estimates

In the context of the previous section, concerning the analysis of boundary terms for the system (9.23), we will consider the following.

For $v_{i,j} \in C(0, T; (H_0^1(M_{i,j}))) \cap C^1(0, T; (L^2(M_{i,j})))$ the weak solution of Problem (9.23), we have

$$\left\langle \nabla_{g_i} v_{i,j}, \nabla_{g_i} d_j \right\rangle_{g_i} = \left\langle \nabla_{g_i} v_{i,j}, \nu \right\rangle_{g_i} \left\langle \nabla_{g_i} d_j, \nu \right\rangle_{g_i},$$

where ν denotes the outward unit normal field along the boundary $\partial M_{i,j}$.

Let τ be a positive parameter, and then we define

$$BT_\tau v_{i,j} \equiv 2\tau \int_0^T \int_{\partial\Omega_j} e^{2\tau\phi_j} \left(\left\langle \nabla_{g_i} v_{i,j}, \nu \right\rangle_{g_i} \right)^2 \left\langle \nabla_{g_i} d_j, \nu \right\rangle_{g_i} \mathrm{d}x\mathrm{d}t, \tag{9.24}$$

with $T > 0$ satisfying (9.17).

Next, we shall investigate the properties on the forces $f_i(\cdot)$ $(i = 1, \cdots, n)$ in the context of the new decomposition. We promptly have from (9.6)–(9.9) that there exists a positive constant C_T such that

$$\int_0^T \int_{\Omega_j} |f_i(v_{1,j}, \cdots, v_{n,j})|_{g_i}^2 \mathrm{d}x\mathrm{d}t \leq C_T \sum_{i=1}^n \int_0^T \int_0^L \mathcal{V}_i(x,t)\mathrm{d}x\mathrm{d}t, \tag{9.25}$$

where

$$\mathcal{V}_i(x,t) \equiv |v_{i,1} + v_{i,2}|^2 + \gamma_i |\partial_x(v_{i,1} + v_{i,2})|^2 + |\partial_t(v_{i,1} + v_{i,2})|^2, \tag{9.26}$$

with

$$v_{i,1} + v_{i,2} = \begin{cases} v_{i,1} \text{ in } [0, L_0] \times [0, T], \\ v_{i,2} \text{ in } [L_0, L] \times [0, T]. \end{cases} \tag{9.27}$$

Note that

$$\mathcal{V}_i|_{\Omega_j \times [0,T]} = |v_{i,j}|^2 + \gamma_i |\partial_x v_{i,j}|^2 + |\partial_t v_{i,j}|^2 \text{ and } \sum_{j=1}^2 \int_{\Omega_j} \mathcal{V}_i|_{\Omega_j \times [0,T]} \mathrm{d}x = \int_0^L \mathcal{V}_i \mathrm{d}x.$$

Now, for any regular solution $v_{i,j}$, we find from (9.25) that

$$\int_0^T \int_{\Omega_j} e^{2\tau\phi_j} |\partial_t^2 v_{i,j} - \Delta_i v_{i,j}|_{g_i}^2 \,\mathrm{d}x\mathrm{d}t \le C_T \sum_{i=1}^n \int_0^T \int_0^L e^{2\tau\phi_j} \mathcal{V}_i(x,t)\mathrm{d}x\mathrm{d}t.$$

In particular, by (9.18) and (9.19), we obtain

$$\int_0^T \int_{[Q(\sigma)]^c} e^{2\tau\phi_j} |\partial_t^2 v_{i,j} - \Delta_i v_{i,j}|_{g_i}^2 \,\mathrm{d}x\mathrm{d}t \le C_T e^{2\tau\sigma} \sum_{i=1}^n \int_{[Q(\sigma)]^c} \mathcal{V}_i(x,t)\mathrm{d}x\mathrm{d}t.$$

Remark 9.3.1 Note that Problem (9.23) satisfies the following compatibility condition:

$$v_{i,j}(L_0,t) = 0 \ \text{in} \ (0,T] \ \text{and} \ v_i^0(L_0) = v_i^1(L_0) = 0.$$

In particular, there exist positive constants k_1, k_2 such that

$$k_1 \sum_{i=1}^n \int_0^L \mathcal{V}_i(x,t)\mathrm{d}x$$

$$\le \sum_{i=1}^n \int_0^L \gamma_i |\partial_x(v_{i,1}+v_{i,2})|^2 + |\partial_t(v_{i,1}+v_{i,2})|^2 \mathrm{d}x \le k_2 \sum_{i=1}^n \int_0^L \mathcal{V}_i(x,t)\mathrm{d}x.$$

Collecting all the above ingredients and proceeding analogously to Theorem 6.1 in [31], we arrive at:

Theorem 9.3.1 (Carleman Estimates) *Let $v_{i,j}$ ($i = 1, \cdots, n$ and $j = 1, 2$) be a regular solution of Problem (9.23) with initial data $(\chi_j(x,0)u_i^0(x), \chi_j(x,0)u_i^1(x))$. Then, for all $\tau > 0$ sufficiently large and $\varepsilon > 0$ small, the following estimate holds true*

$$\sum_{j=1}^2 \sum_{i=1}^n BT_\tau v_{i,j} \ge$$

$$\ge \left[\frac{k_1 e^{2\tau\sigma}(t_1 - t_0)}{2}(\varepsilon\tau(1-c) - 2nC_T)e^{-C_T T}\right] \sum_{i=1}^n \int_0^L \Big[\mathcal{V}_i(x,0) + \mathcal{V}_i(x,T)\Big]\mathrm{d}x$$

$$- \left[\frac{C_{1,T} k_2 e^{2\tau\sigma}}{2k_1} T e^{C_T T} + C_T \tau^3 e^{-2\tau\delta}\right] \sum_{i=1}^n \int_0^L \Big[\mathcal{V}_i(x,0) + \mathcal{V}_i(x,T)\Big]\mathrm{d}x$$

$$\ge k_T \sum_{i=1}^n \Big[\int_0^L \mathcal{V}_i(x,0) + \mathcal{V}_i(x,T)\Big]\mathrm{d}x,$$

where (1) k_1, k_2 *are the positive constants from Remark* 9.3.1*;* (2) σ *is defined in* (ϕ.2)*;* (3) $c \in (0, 1)$ *is given in* (9.17)*;* (4) $C_T, C_{1,T}$ *are positive constants depending only on* T, σ *and* $Q(\sigma)$*;* (5) k_T *is a positive constant depending only on* $\sigma, \phi_1, \phi_2, n$ *and* $C_{1,T}$*. Moreover, the above inequality may be extended to all weak solutions of the system* (9.23) *with data* $(\chi_j(x, 0)u_i^0(x), \chi_j(x, 0)u_i^1(x))$ *in* $H_0^1(M_{i,j}) \times L^2(M_{i,j})$.

Remark 9.3.2 (a) Note that the definitions of σ and $Q(\sigma)$ allow the right-hand term of the inequality in Theorem 9.3.1 to be independent of j. (b) The validity of the inequality extends to weak solutions of the system (9.23) as demonstrated in [31, Theorems 7.1 and 8.1].

9.3.3 A New UCP

Thanks to Theorem 9.3.1, we can now present the main finding of this part of the chapter.

Theorem 9.3.2 *Let* I *be an open interval such that*

$$\omega \equiv I \cap [0, L] \neq \emptyset.$$

Then, under the framework (9.5)–(9.9)*, for* $T > 0$ *large enough, any weak solution*

$$u_i \in C(0, T; H_0^1(0, L)) \cap C^1(0, T; L^2(0, L))$$

of system (9.2)–(9.4) *vanishing in* $\omega \times [0, T]$ *must vanish all over* $[0, L] \times [0, T]$.

Proof Without loss of generality, we can assume

$$\omega = \left(L_0 - \frac{\varepsilon}{2}, L_0 + \frac{\varepsilon}{2}\right) \cap [0, L], \tag{9.28}$$

where $\varepsilon > 0$ and $L_0 \in [0, L]$. The proof of the unique continuation property will be divided into four steps:

Step 1. Equivalent Systems Firstly, we observe that if $(u_1, \cdots, u_n)$ is a weak solution of Problem (9.2)–(9.4) with overdetermined condition (9.20), then $u_{i,j} = \chi_j u_i \in C(0, T; H_0^1(M_{i,j})) \cap C^1(0, T; M_{i,j})$ is a solution of (9.21), where $M_{i,1} = ([0, L_0], g_i)$ and $M_{i,2} = ([L_0, L], g_i)$, for all $i = 1, \cdots, n$ and $j = 1, 2$.

From the Appendix, for $j = 1, 2$, the system (9.23) with the following overdetermined condition

$$(v_{1,j}, \cdots, v_{n,j}) = (0, \cdots, 0) \text{ in } \omega_j \equiv \overline{\Omega}_j \cap \omega \tag{9.29}$$

is well posed and generates a strongly continuous semigroup

$$T_{M_{i,j}} : H_{\omega_j}(M_{i,j}) \to H_{\omega_j}(M_{i,j})$$

in the Hilbert space

$$H_{\omega_j}(M_{i,j}) \equiv \left\{ (v_{1,j}, \cdots, v_{n,j}, w_{1,j}, \cdots, w_{n,j}) \middle| \begin{array}{l} v_{i,j} \in H_0^1(M_{i,j}),\ w_{i,j} \in L^2(M_{i,j}), \\ v_{i,j} = w_{i,j} = 0 \text{ in } \omega_j,\ i = 1, \cdots, n \end{array} \right\}.$$

In particular,

$$\begin{aligned} &T_{M_{i,j}}(\chi_j(0)u_1^0, \cdots, \chi_j(0)u_n^0, \chi_j(0)u_1^1, \cdots, \chi_j(0)u_n^1) \\ &\quad = (u_{1,j}, \cdots, u_{n,j}, \partial_t u_{1,j}, \cdots, \partial_t u_{n,j}) \end{aligned}$$

is the weak solution of (9.23) satisfying (9.29) with initial data

$$(\chi_j(0)u_1^0, \cdots, \chi_j(0)u_n^0, \chi_j(0)u_1^1, \cdots, \chi_j(0)u_n^1) \in H_{\omega_j}(M_{i,j}).$$

Step 2. Carleman Estimate From Step 1 and via Theorem 9.3.1, there exists a positive constant k_T such that

$$\sum_{j=1}^{2}\sum_{i=1}^{n} BT_\tau u_{i,j} \geq k_T \sum_{i=1}^{n}\left[\int_0^L \mathcal{V}_i(x,0) + \mathcal{V}_i(x,T)\mathrm{d}x\right].$$

Next, from (9.8), (9.22), (9.24), (9.26), and (9.27), we find that

$$2\tau \sum_{j=1}^{2}\sum_{i=1}^{n}\int_0^T \int_{\partial\Omega_j} e^{2\tau\phi_j}\left(\left\langle \nabla_{g_i} u_{i,j}, \nu\right\rangle_{g_i}\right)^2 \left\langle \nabla_{g_i} d_j, \nu\right\rangle_{g_i} \mathrm{d}x\mathrm{d}t \geq k_T\left(F_u(0) + F_u(T)\right). \tag{9.30}$$

Step 3. Boundary Estimates The fact that $L_0 \in \omega$ and ω is an open subset of $[0, L]$ and we have

$$\left\langle \nabla_{g_i} u_{i,j}(L_0), \nu(L_0)\right\rangle_{g_i} = 0,\ \forall i = 1, \cdots, n \text{ and } j = 1, 2.$$

Now, recalling Item 5 from Lemma 9.3.1, we obtain

$$\left\langle \nabla_{g_i} d_j, \nu\right\rangle_{g_i} < 0 \text{ in } \{0, L\},\ \forall i = 1, \cdots, n \text{ and } j = 1, 2.$$

Combining the above information with assumption that $\tau > 0$, we infer that

$$2\tau \sum_{j=1}^{2}\sum_{i=1}^{n}\int_0^T \int_{\partial\Omega_j} e^{2\tau\phi_j}\left(\left\langle \nabla_{g_i} u_{i,j}, \nu\right\rangle_{g_i}\right)^2 \left\langle \nabla_{g_i} d_j, \nu\right\rangle_{g_i} \mathrm{d}x\mathrm{d}t \leqslant 0. \tag{9.31}$$

Step 4. Conclusion From inequalities (9.30) and (9.31), we find

$$0 \geq k_T \left(F_u(0) + F_u(T)\right) \geq 0.$$

This last implies that $F_u(0) = 0$. Since (9.2)–(9.4) are well posed, the result is followed.

□

9.4 Dynamics of Locally Damped Bresse Systems

Let us consider the semilinear Bresse system

$$\begin{cases} \rho_1\varphi_{tt} - k(\varphi_x + \psi + lw)_x - k_0 l(w_x - l\varphi) + a_1(x)g_1(\varphi_t) + f_1(\varphi, \psi, w) = 0, \\ \rho_2\psi_{tt} - b\psi_{xx} + k(\varphi_x + \psi + lw) + a_2(x)g_2(\psi_t) + f_2(\varphi, \psi, w) = 0, \\ \rho_1 w_{tt} - k_0(w_x - l\varphi)_x + kl(\varphi_x + \psi + lw) + a_3(x)g_3(w_t) + f_3(\varphi, \psi, w) = 0, \end{cases} \tag{9.32}$$

with Dirichlet boundary conditions

$$\varphi(0,t) = \varphi(L,t) = \psi(0,t) = \psi(L,t) = w(0,t) = w(L,t) = 0, \quad t \in \mathbb{R}^+, \tag{9.33}$$

and with initial condition

$$\varphi(0) = \varphi_0, \ \ \varphi_t(0) = \varphi_1, \ \ \psi(0) = \psi_0, \ \ \psi_t(0) = \psi_1, \ \ w(0) = w_0, \ \ w_t(0) = w_1. \tag{9.34}$$

9.4.1 Well Posedness

In this section, we summarize all the assumptions that will be used to prove the main result. We also introduce the well-posedness result along with some energy inequalities.

Notations Henceforth the symbols $L^p(0, L)$ $(p \geqslant 1)$ and $H^m(0, L)$ $(m \in \mathbb{N})$ denote the Lebesgue and Sobolev spaces, respectively. The norms in $L^p(0, L)$ are indicated by $\|\cdot\|_p$ and $\|\cdot\|_{L^2(0,L)} \equiv \|\cdot\|$. We will also frequently use the inequality

$$\|u\| \leqslant \frac{L}{\pi}\|u_x\|, \ \forall u \in H_0^1(0, L).$$

9.4.1.1 Assumptions for the Nonlinear Bresse System

$(b.1)$ The functions $f_i \in C^1(\mathbb{R})$ $(i = 1, 2, 3)$ are locally Lipschitz, and there exists a potential $F \in C^1(\mathbb{R}^3)$ such that $\nabla F = (f_1, f_2, f_3)$.

$(b.2)$ There exist constants $0 \leq \alpha < \frac{\pi^2}{2\beta L^2}$ and $c_F > 0$ such that

$$F(u, v, w) \geq -\alpha\Big[|u|^2 + |v|^2 + |w|^2\Big] - c_F, \ \forall u, v, w \in \mathbb{R},$$

$$\nabla F(u, v, w)\cdot(u, v, w) \geq F(u, v, w) - \alpha\Big[|u|^2 + |v|^2 + |w|^2\Big] - c_F, \ \forall u, v, w \in \mathbb{R},$$

where $\beta > 0$ is the constant

$$\|\varphi_x\|^2+\|\psi_x\|^2+\|w_x\|^2 \leqslant \beta\Big[b\|\psi_x\|^2+k\|\varphi_x+\psi+lw\|^2+k_0\|w_x-l\varphi\|^2\Big].$$

$(b.3)$ There exists $c_f > 0$ such that

$$|\nabla f_i(u, v, w)|$$
$$\leqslant c_f\Big[1 + |u|^{p-1} + |v|^{p-1} + |w|^{p-1}\Big], \ i = 1, 2, 3, \ p \geqslant 1, \ \forall u, v, w \in \mathbb{R}.$$

$(b.4)$ The damping functions $g_i \in C^1(\mathbb{R})$ $(i = 1, 2, 3)$ are monotone increasing with $g_i(0) = 0$. Moreover, we assume that there exist constants positive constants m and $M > 0$ such that

$$m \leq g_i'(s) \leq M, \ \forall s \in \mathbb{R}.$$

$(b.5)$ The localizing functions $a_i \in L^\infty(0, L)$ $(i = 1, 2, 3)$ are nonnegative, and there exists positive constant a_0 such that

$$a_i(x) \geqslant a_0, \ x \in I_i, \ i = 1, 2, 3,$$

where $I_i \subset [0, L]$ are open intervals with $(L_1, L_2) \equiv \bigcap_i I_i \neq \emptyset$.

9.4.1.2 Dynamical System Generation

Before introducing the well-posedness result, we start with the necessary functional framework. First, the finite energy space H of the well posedness is defined as

$$H = H_0^1(0, L) \times H_0^1(0, L) \times H_0^1(0, L) \times L^2(0, L) \times L^2(0, L) \times L^2(0, L).$$

For $\mathbb{Z} = (\varphi(t), \psi(t), w(t), \tilde{\varphi}(t), \tilde{\psi}, \tilde{w}(t)) \in H$, we define the H norm as

$$\|\mathbb{Z}\|_H^2 = \rho_1\|\tilde{\varphi}\|^2 + \rho_2\|\tilde{\psi}\|^2 + \rho_1\|\tilde{w}\|^2 + b\|\psi_x\|^2 + k\|\varphi_x + \psi + lw\|^2 + k_0\|w_x - l\varphi\|^2. \tag{9.35}$$

Next, let $A\colon D(A) \subset H \to H$ be the differential operator

$$A\mathbb{Z} \equiv \begin{bmatrix} \tilde{\varphi} \\ \tilde{\psi} \\ \tilde{w} \\ k\rho_1{}^{-1}(\varphi_x + \psi + lw)_x + k_0 l\rho_1{}^{-1}(w_x - l\varphi) \\ b\rho_2{}^{-1}\psi_{xx} - k\rho_2{}^{-1}(\varphi_x + \psi + lw) \\ k_0\rho_1{}^{-1}(w_x - l\varphi)_x - kl\rho_1^{-1}(\varphi_x + \psi + lw) \end{bmatrix},$$

with domain

$$D(A) = \left[H^2(0, L) \cap H_0^1(0, L)\right]^3 \times H_0^1(0, L)^3,$$

and define $B : D(B) = H \to H$ and $\mathcal{F} : H \to H$,

$$B\mathbb{Z} \equiv \begin{bmatrix} 0 \\ 0 \\ 0 \\ -\alpha_1(x)\rho_1^{-1}g_1(\tilde{\varphi}) \\ -\alpha_2(x)\rho_2^{-1}g_2(\tilde{\psi}) \\ -\alpha_3(x)\rho_1^{-1}g_3(\tilde{w}) \end{bmatrix} \quad \text{and} \quad \mathcal{F}\mathbb{Z} \equiv \begin{bmatrix} 0 \\ 0 \\ 0 \\ -\rho_1^{-1}f_1(\varphi, \psi, w) \\ -\rho_2^{-1}f_2(\varphi, \psi, w) \\ -\rho_1^{-1}f_3(\varphi, \psi, w) \end{bmatrix},$$

that represent respectively the damping and forcing operators.

Now, using the definitions of operators $A, B, \mathcal{F}$, we can abstract represent the problem as follows:

$$\frac{\mathrm{d}}{\mathrm{d}t}\mathbb{Z}(t) - (A + B)\mathbb{Z}(t) = \mathcal{F}(\mathbb{Z}(t)),\ \ \mathbb{Z}(0) \equiv \mathbb{Z}_0 = (\varphi_0, \psi_0, w_0, \varphi_1, \psi_1, w_1), \tag{9.36}$$

where

$$\mathbb{Z}(t) = (\varphi(t), \psi(t), w(t), \tilde{\varphi}(t), \tilde{\psi}, \tilde{w}(t)) \text{ with } \tilde{\varphi} = \varphi_t,\ \tilde{\psi} = \psi_t,\ \tilde{w} = w_t.$$

We observe that the well posedness of (9.36) induces the well posedness for Problem (9.32)–(9.34). In the following, we present the well posedness for (9.36).

Theorem 9.4.1 (Well Posedness) *Assume that assumptions* $(b.1)$–$(b.4)$ *of Sect. 9.4.1.1 hold. Then for any initial data* $\mathbf{Z}_0 \in H$ *and* $T > 0$*, the Cauchy problem* (9.36) *admits a unique weak solution* $\mathbf{Z} \in C([0, T]; H)$ *that depends continuously on the initial data and satisfies the variation of parameters formula*

$$\mathbf{Z}(t) = e^{(A+B)t}\mathbf{Z}_0 + \int_0^t e^{(A+B)(t-s)}\mathcal{F}(\mathbf{Z}(s))\mathrm{d}s, \quad t \in [0, T]. \tag{9.37}$$

Moreover, if $\mathbf{Z}_0 \in D(A)$*, then the solution is strong.*

The well-posedness result stated above is known and can be found in [20]. Theorem 9.4.1, in particular, implies that the map $H \ni \mathbb{Z}_0 \mapsto \mathbb{Z}(t) = (\varphi, \psi, w, \varphi_t, \psi_t, w_t)$, where $\mathbb{Z}(t)$ solves (9.36), defines a strongly continuous semigroup $\{S(t)\}_{t \geqslant 0}$ on H.

Energy Let $\mathbb{Z}(t) = (\varphi(t), \psi(t), w(t), \varphi_t(t), \psi_t(t), w_t(t))$ be a solution of (9.32)–(9.34). The energy is defined by the following functional:

$$\mathscr{E}_{\mathbb{Z}}(t) \equiv E_{\mathbb{Z}}(t) + \int_0^L F(\varphi, \psi, w)\mathrm{d}x = \|\mathbb{Z}(t)\|_H^2 + \int_0^L F(\varphi, \psi, w)\mathrm{d}x. \tag{9.38}$$

The weak solution $\mathbb{Z}(t) = (\varphi(t), \psi(t), w(t), \varphi_t(t), \psi_t(t), w_t(t))$ satisfies the energy identity

$$\mathscr{E}_{\mathbb{Z}}(t) + \int_s^t \int_0^L \Big[\alpha_1(x)g_1(\varphi_t)\varphi_t + \alpha_2(x)g_2(\psi_t)\psi_t + \alpha_3(x)g_3(w_t)w_t\Big]\mathrm{d}x\mathrm{d}\tau = \mathscr{E}_{\mathbb{Z}}(s), \tag{9.39}$$

for all $0 \leqslant s < t$.

As in [20], the energy $\mathscr{E}_{\mathbb{Z}}(\cdot)$ (9.38) and the norm $\|\cdot\|_H$ (9.35) satisfy

$$C_E\|\mathbb{Z}(t)\|_H^2 - Lc_F \leqslant \mathscr{E}_{\mathbb{Z}}(t) \leqslant \|\mathbb{Z}(t)\|_H^2 + c_E(1 + \|\mathbb{Z}(t)\|_H^{p+1}), \ \forall t \geqslant 0, \tag{9.40}$$

for some positive constants C_E and c_E.

9.4.2 Global Attractors

Abstract Existence Theorems A compilation of fundamental definitions and results from the theory of attractors for gradient systems is provided, with the main reference being the book [9].

Definition 9.4.1 A global attractor for a dynamical system $(H, S(t))$ with evolution operator $\{S(t)\}_{t \geqslant 0}$ on a complete metric space H is defined as a compact set $\mathscr{A} \subset H$ that is fully invariant, that is $S(t)\mathscr{A} = \mathscr{A}$ for all $t \geqslant 0$, and uniformly attracts all bounded subsets of H

$$\lim_{t\to\infty} \sup\Big\{\mathrm{dist}_H(S(t)\mathbb{Z}, \mathscr{A}) \mid \mathbb{Z} \in \mathscr{B}\Big\} = 0, \text{ for any bounded set } \mathscr{B} \subset H.$$

Definition 9.4.2 The fractal dimension of a compact set $\mathscr{A} \subset H$ in a metric space H is defined as

$$\dim_f^H(\mathscr{A}) = \limsup_{\varepsilon \to 0} \frac{\ln N_\varepsilon(\mathscr{A})}{\ln(1/\varepsilon)},$$

where $N_\varepsilon(\mathscr{A})$ is the minimal number of closed balls of radius ε that cover the set A.

To ascertain the existence of a global attractor, we use the concept of gradient and quasi-stable dynamical systems. The global attractors for these systems admit additional structures and properties: (i) The attractor for gradient systems has a regular structure, that is, it is described by the unstable manifold emanating from the set of stationary points, and (ii) quasi-stable systems provide specific properties of attractors, such as finite dimensionality.

Definition 9.4.3 Let $\mathscr{Y} \subset H$ be a forward invariant set of a dynamical system $(H, S(t))$. (i) A continuous functional $\Phi\colon \mathscr{Y} \to \mathbb{R}$ is said to be a Lyapunov function on $\mathscr{Y}$ for $(H, S(t))$ if the map $t \mapsto \Phi(S(t)\mathbb{Z})$ is nonincreasing for any $\mathbb{Z} \in \mathscr{Y}$. (ii) The Lyapunov function is said to be strict on $\mathscr{Y}$ if the equation if $\Phi(S(t)\mathbb{Z}) = \Phi(\mathbb{Z})$ for all $t > 0$ for some $\mathbb{Z} \in \mathscr{Y}$ implies that y is a stationary point of $(H, S(t))$. (iii) The dynamical system $(H, S(t))$ is said to be gradient if there exists a strict Lyapunov function on the whole phase space H.

Definition 9.4.4 Let X, Y be reflexive Banach spaces, X compactly embedded in Y. We consider a dynamical system $(H, S(t))$ with $H = X \times Y$ and evolution operator defined by

$$S(t)\mathbb{Z} = (u(t), u_t(t)), \qquad \mathbb{Z} = (u(0), u_t(0)) \in H, \tag{9.41}$$

where the function u possesses the property

$$u \in C([0, \infty); X) \cap C^1([0, \infty); Y). \tag{9.42}$$

A dynamical system of the form (9.41) with regularity (9.42) is said to be quasi-stable on a set $\mathscr{B} \subset H$, if there exist a compact semi-norm $[\cdot]_X$ on X and nonnegative scalar functions $a(t), b(t), c(t)$, such that (i) $a(t), b(t)$ are locally bounded on $[0, \infty)$, (ii) $b(t) \in L^1(0, \infty)$ with $\lim_{t\to\infty} b(t) = 0$, and (iii) for any $\mathbb{Z}^1, \mathbb{Z}^2 \in \mathscr{B}$ the following estimates hold true

$$\|S(t)\mathbb{Z}^1 - S(t)\mathbb{Z}^2\|_H^2 \leqslant a(t)\|\mathbb{Z}^1 - \mathbb{Z}^2\|_H^2, \tag{9.43}$$

and

$$\|S(t)\mathbb{Z}^1 - S(t)\mathbb{Z}^2\|_H^2 \leqslant b(t)\|\mathbb{Z}^1 - \mathbb{Z}^2\|_H^2 + c(t) \sup_{0<s<t} [u^1(s) - u^2(s)]_X^2, \tag{9.44}$$

where $S(t)\mathbb{Z}^i = (u^i(t), u_t^i(t))$, $i = 1, 2$.

Unifying the abstracts results from [9], we arrive at the following criteria for existence and properties of global attractors.

Theorem 9.4.2 *Let $(H, S(t))$ be a gradient quasi-stable dynamical system. Assume its Lyapunov function $\Phi(\cdot)$ is bounded from above on any bounded subset of H and the set $\Phi(R) = \{\mathbb{Z} \in H \mid \Phi(\mathbb{Z}) \leqslant R\}$ is bounded for every R. If the set $\mathscr{N}$ of stationary points of $(H, S(t))$ is bounded, then $(H, S(t))$ possesses a finite fractal dimensional global attractor $\mathscr{A}$ defined by the unstable manifold emanating from the set of stationary solution. Moreover, any trajectory stabilizes to the set $\mathscr{N}$ of stationary points, that is,*

$$\lim_{t\to+\infty} \mathrm{dist}_H(S(t)\mathbb{Z}, \mathscr{N}) = 0, \ \forall \mathbb{Z} \in H.$$

Main Result We now state the main result of this section.

Theorem 9.4.3 *Under the assumptions* $(b.1)$–$(b.5)$ *the dynamical system* $(H, S(t))$ *generated by the problem* (9.32)–(9.34) *has a global attractor* $\mathbf{A}$*, with finite fractal dimension, characterized by* $\mathbf{A} = \mathbb{M}_+(\mathcal{N})$*, where* $\mathbb{M}_+(\mathcal{N})$ *is the unstable manifold emanating from* $\mathcal{N}$*, the set of stationary points of* $\{S(t)\}_{t\geq 0}$.

To prove Theorem 9.4.3, our strategy is to apply Theorem 9.4.2. We start exhibiting the gradient structure for $(H, S(t))$ and focusing our attention on the strictness property of the Lyapunov function where the new observability result stated in Theorem 9.3.2 plays an essential role in the proof.

Proposition 9.4.1 *Let the assumptions of Theorem* 9.4.3 *be satisfied. Then* $(H, S(t))$ *is a gradient dynamical system.*

Proof The dynamical system $(H, S(t))$ is gradient with the energy $\mathscr{E}_{\mathbb{Z}}(\cdot)$—defined in (9.38)—being the strict Lyapunov function $\Phi(\cdot)$. In fact, from identity (9.39), we find that $t \to \Phi(S(t)\mathbb{Z})$ is a nonincreasing function for any $\mathbb{Z} \in H$.

Next, we suppose that $\Phi(S(t)\mathbb{Z}) = \Phi(\mathbb{Z})$, for all $t > 0$. Then, from identity (9.39), we obtain

$$\int_s^t \int_0^L \Big[\alpha_1(x)g_1(\varphi_t)\varphi_t + \alpha_2(x)g_2(\psi_t)\psi_t + \alpha_3(x)g_3(w_t)w_t\Big]\mathrm{d}x\mathrm{d}\tau = 0.$$

This shows that $\varphi_t = \psi_t = w_t = 0$ a.e. in $(L_1, L_2) \times [0, T]$, where $(L_1, L_2) = \bigcap_i I_i$. Thus, (φ, ψ, w) satisfies the problem

$$\begin{cases} \rho_1\varphi_{tt} - k(\varphi_x + \psi + lw)_x - k_0 l(w_x - l\varphi) + f_1(\varphi, \psi, w) = 0 & \text{in } (0, L)\times[0, T], \\ \rho_2\psi_{tt} - b\psi_{xx} + k(\varphi_x + \psi + lw) + f_2(\varphi, \psi, w) = 0 & \text{in } (0, L)\times[0, T], \\ \rho_1 w_{tt} - k_0(w_x - l\varphi)_x + kl(\varphi_x + \psi + lw) + f_3(\varphi, \psi, w) = 0 & \text{in } (0, L)\times[0, T]. \end{cases} \tag{9.45}$$

Using the notation $u^1 = \varphi_t$, $u^2 = \psi_t$, $u^3 = w_t$ and taking the derivative in the distributional sense of (9.45), we find that (u^1, u^2, u^3) solves the problem

$$\begin{cases} u^1_{tt} - \gamma_1 u^1_{xx} = F^1(u^1, u^2, u^3) & \text{in } (0, L) \times [0, T], \\ u^2_{tt} - \gamma_2 u^2_{xx} = F^2(u^1, u^2, u^3) & \text{in } (0, L) \times [0, T], \\ u^3_{tt} - \gamma_2 u^3_{xx} = F^3(u^1, u^2, u^3) & \text{in } (0, L) \times [0, T], \end{cases}$$

with $\gamma_1 = \frac{k}{\rho_1}$, $\gamma_2 = \frac{b}{\rho_2}$, $\gamma_3 = \frac{k_0}{\rho_1}$ and with forcing $F^i(\cdot)$ defined by

$$F^1(u^1, u^2, u^3) = \gamma_1 u^2_x + \gamma_1 u^3_x + k_0 \rho_1^{-1} l(u^3_x - l u^1) - \rho_1^{-1} \partial_t [f_1(\varphi, \psi, w)],$$

$$F^2(u^1, u^2, u^3) = -k\rho_2^{-1}(u^1_x - u^2 + l u^3) - \rho_2^{-1} \partial_t [f_2(\varphi, \psi, w)],$$

$$F^3(u^1, u^2, u^3) = -\gamma_3 l u^1_x - k\rho_1^{-1} l(u^1_x + u^2 + l u^3) - \rho_1^{-1} \partial_t [f_3(\varphi, \psi, w)].$$

Now, we apply the UCP—Theorem 9.3.2—to conclude that $(\varphi_t, \psi_t, w_t) = (u^1, u^2, u^3) = (0, 0, 0)$. Therefore, the solution $\mathbb{Z} \in H$ must be stationary. This implies that the energy $\mathscr{E}_{\mathbb{Z}}(\cdot)$ is strict on H. □

Our next aim is to show the quasi-stability of $(H, S(t))$. According to Definition 9.4.4, the difference of two trajectories should obey the estimates (9.43) and (9.44). Taking advantage of the locally Lipschitz property of the functions f_i and the variation of parameter formula (9.37), it can be demonstrated that (9.43) holds. Subsequently, employing the multiplier technique, we establish the stabilization inequality (9.44).

Proposition 9.4.2 *Let the assumptions of Theorem* 9.4.3 *be satisfied. Then,* $(H, S(t))$ *is a quasi-stable dynamical system.*

Proof The proof is carried out through several energy estimates. In the text that follows, we use the notations

$$\tilde{v} = v^1 - v^2, \;\; G(v) = g_i(v^1) - g_i(v^2) \text{ and } F_i(v) = f_i(v^1) - f_i(v^2), \;\; i = 1, 2, 3.$$

First, we shall consider the difference of two trajectories with initial data $\mathbb{Z}^1_0, \mathbb{Z}^2_0 \in \mathscr{B}$, where $\mathscr{B}$ is a bounded subset of H. The corresponding solution $S(t)(\mathbb{Z}^1_0 - \mathbb{Z}^2_0) \equiv \mathbb{Z}^1 - \mathbb{Z}^2 = (\tilde{\varphi}, \tilde{\psi}, \tilde{w}, \tilde{\varphi}_t, \tilde{\psi}_t, \tilde{w}_t)$ verifies the following problem:

$$\begin{cases} \rho_1 \tilde{\varphi}_{tt} - k(\tilde{\varphi}_x + \tilde{\psi} + l\tilde{w})_x - k_0 l(\tilde{w}_x - l\tilde{\varphi}) = -a_1(x) G_1(\tilde{\varphi}_t) - F_1(\tilde{\varphi}, \tilde{\psi}, \tilde{w}), \\ \rho_2 \tilde{\psi}_{tt} - b\tilde{\psi}_{xx} + k(\tilde{\varphi}_x + \tilde{\psi} + l\tilde{w}) = -a_2(x) G_2(\tilde{\psi}_t) - F_2(\tilde{\varphi}, \tilde{\psi}, \tilde{w}), \\ \rho_1 \tilde{w}_{tt} - k_0(\tilde{w}_x - l\tilde{\varphi})_x + kl(\tilde{\varphi}_x + \tilde{\psi} + l\tilde{w}) = -a_3(x) G_3(\tilde{w}_t) - F_3(\tilde{\varphi}, \tilde{\psi}, \tilde{w}), \end{cases} \tag{9.46}$$

with zero Dirichlet boundary conditions and initial conditions $\mathbb{Z}^1_0 - \mathbb{Z}^2_0$.

Second, let ϵ_0 be a positive real number, such that $\epsilon_0 \leqslant (L_2 - L_1)$, where $(L_1, L_2) = \bigcap_i I_i$. We consider the following real function $\xi(\cdot)$ defined as follows:

$$\xi(x)=\begin{cases}(\lambda-1)x, \text{ if } x\in[0, L_1+\epsilon_0),\\ \lambda(x-L_1-\epsilon_0)+(L_1-L_2+2\epsilon_0)(L_1+\epsilon_0)L^{-1}, \text{ if } x\in(L_1+\epsilon_0, L_2-\epsilon_0],\\ (\lambda-1)(x-L), \text{ if } x\in[L_2-\epsilon_0, L],\end{cases}$$

where $\lambda = \frac{L-(L_2-L_1-2\epsilon_0)}{L} \in (0, 1)$.

Now, we take $\tilde{\varphi}_x\xi$, $\tilde{\psi}_x\xi$, $\tilde{w}_x\xi$ as multipliers for (9.46). Thus, we find

$$\begin{aligned}
&\frac{1}{2}\int_0^T\int_0^L(\lambda-1)E(t)\,\mathrm{d}x\mathrm{d}t\\
&=-\int_0^L\Big[\rho_1\tilde{\varphi}_t\tilde{\varphi}_x+\rho_2\tilde{\psi}_t\tilde{\psi}_x+\rho_1\tilde{w}_t\tilde{w}_x\Big]\xi\,\mathrm{d}x\Big|_0^T-\frac{1}{2}\int_0^T\int_{L_1+\epsilon_0}^{L_2-\epsilon_0}E(t)\,\mathrm{d}x\mathrm{d}t\\
&\quad-\int_0^T\int_0^L\Big[k(\tilde{\varphi}_x-\tilde{\psi}+l\tilde{w})(\tilde{\psi}+l\tilde{w})+k_0l(\tilde{w}_x-l\tilde{\varphi})\tilde{\varphi}\Big]\xi'\,\mathrm{d}x\mathrm{d}t\\
&\quad-\int_0^T\int_0^L\Big[a_1G_1(\tilde{\varphi}_t)\tilde{\varphi}_x+a_2G_2(\tilde{\psi}_t)\tilde{\psi}_x+a_3G_3(\tilde{w}_t)\tilde{w}_x\Big]\xi\,\mathrm{d}x\mathrm{d}t\\
&\quad-\int_0^T\int_0^L\Big[F_1(\tilde{\varphi},\tilde{\psi},\tilde{w})\tilde{\varphi}_x+F_2(\tilde{\varphi},\tilde{\psi},\tilde{w})\tilde{\psi}_x+F_3(\tilde{\varphi},\tilde{\psi},\tilde{w})\tilde{w}_x\Big]\xi\,\mathrm{d}x\mathrm{d}t,
\end{aligned}\tag{9.47}$$

where

$$E(t)=\rho_1|\tilde{\varphi}_t|^2+\rho_2|\tilde{\psi}_t|^2+\rho_1|\tilde{w}_t|^2+b|\tilde{\psi}_x|^2+k|\tilde{\varphi}_x+\tilde{\psi}+l\tilde{w}|^2+k_0|\tilde{w}_x-l\tilde{\varphi}|^2. \tag{9.48}$$

Note that from the definitions of the energy (9.38) and (9.48), we find

$$E_{\mathbb{Z}}(t)=\int_0^L E(t)\mathrm{d}t=\|\mathbb{Z}\|_H^2.$$

Let us estimate the left-hand side of (9.47). From the definition (9.38) we can show that there exists $c>0$ satisfying

$$\left|\int_0^L\Big[\rho_1\tilde{\varphi}_t\tilde{\varphi}_x+\rho_2\tilde{\psi}_t\tilde{\psi}_x+\rho_1\tilde{w}_t\tilde{w}_x\Big]\xi\mathrm{d}x\right|\leqslant c\sup_{x\in[0,L]}\{\xi(x)\}E_{\mathbb{Z}}(t).$$

This last inequality implies that

$$\left|\int_0^L\Big[\rho_1\tilde{\varphi}_t\tilde{\varphi}_x+\rho_2\tilde{\psi}_t\tilde{\psi}_x+\rho_1\tilde{w}_t\tilde{w}_x\Big]\xi\mathrm{d}x\Big|_0^T\right|\leqslant c\sup_{x\in[0,L]}\{\xi(x)\}(E_{\mathbb{Z}}(T)+E_{\mathbb{Z}}(0)). \tag{9.49}$$

Again, using the definition of $E_{\mathbb{Z}}(t)$, one obtains

$$\left| \int_0^T \int_0^L \Big[k(\tilde{\varphi}_x - \tilde{\psi} + l\tilde{w})(\tilde{\psi} + l\tilde{w}) + k_0 l(\tilde{w}_x - l\tilde{\varphi})\tilde{\varphi} \Big] \xi' \, \mathrm{d}x\mathrm{d}t \right| \leqslant \epsilon \int_0^T E_{\mathbb{Z}}(t)\mathrm{d}t + c_\epsilon \mathrm{l.o.t}(\tilde{\varphi}, \tilde{\psi}, \tilde{w}),$$

with lower order terms defined by

$$\mathrm{l.o.t}(\tilde{\varphi}, \tilde{\psi}, \tilde{w}) \equiv \sup_{\sigma \in [0,T]} \Big[\|\tilde{\varphi}(\sigma)\|_{2p}^2 + \|\tilde{\psi}(\sigma)\|_{2p}^2 + \|\tilde{w}(\sigma)\|_{2p}^2 \Big].$$

To estimate the damping terms, we use Assumption $(b.4)$ to obtain

$$\left| \int_0^T \int_0^L a_1 G_1(\tilde{\varphi}_t)\tilde{\varphi}_x \xi \, \mathrm{d}x\mathrm{d}t \right| \leqslant \sup_{x \in [0,L]} \{\xi(x)\} \int_0^T \int_0^L a_1 M |\varphi_t^1 - \varphi_t^2| |\tilde{\varphi}_x| \, \mathrm{d}x\mathrm{d}t \leqslant \frac{\epsilon}{3} \int_0^T E_{\mathbb{Z}}(t)\mathrm{d}t + c_\epsilon \int_0^T \int_0^L a_1 \tilde{\varphi}_t^2 \, \mathrm{d}x\mathrm{d}t.$$

This allows us to conclude the following estimate:

$$\begin{aligned} &\left| \int_0^T \int_0^L \Big[a_1 G_1(\tilde{\varphi}_t)\tilde{\varphi}_x + a_2 G_2(\tilde{\psi}_t)\tilde{\psi}_x + a_3 G_3(\tilde{w}_t)\tilde{w}_x \Big] \xi \, \mathrm{d}x\mathrm{d}t \right| \\ &\quad \leqslant \epsilon \int_0^T E_{\mathbb{Z}}(t)\mathrm{d}t + c_\epsilon \int_0^T \int_0^L \Big[a_1 \tilde{\varphi}_t^2 + a_2 \tilde{\psi}_t^2 + a_3 \tilde{w}_t^2 \Big] \mathrm{d}x\mathrm{d}t. \end{aligned} \tag{9.50}$$

Let us estimate the kinetic energy in (9.50). Assumption $(b.4)$ implies that

$$\int_0^T \int_0^L \Big[a_1 \tilde{\varphi}_t^2 + a_2 \tilde{\psi}_t^2 + a_3 \tilde{w}_t^2 \Big] \mathrm{d}x\mathrm{d}t \leqslant c \int_0^T \int_0^L \Big[a_1 G_1(\tilde{\varphi}_t)\tilde{\varphi}_t + a_2 G_2(\tilde{\psi}_t)\tilde{\psi}_t + a_3 G_3(\tilde{w}_t)\tilde{w}_t \Big] \mathrm{d}x\mathrm{d}t.$$

Next, we estimate the source terms. Invoking Assumption $(b.3)$, we find a positive constant c, such that

$$\left| \int_0^T \int_0^L F_1(\tilde{\varphi}, \tilde{\psi}, \tilde{w})\tilde{\varphi}_x \xi \, \mathrm{d}x\mathrm{d}t \right| \leqslant c \int_0^T \int_0^L c(\nabla f_1)(|\varphi| + |\psi| + |w|)|\varphi_x| \, \mathrm{d}x\mathrm{d}t \leqslant \epsilon \int_0^T E_{\mathbb{Z}}(t)\mathrm{d}t + c_{\epsilon,\mathscr{B}} \mathrm{l.o.t}(\tilde{\varphi}, \tilde{\psi}, \tilde{w}),$$

where

$$c(\nabla f_1) = 1 + |\varphi^1|^{p-1} + |\varphi^2|^{p-1} + |\psi^1|^{p-1} + |\psi^2|^{p-1} + |w^1|^{p-1} + |w^2|^{p-1}.$$

The above implies that

$$\begin{aligned}\left|\int_0^T \int_0^L \Big[F_1(\tilde{\varphi}, \tilde{\psi}, \tilde{w})\tilde{\varphi}_x + F_2(\tilde{\varphi}, \tilde{\psi}, \tilde{w})\tilde{\psi}_x + F_3(\tilde{\varphi}, \tilde{\psi}, \tilde{w})\tilde{w}_x\Big]\xi \, \mathrm{d}x\mathrm{d}t\right| \\ \leqslant \epsilon \int_0^T E_{\mathbb{Z}}(t)\mathrm{d}t + c_{\epsilon,\mathscr{B}}\mathrm{l.o.t}(\tilde{\varphi}, \tilde{\psi}, \tilde{w}).\end{aligned} \tag{9.51}$$

Now, we combine (9.49)–(9.51) with (9.47). For sufficiently small $\epsilon > 0$, we obtain

$$\begin{aligned}\int_0^T E_{\mathbb{Z}}(t)\mathrm{d}t \leqslant c\big[E_{\mathbb{Z}}(T) + E_{\mathbb{Z}}(0)\big] + \frac{1}{2}\int_0^T \int_{L_1+\epsilon_0}^{L_2-\epsilon_0} E(t) \, \mathrm{d}x\mathrm{d}t + c_{\mathscr{B}}\mathrm{l.o.t}(\tilde{\varphi}, \tilde{\psi}, \tilde{w}) \\ + c_{\mathscr{B}}\int_0^T \int_0^L \Big[a_1 G_1(\tilde{\varphi}_t)\tilde{\varphi}_t + a_2 G_2(\tilde{\psi}_t)\tilde{\psi}_t + a_3 G_3(\tilde{w}_t)\tilde{w}_t\Big]\mathrm{d}x\mathrm{d}t.\end{aligned} \tag{9.52}$$

The next step is to estimate the integral of $E(\cdot)$ over the interval $[L_1 + \epsilon, L_2 + \epsilon]$. To this end, we consider the function $[0, 1] \ni \eta \in C^\infty(0, L)$ defined as follows:

$$\eta(x) = \begin{cases} \eta(x) = 0, & \text{if } x \in (0, L_1) \cup (L_2, L), \\ \eta(x) = 1, & \text{if } x \in (L_1 + \epsilon_0, L_2 - \epsilon_0). \end{cases}.$$

We start multiplying Eqs. (9.46) by $\tilde{\varphi}\eta$, $\tilde{\psi}\eta$, and $\tilde{w}\eta$, respectively, and after integrate over $[0, T] \times [0, L]$, we add the kinetic energy $\int_0^T \int_0^L \big[\rho_1\tilde{\varphi}^2 + \rho_2\tilde{\psi}_t^2 + \rho_1\tilde{w}_t^2\big]\eta \, \mathrm{d}x\mathrm{d}t$ to obtain

$$\begin{aligned}&\int_0^T \int_0^L E(t)\eta \, \mathrm{d}x\mathrm{d}t \\ &= -\int_0^L \Big[\rho_1\tilde{\varphi}_t\tilde{\varphi} + \rho_2\tilde{\psi}_t\tilde{\psi} + \rho_1\tilde{w}_t\tilde{w}\Big]\eta \, \mathrm{d}x\Big|_0^T + 2\int_0^T \int_0^L \Big[\rho_1\tilde{\varphi}^2 + \rho_2\tilde{\psi}_t^2 + \rho_1\tilde{w}_t^2\Big]\eta \, \mathrm{d}x\mathrm{d}t \\ &\quad - \int_0^T \int_0^L \Big[k(\tilde{\varphi}_x + \tilde{\psi} + l\tilde{w})\tilde{\varphi} + b\tilde{\psi}_x\tilde{\psi} + k_0 l(\tilde{w}_x - l\tilde{\varphi})\tilde{w}\Big]\eta' \, \mathrm{d}x\mathrm{d}t \\ &\quad - \int_0^T \int_0^L \Big[a_1 G_1(\tilde{\varphi}_t)\tilde{\varphi} + a_2 G_2(\tilde{\psi}_t)\tilde{\psi} + a_3 G_3(\tilde{w}_t)\tilde{w}\Big]\eta \, \mathrm{d}x\mathrm{d}t \\ &\quad - \int_0^T \int_0^L \Big[F_1(\tilde{\varphi}, \tilde{\psi}, \tilde{w})\tilde{\varphi} + F_2(\tilde{\varphi}, \tilde{\psi}, \tilde{w})\tilde{\psi} + F_3(\tilde{\varphi}, \tilde{\psi}, \tilde{w})\tilde{w}\Big]\eta \, \mathrm{d}x\mathrm{d}t.\end{aligned} \tag{9.53}$$

We shall estimate the right-hand side of (9.53). To this end, we repeat the pattern of estimates (9.49)–(9.51) to find

$$
\begin{aligned}
\int_0^T \int_0^L E(t)\eta \, \mathrm{d}x\mathrm{d}t &\leqslant c\big[E_{\mathbb{Z}}(T) + E_{\mathbb{Z}}(0)\big] + c_{\mathscr{B}}\mathrm{l.o.t}(\tilde{\varphi}, \tilde{\psi}, \tilde{w}) \\
&\quad + c_{\mathscr{B}} \int_0^T \int_0^L \Big[a_1 G_1(\tilde{\varphi}_t)\tilde{\varphi}_t + a_2 G_2(\tilde{\psi}_t)\tilde{\psi}_t + a_3 G_3(\tilde{w}_t)\tilde{w}_t\Big]\mathrm{d}x\mathrm{d}t.
\end{aligned} \tag{9.54}
$$

Applying the estimate (9.54) in (9.52), we obtain

$$
\begin{aligned}
\int_0^T E_{\mathbb{Z}}(t)\mathrm{d}t &\leqslant c\big[E_{\mathbb{Z}}(T) + E_{\mathbb{Z}}(0)\big] + c_{\mathscr{B}}\mathrm{l.o.t}(\tilde{\varphi}, \tilde{\psi}, \tilde{w}) \\
&\quad + c_{\mathscr{B}} \int_0^T \int_0^L \Big[a_1 G_1(\tilde{\varphi}_t)\tilde{\varphi}_t + a_2 G_2(\tilde{\psi}_t)\tilde{\psi}_t + a_3 G_3(\tilde{w}_t)\tilde{w}_t\Big]\mathrm{d}x\mathrm{d}t.
\end{aligned} \tag{9.55}
$$

Next, we estimate damping terms on the right-hand side of (9.55). Multiply Eqs. (9.46) by φ_t, ψ_t, w_t, respectively. Then we find that

$$
\begin{aligned}
&\int_0^T \int_0^L \Big[a_1 G_1(\tilde{\varphi}_t)\tilde{\varphi}_t + a_2 G_2(\tilde{\psi}_t)\tilde{\psi}_t + a_3 G_3(\tilde{w}_t)\tilde{w}_t\Big]\mathrm{d}x\mathrm{d}t - E_{\mathbb{Z}}(0) + E_{\mathbb{Z}}(T) \\
&\quad = -\int_0^T \int_0^L \Big[F_1(\tilde{\varphi}, \tilde{\psi}, \tilde{w})\tilde{\varphi}_t + F_2(\tilde{\varphi}, \tilde{\psi}, \tilde{w})\tilde{\psi}_t + F_3(\tilde{\varphi}, \tilde{\psi}, \tilde{w})\tilde{w}_t\Big]\mathrm{d}x\mathrm{d}t.
\end{aligned} \tag{9.56}
$$

Based on estimate (9.51), we obtain

$$
\begin{aligned}
&\left|\int_0^T \int_0^L \Big[F_1(\tilde{\varphi}, \tilde{\psi}, \tilde{w})\tilde{\varphi}_t + F_2(\tilde{\varphi}, \tilde{\psi}, \tilde{w})\tilde{\psi}_t + F_3(\tilde{\varphi}, \tilde{\psi}, \tilde{w})\tilde{w}_t\Big]\mathrm{d}x\mathrm{d}t\right| \\
&\qquad \leqslant \epsilon \int_0^T E_{\mathbb{Z}}(t)\mathrm{d}t + c_{\epsilon,\mathscr{B}}\mathrm{l.o.t}(\tilde{\varphi}, \tilde{\psi}, \tilde{w}).
\end{aligned} \tag{9.57}
$$

Using both (9.56) and (9.57), we find

$$
\begin{aligned}
&\int_0^T \int_0^L \Big[a_1 G_1(\tilde{\varphi}_t)\tilde{\varphi}_t + a_2 G_2(\tilde{\psi}_t)\tilde{\psi}_t + a_3 G_3(\tilde{w}_t)\tilde{w}_t\Big]\mathrm{d}x\mathrm{d}t \\
&\leqslant E_{\mathbb{Z}}(0) - E_{\mathbb{Z}}(T) + \epsilon \int_0^T E_{\mathbb{Z}}(t)\mathrm{d}t + c_{\epsilon,\mathscr{B}}\mathrm{l.o.t}(\tilde{\varphi}, \tilde{\psi}, \tilde{w}).
\end{aligned} \tag{9.58}
$$

We return to (9.55) and obtain by use of (9.58), with $\epsilon > 0$ small enough, the following estimate

$$\int_0^T E_{\mathbb{Z}}(t)\mathrm{d}t \leqslant (c - c_{\mathscr{B}})E_{\mathbb{Z}}(T) + (c + c_{\mathscr{B}})E_{\mathbb{Z}}(0) + c_{\mathscr{B}}\text{l.o.t}(\tilde{\varphi}, \tilde{\psi}, \tilde{w}). \tag{9.59}$$

The next step is to estimate the energy $E_{\mathbb{Z}}(\cdot)$. To this end, we use the multipliers φ_t, ψ_t, $\tilde{w}_t$ for (9.46). Then, after integration, we find

$$\begin{aligned} TE_{\mathbb{Z}}(T) = \\ = \int_0^T E_{\mathbb{Z}}(t)\mathrm{d}t - \int_0^T\int_s^T\int_0^L \Big[a_1 G_1(\tilde{\varphi}_t)\tilde{\varphi}_t + a_2 G_2(\tilde{\psi}_t)\tilde{\psi}_t + a_3 G_3(\tilde{w}_t)\tilde{w}_t\Big]\mathrm{d}x\mathrm{d}t\mathrm{d}s \\ - \int_0^T\int_s^T\int_0^L \Big[F_1(\tilde{\varphi}, \tilde{\psi}, \tilde{w})\tilde{\varphi}_t + F_2(\tilde{\varphi}, \tilde{\psi}, \tilde{w})\tilde{\psi}_t + F_3(\tilde{\varphi}, \tilde{\psi}, \tilde{w})\tilde{w}_t\Big]\mathrm{d}x\mathrm{d}t\mathrm{d}s. \end{aligned} \tag{9.60}$$

Now, the forcing assumptions give

$$\begin{aligned} \int_0^L \Big[F_1(\tilde{\varphi}, \tilde{\psi}, \tilde{w})\tilde{\varphi}_t + F_2(\tilde{\varphi}, \tilde{\psi}, \tilde{w})\tilde{\psi}_t + F_3(\tilde{\varphi}, \tilde{\psi}, \tilde{w})\tilde{w}_t\Big]\mathrm{d}x \\ \leqslant \frac{1}{T}E_{\mathbb{Z}}(t) + c_{T,\mathscr{B}}\text{l.o.t}(\tilde{\varphi}, \tilde{\psi}, \tilde{w}). \end{aligned}$$

The above implies

$$\begin{aligned} \int_0^T\int_s^T\int_0^L \Big[F_1(\tilde{\varphi}, \tilde{\psi}, \tilde{w})\tilde{\varphi}_t + F_2(\tilde{\varphi}, \tilde{\psi}, \tilde{w})\tilde{\psi}_t + F_3(\tilde{\varphi}, \tilde{\psi}, \tilde{w})\tilde{w}_t\Big]\mathrm{d}x\mathrm{d}t\mathrm{d}s \\ \leqslant \int_0^T E_{\mathbb{Z}}(t)\mathrm{d}t + c_{T,\mathscr{B}}\text{l.o.t}(\tilde{\varphi}, \tilde{\psi}, \tilde{w}). \end{aligned}$$

We combine the above estimates with (9.60)

$$TE_{\mathbb{Z}}(T) \leqslant 2\int_0^T E_{\mathbb{Z}}(t)\mathrm{d}t + c_{T,\mathscr{B}}\text{l.o.t}(\tilde{\varphi}, \tilde{\psi}, \tilde{w}). \tag{9.61}$$

Next, using estimate (9.59) in (9.61), we arrive at

$$TE_{\mathbb{Z}}(T) \leq 2(c - c_{\mathscr{B}})E_{\mathbb{Z}}(T) + 2(c + c_{\mathscr{B}})E_{\mathbb{Z}}(0) + c_{T,\mathscr{B}}\text{l.o.t}(\tilde{\varphi}, \tilde{\psi}, \tilde{w}).$$

Taking $T > 4c$, we find

$$E_{\mathbb{Z}}(T) \leq \frac{2(c + c_{\mathscr{B}})}{T - 2(c - c_{\mathscr{B}})}E_{\mathbb{Z}}(0) + c_{T,\mathscr{B}}\text{l.o.t}(\tilde{\varphi}, \tilde{\psi}, \tilde{w}).$$

Using standard stabilization arguments, we obtain the existence of positive constants $c_1 = c_{\mathscr{B}}$ and $\omega = \omega_{\mathscr{B}}$ such that

$$\|\mathbb{Z}(t)\|_H^2 \le c_1 \|\mathbb{Z}(0)\|_H^2 e^{-\omega t} + c_1 \sup_{\sigma \in [0,t]} \Big[\|\tilde{\varphi}(\sigma)\|_{2p}^2 + \|\tilde{\psi}(\sigma)\|_{2p}^2 + \|\tilde{w}(\sigma)\|_{2p}^2 \Big].$$

Now, the inequality (9.44) holds with $X = [H_0^1(0, L)]^3$, $Y = [L^2(0, L)]^3$, $b(t) = c_1 e^{-\omega t}$, $c(t) = c_1$ and with compact semi-norm

$$[(\tilde{\varphi}, \tilde{\psi}, \tilde{w})]_X^2 = \|\tilde{\varphi}\|_{2p}^2 + \|\tilde{\psi}\|_{2p}^2 + \|\tilde{w}\|_{2p}^2.$$

Therefore, $(H, S(t))$ is quasi-stable. □

Completion of the Proof of Theorem 9.4.3 With these preparations completed, we are now ready to present the proof of our main result.

Proof of Theorem 9.4.3 From Propositions 9.4.1 and 9.4.2, we deduce that $(H, S(t))$ is a gradient quasi-stable system. Furthermore, through inequality (9.40), it becomes evident that the Lyapunov function defined as the energy $\mathscr{E}_{\mathbb{Z}}(\cdot)$ satisfies the following conditions: (i) $\Phi(\cdot)$ is bounded from above on any bounded subset of H and (ii) the set $\Phi(R) = \{\mathbf{Z} \in H \mid \Phi(\mathbf{Z}) \le R\}$ is bounded for every R. To conclude the proof, we establish that the set of stationary solutions is bounded. Notably, if $\mathbf{Z} \in \mathscr{N}$, then $\mathbf{Z} = (\varphi, \psi, w, 0, 0, 0)$ solves the stationary problem

$$\begin{cases} -k(\varphi_x + \psi + lw)_x - k_0 l(w_x - l\varphi) + f_1(\varphi, \psi, w) = 0, \\ -b\psi_{xx} + k(\varphi_x + \psi + lw) + f_2(\varphi, \psi, w) = 0, \\ -k_0(w_x - l\varphi)_x + kl(\varphi_x + \psi + lw) + f_3(\varphi, \psi, w) = 0. \end{cases} \tag{9.62}$$

Now, multiplying in $L^2(0, L)$ the equations in (9.62) by (φ, ψ, w), we find

$$b\|\psi_x\|^2 + k\|\varphi_x + \psi + lw\|^2 + k_0\|w_x + l\varphi\|^2 = -\int_0^L \nabla F(u, v, w)\cdot(u, v, w)dx.$$

Now, we use Assumption (b.1) to show

$$\left[1 - \frac{2\alpha\beta L^2}{\pi^2}\right]\left[\|\varphi_x\|^2 + \|\psi_x\|^2 + \|w_x\|^2\right] \leqslant 2\beta c_F L.$$

Therefore, the set of stationary solutions $\mathscr{N}$ is bounded. This completes the proof. □

9.5 Conclusions

In the present chapter, we have presented a useful Unique Continuation Property (Theorem 9.3.2) for systems of 1D wave equations with potentials in L^2. Then this was used to prove gradient like property for a Bresse system with locally distributed damping (Proposition 9.4.1). It is worth mentioning that our UCP was recently used in some other locally dissipative models, for instance, in [16], [15], and [27].

Appendix: Well Posedness for Overdetermined Wave Equations

In this appendix we will guarantee the well posedness for the system presented in (9.2) with overdetermined condition.

Theorem 9.A.1 *Let $L > 0$ and $T > 0$ be sufficiently large. If Problem* (9.2)–(9.4) *satisfies the conditions* (9.5)–(9.9) *along with the supplementary condition*

$$(u_1, \cdots, u_n) = (0, \cdots, 0) \ in \ \omega \times [0, T], \tag{A.1}$$

where $\omega \subset [0, L_0]$ as defined in (9.28)*, then the overdetermined problem is well posed and generates a strongly continuous semigroup over the Hilbert space*

$$H_{\omega_j}(M_i) \equiv \left\{ (u_1, \cdots, u_n, v_1, \cdots, v_n) \,\middle|\, \begin{array}{l} u_i \in H_0^1(M_i), \ v_i \in L^2(M_i), \\ u_i = v_i = 0 \ in \ \omega, i = 1, \cdots, n \end{array} \right\},$$

where $M_i = ([0, L], g_i)$.

Proof First, for the state vector $\mathbb{Z}(t) \equiv (u_1, \cdots, u_n, \partial_t u_1, \cdots, \partial_t u_n)^\top$, Problem (9.2)–(9.4) is equivalent to the following vectorial Cauchy problem:

$$\partial_t \mathbb{Z}(t) + A\mathbb{Z}(t) = \mathcal{F}(\mathbb{Z}(t)), \ \ \mathbb{Z}(0) = (u_1^0, \cdots, u_n^0, u_1^1, \cdots, u_n^1)^\top, \tag{A.2}$$

with operators defined by

$$A = \begin{bmatrix} 0 & -I|_{V_2(M_i)} \\ B|_{V_1(M_i)} & 0 \end{bmatrix}, \quad B = \begin{bmatrix} -\Delta_1 & 0 & \cdots & 0 \\ 0 & -\Delta_2 & \cdots & 0 \\ \vdots & \vdots & \cdots & \vdots \\ 0 & 0 & 0 & -\Delta_n \end{bmatrix},$$

$$I = (\delta_{ij})_{n\times n} \quad \text{and} \quad \mathscr{F}(\mathbb{Z}) = \begin{bmatrix} 0 \\ \vdots \\ 0 \\ f_1(u_1,\cdots,u_n) \\ \vdots \\ f_n(u_1,\cdots,u_n) \end{bmatrix}.$$

The domain of operator A is defined by $D(A) = [V_1(M_i)]^n \times [V_2(M_i)]^n$ where

$$V_1(M_i) \equiv \big\{(u_1,\cdots,u_n) \in D(B) \mid (u_1,\cdots,u_n) = (0,\cdots,0) \text{ in } \omega\big\},$$
$$V_2(M_i) \equiv \big\{(v_1,\cdots,v_n) \in D(B^{\frac{1}{2}}) \mid (v_1,\cdots,v_n) = (0,\cdots,0) \text{ in } \omega\big\}$$

and

$$D(B) = D(-\Delta_1)\times\cdots\times D(-\Delta_n) = H^2(M_1)\cap H_0^1(M_1)\times\cdots H^2(M_n)\cap H_0^1(M_n).$$

The finite energy space for (A.2) is the Hilbert space defined by

$$H_\omega(M_i) \equiv \big\{(u_1,\cdots,u_n,v_1,\cdots,v_n) \in H(M_i) \mid u_i = v_i = 0 \text{ in } \omega,\ \forall i = 1,\cdots,n\big\},$$

where

$$H(M_i) \equiv [D(B^{\frac{1}{2}})]^n \times L(M_i) \text{ and } L(M_i) \equiv L^2(M_1)\times\cdots\times L^2(M_n).$$

Using classical semigroup theory, one can establish existence and uniqueness of a solution to the Cauchy problem (A.2). Moreover, the solution operator generates a strongly continuous semigroup

$$T_{M_i}(t) : H_w(M_i) \to H_w(M_i),$$

defined by

$$(u_1^0,\cdots,u_n^0,u_1^1,\cdots,u_n^1) \mapsto (u_1(t),\cdots,u_n(t),\partial_t u_1(t),\cdots,\partial_t u_n(t)), \quad t \geq 0,$$

where $(u_1,\cdots,u_n,\partial_t u_1,\cdots,\partial_t u_n)$ is the weak solution corresponding to the initial data

$$(u_1^0,\cdots,u_n^0,u_1^1,\cdots,u_n^1).$$

In addition, $\{T_{M_i}(t)\}_{t\geq 0}$ is also strongly continuous semigroup on $H_\omega(M_i)$ satisfying the compatibility condition (A.1). □

Remark 9.A.1 It is not difficult to show that if (A.1) is fulfilled, then we also have

$$(\partial_t u_1, \cdots, \partial_t u_n) = (0, \cdots, 0) \text{ in } \omega \times [0, T].$$

Funding T. F. Ma and R. N. Monteiro were supported by FAPDF grant 0193.00001821/2022-21. T. F. Ma and P. N. Seminario-Huertas were supported by Ministerio de Ciencia e Innovación (Spain) and FEDER (European Community) under grant PID2021-122991NB-C21. T. F. Ma was also supported by CNPq grant 315165/2021-9.

References

1. F. Alabau Boussouira, J.E. Muñoz Rivera, D.S. Almeida Júnior, Stability to weak dissipative Bresse system. J. Math. Anal. Appl. **374**, 481–498 (2011)
2. C. Bardos, G. Lebeau, J. Rauch, Sharp sufficient conditions for the observation, control, and stabilization of waves from the boundary. SIAM J. Control Optim. **30**, 1024–1065 (1992)
3. G.E. Bittencourt Moraes, M.A. Jorge Silva, Arched beams of Bresse type: observability and application in thermoelasticity. Nonl. Dyn. **103**, 2365–2390 (2021)
4. J.A.C. Bresse, *Cours de Méchanique Appliquée* (Mallet Bachelier, Paris, 1859)
5. M.M. Cavalcanti, V.N. Domingos Cavalcanti, R. Fukuoka, J.A. Soriano, Uniform stabilization of the wave equation on compact surfaces and locally distributed damping - a sharp result. Trans. Amer. Math. Soc. **361**, 4561–4580 (2009)
6. M.M. Cavalcanti, V.N. Domingos Cavalcanti, R. Fukuoka, J.A. Soriano, Asymptotic stability of the wave equation on compact manifolds and locally distributed damping: a sharp result. Arch. Ration. Mech. Anal. **197**, 925–964 (2010)
7. M.M. Cavalcanti, T.F. Ma, P. Marín-Rubio, P.N. Seminario Huertas, Dynamics of Riemann waves with sharp measure-controlled damping, arXiv:1908.04814 (2019)
8. W. Charles, J.A. Soriano, F.A. Falcão Nascimento, J.H. Rodrigues, Decay rates for Bresse system with arbitrary nonlinear localized damping. J. Differ. Equ. **255**, 2267–2290 (2013)
9. I. Chueshov, I. Lasiecka, *Von Karman Evolution Equations. Well-posedness and Long Time Dynamics*, Springer Monographs in Mathematics (Springer, New York, 2010)
10. I. Chueshov, I. Lasiecka, D. Toundykov, Long-term dynamics of semilinear wave equation with nonlinear localized interior damping and a source term of critical exponent. Discrete Contin. Dyn. Syst. **20**, 459–509 (2008)
11. F. Dell'Oro, Asymptotic stability of thermoelastic systems of Bresse type. J. Differ. Equ. **258**, 3902–3927 (2015)
12. L.H. Fatori, R.N. Monteiro, The optimal decay rate for a weak dissipative Bresse system. Appl. Math. Lett. **25**, 600–604 (2012)
13. L.H. Fatori, J.E. Muñoz Rivera, Rates of decay to weak thermoelastic Bresse system. IMA J. Appl. Math. **75**, 881–904 (2010)
14. E. Feireisl, E. Zuazua, Global attractors for semilinear wave equations with locally distributed nonlinear damping and critical exponent. Comm. Partial Differ. Equ. **18**, 1539–1555 (1993)
15. M.M. Freitas, R.Q. Caljaro, A.J.A. Ramos, H.C.M. Rodrigues, Long-time dynamics of ternary mixtures with localized dissipation. J. Math. Phys. **63**(12), article 121508, 28 pp. (2022)
16. M.M. Freitas, D.S. Almeida Junior, M.L. Santos, A.J.A. Ramos, R.Q. Caljaro, Dynamics of locally damped Timoshenko systems. Math. Mech. Solids **28**, 1012–1034 (2023)
17. J.E. Lagnese, G. Leugering, E.J.P.G. Schmidt, *Modeling, Analysis and Control of Dynamic Elastic Multi-Link Structures* (Birkhaüser, Boston, 1994)

18. I. Lasiecka, R. Triggiani, X. Zhang, Nonconservative wave equations with unobserved Neumann B.C.: global uniqueness and observability in one shot, in *Differential Geometric Methods in the Control of Partial Differential Equations (Boulder, CO, 1999)*, Contemp. Math., vol. 268 (Amer. Math. Soc., Providence, RI, 2000), pp. 227–325
19. Z. Liu, B. Rao, Energy decay rate of the thermoelastic Bresse system. Z. Angew. Math. Phys. **60**, 54–69 (2009)
20. T.F. Ma, R.N. Monteiro, Singular limit and long-time dynamics of Bresse systems. SIAM J. Math. Anal. **49**, 2468–2495 (2017)
21. T.F. Ma, P.N. Seminario-Huertas, Attractors for semilinear wave equations with localized damping and external forces. Comm. Pure Appl. Anal. **19**, 2219–2233 (2020)
22. P. Olsson, G. Kristensson, Wave splitting of the Timoshenko beam equation in the time domain. Z. Angew. Math. Phys. **45**, 866–881 (1994)
23. J. Rauch, M. Taylor, Exponential decay of solutions to hyperbolic equations in bounded domains. Indiana Univ. Math. J. **24**, 79–86 (1974)
24. J. Rauch, M. Taylor, Decay of solutions to nondissipative hyperbolic systems on compact manifolds. Comm. Pure Appl. Math. **28**, 501–523 (1975)
25. A. Ruiz, Unique continuation for weak solutions of the wave equation plus a potential. J. Math. Pures Appl. (9) **71**(5), 455–467 (1992)
26. M.L. Santos, A. Soufyane, D. Almeida Júnior, Asymptotic behavior to Bresse system with past history. Q. Appl. Math. **73**, 23–54 (2015)
27. M.L. Santos, D.S. Almeida Junior, S.M.S. Cordeiro, Energy decay for a porous-elastic system with nonlinear localized damping. Z. Angew. Math. Phys. **73**(1), 7–21 (2022)
28. A. Soufyane, Stabilisation de la poutre de Timoshenko. C. R. Acad. Sci. Paris Ser. I Math. **328**(8), 731–734 (1999)
29. A. Soufyane, B. Said-Houari, The effect of the wave speeds and the frictional damping terms on the decay rate of the Bresse system. Evol. Equ. Control Theory **3**, 713–738 (2014)
30. S. Timoshenko, On the correction for shear of the differential equation for transverse vibrations of prismatic bars. Philos. Mag. **41**, 744–746 (1921)
31. R. Triggiani, P.F. Yao, Carleman estimates with no lower-order terms for general Riemann wave equations. Global uniqueness and observability in one shot. Special issue dedicated to the memory of Jacques-Louis Lions. Appl. Math. Optim. **46**(2–3), 331–375 (2002)
32. A. Wehbe, W. Youssef, Exponential and polynomial stability of an elastic Bresse system with two locally distributed feedbacks. J. Math. Phys **51**, article 103523 (2010)

Chapter 10
Spectral Decomposition of $H^1(\mathbb{R}^n)$

César Augusto Rubim and Eduard Toon

Abstract In this chapter, we are interested in presenting a spectral decomposition of the Sobolev Space $H^1(\mathbb{R}^n)$. To this end, we provide a thorough construction stating the foundations of functional calculus for operators and present a proof of the Spectral Theorem for unbounded self-adjoint operators. Despite this last mentioned result is a classical result in functional analysis, there are few references that build all arguments from scratch, which we make here.

Keywords Schrödinger equation · Spectral theory · Sobolev spaces · Spectral decomposition · Spectral measures

MSC 2020 35J10, 35P05, 46E35, 81Q10

10.1 Introduction

The Schrödinger equation

$$-\Delta\psi(x,t)+V(x,t)\psi=-i\frac{\partial(x,t)\psi}{\partial t}\quad\text{for } (x,t)\in\mathbb{R}\times[0,+\infty) \tag{10.1}$$

is one of the most important, if not the most important, equations in quantum theory, describing the behavior of particles under the action of a potential $V(x,t)$. When the potential V does not vary with time, we can apply the technique of separation of variables and look for solutions $\psi(x,t)=\phi(t)u(x)$, with $\phi(t)=\exp(-iEt)$. Accordingly, the spatial part of (10.1) can be written as

$$-\Delta u+(V(x)-E)u=0, \tag{10.2}$$

C. A. Rubim · E. Toon (✉)
Departamento de Matemática, Universidade Federal de Juiz de Fora, Minas Gerais, Brazil
e-mail: rubim@ice.ufjf.br; eduard.toon@ufjf.br

P. Benevieri, J. G. Mesquita (eds.), *Functional Differential Equations and Dynamic Equations on Time Scales*, Advances in Continuum Mechanics 52,
https://doi.org/10.1007/978-3-031-83327-4_10

where E is understood as the energy of the particle. The association of nonlinear terms of the form $f(x, u)$ to (10.2) constitutes a vast field of study that is interesting in itself and has various applications, such as condensed matter physics, plasma physics, among others, see [2, 25]. Thus, equations of the form

$$-\Delta u + V(x)u = f(x, u(x)), \quad x \in \mathbb{R}^n, \tag{10.3}$$

have been extensively studied through various approaches such as formulations of the mountain pass theorem, generalized linking theorems, to mention a few, see [4, 9, 18, 21–24, 26–28]. In these techniques, it is common to employ the study of the energy functional associated with (10.3)

$$I(u) = \int_{\mathbb{R}^n} |\nabla u(x)|^2 + V(x)u(x)^2 dx - \int_{\mathbb{R}^n} F(x, u(x))dx, \, u \in H^1(\mathbb{R}^n), \tag{10.4}$$

with $F(x,t) = \int_0^t f(x,s)ds$, where commonly we seek critical points of I. The potential V plays a fundamental role in these studies, and, furthermore, the spectrum of the Schrödinger operator $S = -\Delta + V$ with $D(S) = H^2(\mathbb{R}^n)$ is crucial in the possibility of existence of solutions to (10.3), see [5–7, 11–13, 20] and the references therein. More than just decomposing the spectrum of S, it is possible to decompose the domain of I (that is, decompose $H^1(\mathbb{R}^n)$) by applying the Spectral Theorem for unbounded self-adjoint operators to S.

We will then be interested in decomposing $H^1(\mathbb{R}^n)$ through the use of the Spectral Theorem. However, despite spectral theory being an extremely consolidated area, there are few references that build all arguments from scratch and still in a nearly self-contained manner. Our goal then is to try to be as simple, self-contained, but in a way that we are able to prove the Spectral Theorem for self-adjoint operators. Finally, following the developments of [5], we will obtain a decomposition of $H^1(\mathbb{R}^n)$ from the Spectral Theorem, and furthermore, obtain a simpler way to express the first integral of the functional (10.4) through a norm equivalent to the usual norm of $H^1(\mathbb{R}^n)$, which will also be constructed with the aid of spectral theory.

This chapter is divided as follows: In the second section, we will construct the spectral measures and obtain some properties of these objects; in the third section, we will construct the proof of the Spectral Theorem, in two versions, first for bounded operators and then for unbounded operators; finally, in the last section, we will apply the Spectral Theorem to the Schrödinger operator S and obtain the mentioned decomposition of $H^1(\mathbb{R}^n)$.

10.2 Spectral Integral

Let $\mathcal{U}$ be a σ-algebra of subsets of a general set Ω and $\mathcal{H}$ be a complex Hilbert space. We will denote by $\mathcal{P}(\mathcal{H})$ the set of orthogonal projections of $\mathcal{H}$.

Definition 10.1 A spectral measure on $\mathcal{U}$ is a mapping $E\colon \mathcal{U} \to \mathcal{P}(\mathcal{H})$ that satisfies:

(i) $E(\Omega) = I$.
(ii) If $(M_n)_{n\in\mathbb{N}}$ is a sequence of sets from $\mathcal{U}$ that are pairwise disjoint, and if $\bigcup_{n\in\mathbb{N}} M_n \in \mathcal{U}$, then $E\left(\bigcup_{n\in\mathbb{N}} M_n\right) = \sum_{n\in\mathbb{N}} E(M_n)$.

An important observation is that, in item (ii) of the definition, the infinite sum appearing in the equation should be understood as follows:

$$E\left(\bigcup_{n\in\mathbb{N}} M_n\right)x = \lim_{k\to\infty} \sum_{n=1}^{k} E(M_n)x, \quad \forall x \in \mathcal{H}.$$

Furthermore, if we take $M_n = \emptyset$ for all n, we obtain that $E(\emptyset) = 0$. Now, if we take a finite sequence $(M_n) \subset \mathcal{U}$, with pairwise disjoint terms, it follows from (ii) and the previous consideration that

$$E(M_1 \cup \cdots \cup M_k) = E(M_1) + \cdots + E(M_k).$$

In the rest of this chapter, we shall denote $R(T)$ as the range of a linear transformation $T : D(T) \subset \mathcal{H}_1 \to \mathcal{H}_2$, $\mathcal{N}(T)$ as the kernel of T, and if T is an extension of a linear transformation S, we will write $S \subset T$. More than that, we will say that a linear transformation $T : D(T) \subset \mathcal{H}_1 \to \mathcal{H}_2$ is closed if the graph of T is closed in $\mathcal{H}_1 \times \mathcal{H}_2$, and closable if it exists an extension of the linear transformation which is closed. For a linear transformation T, a linear subspace $D \subset D(T)$ is said to be core of T if, for all $x \in D(T)$ there exists a sequence $(x_n)_{n\in\mathbb{N}}$ converging to x and $(Tx_n)_{n\in\mathbb{N}}$ converging to Tx.

Let us define some fundamental concepts related to spectral theory. Let $T : D(T) \subset \mathcal{H} \to \mathcal{H}$ be a linear operator. The set

$$\rho(T) = \{\lambda \in \mathbb{C} : T - \lambda I : D(T) \subset \mathcal{H} \to \mathcal{H} \text{ is an isomorphism }\} \tag{10.5}$$

is called the resolvent set of T. The complement of $\rho(T)$, that is, $\sigma(T) = \mathbb{C} \setminus \rho(T)$, is called the spectrum of T.

It is interesting to note that, if $\lambda \in \mathbb{C}$ is such that $\mathcal{N}(T - \lambda I) \neq \{0\}$, then $T - \lambda I$ is not injective, from which $\lambda \notin \rho(T)$, that is, $\lambda \in \sigma(T)$. Furthermore, this means there exists a nonzero $v \in D(T)$ satisfying $Tv = \lambda v$, i.e., λ is an eigenvalue of T. It is also interesting to define some subsets of $\sigma(T)$,

$$\sigma_p(T) = \{\lambda \in \mathbb{C} : \mathcal{N}(T - \lambda I) \neq \{0\}\} \subset \sigma(T), \tag{10.6}$$

called the point spectrum of T,

$$\sigma_d(T) = \{\lambda \in \sigma_p(T) : \dim \mathcal{N}(T - \lambda I) < \infty \text{ and } \lambda \text{ is an isolated point of } \sigma(T)\}, \tag{10.7}$$

called the discrete spectrum of T, and

$$\sigma_{ess}(T) = \sigma(T) \setminus \sigma_d(T)$$

called the essential spectrum of T.

Lemma 10.1 *Let T be a linear operator on $\mathcal{H}$ such that $D(T)$ is dense in $\mathcal{H}$. If $\langle Tx, x\rangle = 0$ for all $x \in D(T)$, then $Tx = 0$ for all $x \in D(T)$.*

Proof We first prove the following polarization identity:

$$\begin{aligned} 4\langle Tx, y\rangle = \langle T(x+y), x+y\rangle - \langle T(x-y), x-y\rangle + i\langle T(x+iy), x+iy\rangle \\ - i\langle T(x-iy), x-iy\rangle. \end{aligned} \tag{10.8}$$

Note that

$$\begin{aligned} \langle T(x+y), x+y\rangle &= \langle Tx, x\rangle + \langle Tx, y\rangle + \langle Ty, x\rangle + \langle Ty, y\rangle, \\ -\langle T(x-y), x-y\rangle &= -\langle Tx, x\rangle + \langle Tx, y\rangle + \langle Ty, x\rangle - \langle Ty, y\rangle, \\ i\langle T(x+iy), x+iy\rangle &= i\langle Tx, x\rangle + \langle Tx, y\rangle - \langle Ty, x\rangle + i\langle Ty, y\rangle, \\ -i\langle T(x-iy), x-iy\rangle &= -i\langle Tx, x\rangle + \langle Tx, y\rangle - \langle Ty, x\rangle - i\langle Ty, y\rangle, \end{aligned}$$

and hence, by adding the equations, the identity follows.

From (10.8), it follows that, if $x, y \in D(T)$, then $\langle Tx, y\rangle = 0$. Now, by the density of $D(T)$, we can construct a sequence $(y_n)_{n\in\mathbb{N}} \subset D(T)$ such that $y_n \to Tx$. Using the continuity of the inner product, we obtain

$$\langle Tx, Tx\rangle = \langle Tx, \lim_{n\to\infty} y_n\rangle = \lim_{n\to\infty} \langle Tx, y_n\rangle = 0,$$

and it follows that $Tx = 0$, for all $x \in D(T)$. □

Lemma 10.2 *If P_1 and P_2 are orthogonal projections and $P_1 + P_2$ is also an orthogonal projection, then $P_1P_2 = 0$.*

Proof Let $x \in R(P_2)$, and then $x = P_2x$. We will show that $x \in N(P_1)$, which implies the statement.

$$\begin{aligned} ||x||^2 = ||P_2x||^2 &\le ||P_2x||^2 + ||P_1x||^2 \\ &= \langle P_2x, P_2x\rangle + \langle P_1x, P_1x\rangle \\ &= \langle P_2^2x, x\rangle + \langle P_1^2x, x\rangle \\ &= \langle P_2x, x\rangle + \langle P_1x, x\rangle \\ &= \langle (P_1 + P_2)x, x\rangle \\ &\le ||(P_1 + P_2)||\,||x||^2 \\ &\le ||x||^2, \end{aligned}$$

in other words, we can rewrite

$$\begin{aligned} ||x||^2 &= ||P_2x||^2 \leq ||P_2x||^2 + ||P_1x||^2 \leq ||x||^2 \\ \Longrightarrow\ ||P_2x||^2 &= ||P_2x||^2 + ||P_1x||^2 \end{aligned}$$

from which we must have $P_1x = 0$, and it follows that $R(P_2) \subset \mathcal{N}(P_1)$. □

Let us now observe how the product of projections correlates with the intersection between sets.

Lemma 10.3 *If $E : \mathcal{U} \to \mathcal{P}(\mathcal{H})$ is a spectral measure, it holds that*

$$E(M)E(N) = E(M \cap N), \quad M, N \in \mathcal{U}. \tag{10.9}$$

In particular, $E(M)E(N) = 0$ if M and N are disjoint.

Proof Let us initially prove the particular case. Take M and N in $\mathcal{U}$, disjoint. Then $E(M) + E(N) = E(M \cup N) \in \mathcal{P}(\mathcal{H})$, that is, $E(M) + E(N)$ is an orthogonal projection, from which we can apply Lemma 10.2, and conclude that $E(M)E(N) = 0$.

Now, for any $M, N \in \mathcal{U}$, we define $M_0 = M \cap N$, $M_1 = M \setminus M_0$, $M_2 = N \setminus M_0$. Since $M_1 \cap M_0 = M_2 \cap M_0 = M_1 \cap M_2 = \emptyset$, we have

$$E(M_1)E(M_0) = E(M_2)E(M_0) = E(M_1)E(M_2) = 0.$$

Furthermore, from $M = M_0 \cup M_1$ and $N = M_0 \cup M_2$, it follows that

$$\begin{aligned} E(M)E(N) &= (E(M_1) + E(M_0))(E(M_2) + E(M_0)) = E(M_0)^2 \\ &= E(M_0) = E(M \cap N). \end{aligned}$$

□

Let us then observe that, from equation (10.9), any two projections $E(M)$ and $E(N)$ commute. Moreover, if $N, M \in \mathcal{U}$ and $N \subset M$, it follows that $E(N) \subset E(M)$, in the sense of extension of operators. Before we proceed, let us use the notation E-a.e. to denote that a certain property holds, except on sets where $E(N) = 0$, for $N \in \mathcal{U}$. Let us now see how we can relate spectral measures to scalar measures.

Lemma 10.4 *Let $\mathcal{U}$ be a σ-algebra of Ω. An application $E : \mathcal{U} \to \mathcal{P}(\mathcal{H})$ is a spectral measure if and only if $E(\Omega) = I$ and, for each $x \in \mathcal{H}$, the function $E_x(\cdot) = \langle E(\cdot)x, x\rangle$ defined on $\mathcal{U}$ is a measure.*

Proof If E is a spectral measure, it immediately follows that $E(\Omega) = I$. Moreover, if x is an element of $\mathcal{H}$ and $(M_n)_{n\in\mathbb{N}}$ is a sequence of pairwise disjoint sets in $\mathcal{U}$, then

$$E_x\left(\bigcup_{n=1}^{\infty} M_n\right) = \left\langle E\left(\bigcup_{n=1}^{\infty} M_n\right)x, x\right\rangle = \left\langle \sum_{n=1}^{\infty} E(M_n)x, x\right\rangle = \sum_{n=1}^{\infty} E_x(M_n).$$

Conversely, let $(M_n)_{n\in\mathbb{N}}$ be a sequence of pairwise disjoint sets from $\mathcal{U}$, such that $M = \cup_{n=1}^{\infty} M_n \in \mathcal{U}$. Noting that as E_x is finitely additive, it follows that E is as well. From Lemma 10.3, it follows that $(E(M_n))_{n\in\mathbb{N}}$ is a sequence of pairwise disjoint orthogonal projections, and more than that, it follows that the sequence $(\sum_{k=1}^{n} E(M_k))x)_{n\in\mathbb{N}}$ converges, to any $x \in \mathcal{H}$, and then

$$\langle E(M)x, x\rangle = E_x(M) = \sum_{n=1}^{\infty} E_x(M_n) = \sum_{n=1}^{\infty} \langle E(M_n)x, x\rangle = \left\langle \sum_{n=1}^{\infty} E(M_n)x, x\right\rangle.$$

Thus, by Lemma 10.1, it follows that $E(M) = \sum_{n=1}^{\infty} E(M_n)$, which proves that E is a spectral measure. □

As in the above result, given a spectral measure E over the σ-algebra $\mathcal{U}$ and $x \in \mathcal{H}$, we can define a scalar measure $E_x : \mathcal{U} \to \mathbb{R}$ by

$$E_x(M) = \langle E(M)x, x\rangle = \langle E(M)^2 x, x\rangle = \langle E(M)x, E(M)x\rangle = ||E(M)x||^2, \ \ M \in \mathcal{U}.$$

Note that such measure is finite, since $E_x(\Omega) = ||E(\Omega)x||^2 = ||x||^2$. We can generalize a bit further and, for $x, y \in \mathcal{H}$, also define a complex measure $E_{x,y}$ over $\mathcal{U}$ given by $E_{x,y}(M) = \langle E(M)x, y\rangle$. It is possible, through direct calculation, to verify that the measure $E_{x,y}$ satisfies a type of polarization identity, namely

$$E_{x,y} = \frac{1}{4}(E_{x+y} - E_{x-y} + iE_{x+iy} - iE_{x-iy}).$$

Moreover, with an entirely analogous proof to the scalar case (and therefore we omit such proof, but it can be found in [1], Lemma 3.4, page 21), it is possible to show the following result.

Lemma 10.5 *Let E be a spectral measure in $\mathcal{U}$. Suppose that $(M_n)_{n\in\mathbb{N}}$ is a decreasing sequence and $(N_n)_{n\in\mathbb{N}}$ is an increasing sequence in $\mathcal{U}$, such that $M = \cap_n M_n$ and $N = \cup_n N_n$ are in $\mathcal{U}$. Then*

$$E(M)x = \lim_{n\to\infty} E(M_n)x, \ \ E(N)x = \lim_{n\to\infty} E(N_n)x, \ \ \forall x \in \mathcal{H}.$$

Let us recall that the total variation $|\mu|$ of a complex measure μ over $\mathcal{U}$ is a positive measure such that $|\mu|(M)$ is the supremum of the sums $\sum_{k=1}^{n} |\mu(M_k)|$ taken over all disjoint unions $M = \cup_{k=1}^{n} M_k$, with $M_k \in \mathcal{U}$. From this definition, it immediately follows that $|\mu(M)| \le |\mu|(M)$. We can then state the following lemma, which makes use of the total variation.

Lemma 10.6 *Let E be a spectral measure over $(\Omega, \mathcal{U})$, in a Hilbert space $\mathcal{H}$. The following hold:*

(i) $|E_{x,y}|(M) \leq E_x(M)^{1/2}E_y(M)^{1/2}$, *for* $x, y \in \mathcal{H}$ *and* $M \in \mathcal{U}$.
(ii) *If* $f \in L^2(\Omega, E_x)$, $g \in L^2(\Omega, E_y)$, *then*

$$\left|\int_\Omega fg dE_{x,y}\right| \leq \int_\Omega |fg| d|E_{x,y}| \leq ||f||_{L^2(\Omega, E_x)}||g||_{L^2(\Omega, E_y)}. \tag{10.10}$$

Proof

(i) Let $M = \cup_{k=1}^n M_n \in \mathcal{U}$, with $M_k \in \mathcal{U}$, for $k = 1, \ldots, n$ a disjoint union of M. Note that, as $E(M)$ is an orthogonal projection, we obtain

$$\begin{aligned}
|E_{x,y}(M_k)| &= |\langle E(M_k)x, y\rangle| \\
&= |\langle E(M_k)x, E(M_k)y\rangle| \\
&\leq ||E(M_k)x||||E(M_k)y|| \\
&= E_x(M_k)^{1/2}E_y(M_k)^{1/2}.
\end{aligned}$$

We can use this estimate to obtain

$$\begin{aligned}
\sum_{k=1}^n |E_{x,y}(M_k)| &\leq \sum_{k=1}^n E_x(M_k)^{1/2}E_y(M_k)^{1/2} \\
&\leq \left(\sum_{k=1}^n E_x(M_k)\right)^{1/2}\left(\sum_{k=1}^n E_y(M_k)\right)^{1/2} \\
&= E_x(M)^{1/2}E_y(M)^{1/2}.
\end{aligned}$$

Taking the supremum over all disjoint unions of M, the statement follows.
(ii) Let $f = \sum_{k=1}^n a_k\chi_{M_k}$ and $g = \sum_{k=1}^n b_k\chi_{M_k}$ be simple functions, with $M_k \in \mathcal{U}$. Then, $fg = \sum_{k=1}^n a_k b_k \chi_{M_k}$ and

$$\begin{aligned}
\left|\sum_{k=1}^n a_k b_k E_{x,y}(M_k)\right| &\leq \sum_{k=1}^n |a_k b_k||E_{x,y}|(M) \\
&\leq \left(\sum_{k=1}^n |a_k|^2 E_x(M_k)\right)^{1/2}\left(\sum_{k=1}^n |b_k|^2 E_y(M_k)\right)^{1/2},
\end{aligned}$$

which is exactly the inequality from item (ii) for simple functions. From the density of simple functions in $L^2(\Omega, E_x)$, the result follows immediately.

□

Let us denote by $\mathcal{B} = \mathcal{B}(\Omega, \mathcal{U})$ the Banach space of $\mathcal{U}$-measurable and bounded complex valued functions defined on Ω with the norm $||f||_\Omega = \sup\{|f(t)| : t \in \Omega\}$. In this line of thought, $\mathcal{B}_s$ will denote the subset of $\mathcal{B}$ of simple functions (i.e., functions that assume a finite number of values). We know that each simple function can be written in the form

$$f = \sum_{k=1}^{n} c_k \chi_{M_k}, \tag{10.11}$$

where $c_1, \ldots, c_n \in \mathbb{C}$ and $M_1, \ldots, M_n$ are disjoint sets from $\mathcal{U}$.

Definition 10.2 Let f be a function in $\mathcal{B}_s$, i.e., a function that can be written as (10.11). We define the spectral integral of f in the spectral measure E as the operator

$$\mathbb{I}(f) = \int_\Omega f dE = \sum_{k=1}^{n} c_k E(M_k). \tag{10.12}$$

From the additivity of E, it follows that $\mathbb{I}$ does not depend on the representation of f.

We will now seek to extend the previous definition to functions in $\mathcal{B}$. Let us begin this endeavor with the following result.

Lemma 10.7 *For $f \in \mathcal{B}_s$, it holds that $||\mathbb{I}(f)|| \leq ||f||_\Omega$.*

Proof Let f be a simple function, as in Eq. (10.11), maintaining the notion that the sets $M_1, \ldots, M_n$ are pairwise disjoint. Therefore, $E(M_k)\mathcal{H}$ and $E(M_l)\mathcal{H}$ are orthogonal sets if $l \neq k$, from which we can derive

$$\begin{aligned}||\mathbb{I}(f)x||^2 &= \left|\left|\sum_{k=1}^{n} c_k E(M_k)x\right|\right|^2 = \sum_{k=1}^{n} |c_k|^2 ||E(M_k)x||^2 \leq \sum_{k=1}^{n} ||f||_\Omega^2 ||E(M_k)x||^2 \\ &= ||f||_\Omega^2 \left|\left|\sum_{k=1}^{n} E(M_k)x\right|\right|^2 \leq ||f||_\Omega^2 ||x||^2.\end{aligned}$$

□

We will now construct the spectral integral for functions in $\mathcal{B}$. Given $f \in \mathcal{B}$, from [16], Theorem 8.8, page 61, $\mathcal{B}_s$ is dense in $\mathcal{B}$ (in the norm $||\cdot||_\Omega$), and it follows that there exists a sequence $(f_n)_{n\in\mathbb{N}} \subset \mathcal{B}_s$ that converges to f. Therefore, $(f_n)_{n\in\mathbb{N}}$ is a Cauchy sequence (in the supremum norm), from which it follows from Lemma 10.7 that $(\mathbb{I}(f_n))_{n\in\mathbb{N}}$ is a Cauchy sequence in the operator norm. Thus, there exists a bounded operator $\mathbb{I}(f) : \mathcal{H} \to \mathcal{H}$ that is the limit of the constructed sequence of operators.

Proposition 10.1 *For $f, g \in \mathcal{B}(\Omega, \mathcal{U})$, $\alpha, \beta \in \mathbb{C}$, and $x, y \in \mathcal{H}$, we have:*

(i) $\mathbb{I}(\overline{f}) = \mathbb{I}(f)^*$, $\mathbb{I}(\alpha f + \beta g) = \alpha\mathbb{I}(f) + \beta\mathbb{I}(g)$, *and* $\mathbb{I}(fg) = \mathbb{I}(f)\mathbb{I}(g)$.
(ii) $\langle \mathbb{I}(f)x, y\rangle = \int_\Omega f dE_{x,y}$.
(iii) $||\mathbb{I}(f)x||^2 = \int_\Omega |f|^2 dE_x$.
(iv) $||\mathbb{I}(f)|| \le ||f||_\Omega$.
(v) *Let $(f_n)_{n\in\mathbb{N}} \subset \mathcal{B}$ be a sequence converging pointwise to f E-a.e. in Ω and a constant c such that $|f_n(t)| \le c$ for $n \in \mathbb{N}$ and $t \in \Omega$. Then $\mathbb{I}(f_n)x \to \mathbb{I}(f)x$ for every $x \in \mathcal{H}$.*

Proof (i) Let $f = \sum_k c_k \chi_{M_k} \in \mathcal{B}_s$, with $M_1, \ldots, M_n$ disjoint. Then $\overline{f} = \sum_k \overline{c_k}\chi_{M_k}$ and

$$\mathbb{I}(\overline{f}) = \sum_k \overline{c_k} E(M_k) = \mathbb{I}(f)^*.$$

Now, if $f \in \mathcal{B}$, then taking a sequence $(f_n)_{n\in\mathbb{N}} \subset \mathcal{B}_s$ converging to f, it follows that $(\overline{f_n})_{n\in\mathbb{N}}$ converges to $\overline{f}$. Thus, $\mathbb{I}(f_n)^* = \mathbb{I}(\overline{f_n}) \to \mathbb{I}(\overline{f})$. Let us finally show that $\mathbb{I}(f_n)^*$ converges to $\mathbb{I}(f)^*$. For this, take any $x, y \in \mathcal{H}$

$$\langle \mathbb{I}(f)^* x, y\rangle = \langle x, \mathbb{I}(f)y\rangle = \langle x, \lim_{n\to\infty} \mathbb{I}(f_n)y\rangle = \langle \lim_{n\to\infty} \mathbb{I}(f_n)^* x, y\rangle,$$

from which it follows that $\mathbb{I}(f_n)^*$ converges strongly to $\mathbb{I}(f)^*$, but as such sequence of operators already converges in the operator norm topology to $\mathbb{I}(\overline{f})$, it follows that $\mathbb{I}(\overline{f}) = \mathbb{I}(f)^*$.

The second equality of item (i) is immediate for $f, g \in \mathcal{B}_s$, and for $f, g \in \mathcal{B}$, it suffices to take sequences $(f_n)_{n\in\mathbb{N}}$ and $(g_n)_{n\in\mathbb{N}}$ in $\mathcal{B}_s$ such that $f_n \to f$ and $g_n \to g$, and apply the previous case.

For the equality $\mathbb{I}(fg) = \mathbb{I}(f)\mathbb{I}(g)$, let $f = \sum_k a_k \chi_{M_k}$ and $g = \sum_l b_l \chi_{N_l}$, where the sets M_k and N_l are pairwise disjoint. We can then write $fg = \sum_{k,l} a_k b_l \chi_{M_k \cap N_l}$. Thus

$$\mathbb{I}(fg) = \sum_{k,l} a_k b_l E(M_k \cap N_l) = \sum_{k,l} a_k b_l E(M_k)E(N_l)$$
$$= \left(\sum_k a_k E(M_k)\right)\left(\sum_l b_l E(N_l)\right) = \mathbb{I}(f)\mathbb{I}(g).$$

Now, for $f, g \in \mathcal{B}$, taking sequences (f_n) and $(g_n) \in \mathcal{B}_s$ such that f_n converges to f and g_n converges to g, it follows that $f_n g_n$ converges pointwise to fg, from which, in the operator norm topology, $\mathbb{I}(f_n g_n)$ converges to $\mathbb{I}(fg)$, but from the property proved for simple functions, it follows that $\mathbb{I}(f_n g_n)$ converges to $\mathbb{I}(f)\mathbb{I}(g)$, and the result is proved.

(ii) Let us initially prove the assertion in the case when $f \in \mathcal{B}_s$, and then use it to prove the general case.

Assume f is in $\mathcal{B}_s$, written as $f = \sum_k c_k \chi_{M_k}$, with the sets M_k pairwise disjoint. Then, for $x, y \in \mathcal{H}$,

$$\langle \mathbb{I}(f)x, y\rangle = \left\langle \sum_k c_k E(M_k)x, y \right\rangle = \sum_k c_k \langle E(M_k)x, y\rangle$$
$$= \sum_k c_k E_{x,y}(M_k) = \int_\Omega f dE_{x,y}.$$

Now, for $f \in \mathcal{B}$, from [16], Theorem 8.8, page 61, we know that we can construct an increasing sequence f_n of simple functions that converges to f, with $|f_n| \leq |f| \leq K$ (since f is bounded). Thus, from the Dominated Convergence Theorem

$$\langle \mathbb{I}(f)x, y\rangle = \lim_{n\to\infty} \langle \mathbb{I}(f_n)x, y\rangle = \lim_{n\to\infty} \int_\Omega f_n dE_{x,y} = \int_\Omega f dE_{x,y}.$$

Item (iii) follows from a calculation entirely analogous to the calculation developed in item (ii). Item (iv) follows from Lemma 10.7, taking a sequence of simple functions in $\mathcal{B}_s$ converging to f and using the continuity of the operator norm.

(v) Given $x \in \mathcal{H}$, the constant function $g(t) = c$ is in $L^2(\Omega, E_x)$, since the measure E_x is finite. Thus, from the Dominated Convergence Theorem

$$\lim_{n\to\infty} ||\mathbb{I}(f_n)x - \mathbb{I}(f)x||^2 = \lim_{n\to\infty} ||\mathbb{I}(f_n - f)x||^2 = \lim_{n\to\infty} \int_\Omega |f_n - f|^2 dE_x$$
$$= \int_\Omega \lim_{n\to\infty} |f_n - f|^2 dE_x = 0$$

from which $\mathbb{I}(f_n)x \to \mathbb{I}(f)x$. □

In what follows, we aim to extend the definition of spectral integral to a larger set of functions. Let $\mathcal{S} = \mathcal{S}(\Omega, \mathcal{U}, E)$ be the set of $\mathcal{U}$-measurable functions defined on Ω such that $E(\{t \in \Omega : f(t) = \infty\}) = 0$.

Definition 10.3 A sequence $(M_n)_{n\in\mathbb{N}}$ of sets in $\mathcal{U}$ is called a limiting sequence of $\mathcal{F} \subset \mathcal{S}$ if every function $f \in \mathcal{F}$ is bounded on M_n, $M_n \subset M_{n+1}$, and $E(\cup_n M_n) = I$.

Note that $E(M_n) \subset E(M_{n+1})$, $\lim_{n\to\infty} E(M_n)x = x$ for every $x \in \mathcal{H}$. In other words, given $x \in \mathcal{H}$, we construct a sequence $(x_n)_{n\in\mathbb{N}} \subset \cup_n E(M_n)\mathcal{H}$ converging to x, which means that $\cup_n E(M_n)\mathcal{H}$ is dense in $\mathcal{H}$. Another useful remark is that, given a finite set $\{f_1, \ldots, f_r\} \subset \mathcal{S}$, there always exists a limiting sequence, given by the sequence of sets of the form $M_n = \{t \in \Omega; |f_j(t)| \leq n,\ j = 1, \ldots, r\}$.

Theorem 10.1 *Assume that $f \in \mathcal{S}$. Define*

$$D(\mathbb{I}(f)) = \left\{ x \in \mathcal{H} : \int_\Omega |f(t)|^2 dE_x < \infty \right\}, \tag{10.13}$$

and let $(M_n)_{n\in\mathbb{N}}$ be a limiting sequence for f. We have:

(i) $x \in \mathcal{H}$ belongs to $D(\mathbb{I}(f))$ if and only if $(\mathbb{I}(f\chi_{M_n}x))_{n\in\mathbb{N}}$ converges in $\mathcal{H}$, or equivalently, $\sup_{n\in\mathbb{N}} ||\mathbb{I}(f\chi_{M_n})|| < \infty$.
(ii) The limit of $(\mathbb{I}(f\chi_{M_n})x)$ does not depend on the limiting sequence (M_n). Thus, there exists a linear operator $\mathbb{I}(f)$ defined on $D(\mathbb{I}(f))$ satisfying

$$\mathbb{I}(f)x = \lim_{n\to\infty} \mathbb{I}(f\chi_{M_n})x, \quad x \in D(\mathbb{I}(f)). \tag{10.14}$$

(iii) $\cup_{n\in\mathbb{N}} E(M_n)\mathcal{H} \subset D(\mathbb{I}(f))$ and is dense in $\mathbb{I}(f)$. Furthermore, it holds that

$$E(M_n)\mathbb{I}(f) \subset \mathbb{I}(f)E(M_n) = \mathbb{I}(f\chi_{M_n}),\ n \in \mathbb{N}. \tag{10.15}$$

Proof (i) Take $x \in D(\mathbb{I}(f))$. Since f is bounded on each M_n, it follows that $f\chi_{M_n} \in \mathcal{B}$, where the operator $\mathbb{I}(f\chi_{M_n})$ is well defined. From Proposition 10.1 (iii), it follows that

$$\begin{aligned} ||\mathbb{I}(f\chi_{M_k})x - \mathbb{I}(f\chi_{M_n})x||^2 = ||\mathbb{I}(f\chi_{M_k} - f\chi_{M_n})x||^2 &= \int_\Omega |f\chi_{M_k} - f\chi_{M_n}|^2 dE_x \\ &= ||f\chi_{M_k} - f\chi_{M_n}||^2_{L^2(\Omega, E_x)}. \end{aligned} \tag{10.16}$$

Now, as $f \in L^2(\Omega, E_x)$ and from the fact that $f\chi_{M_n} \to f$ pointwise and $|f\chi_{M_n}| \leq |f|$, it follows that $(f\chi_{M_n})_{n\in\mathbb{N}}$ is a Cauchy sequence. From Eq. (10.16), it follows that $(\mathbb{I}(f\chi_{M_n})x)_{n\in\mathbb{N}}$ is also a Cauchy sequence in $\mathcal{H}$, and thus $(\mathbb{I}(f\chi_{M_n})x)_{n\in\mathbb{N}}$ is convergent.

Notice that, if $(\mathbb{I}(f\chi_{M_n}))_{n\in\mathbb{N}}$ converges, then $\{\|(\mathbb{I}(f\chi_{M_n}))x\|\}$ is bounded.

Conversely, suppose that

$$(\|(\mathbb{I}(f\chi_{M_n}))x\|)_{n\in\mathbb{N}}$$

is bounded, i.e., $c = \sup_n \|(\mathbb{I}(f\chi_{M_n}))x\| < \infty$. As $(|f\chi_{M_n}(t)|^2)_{n\in\mathbb{N}}$ monotonically converges to $|f(t)|^2$ E_x-a.e. in Ω, from the Monotone Convergence Theorem, we have

$$\int_\Omega |f|^2 dE_x = \lim_{n\to\infty} \int_\Omega |f\chi_{M_n}|^2 dE_x = \lim_{n\to\infty} \|\mathbb{I}(f\chi_{M_n})x\|^2 \leq c^2 < \infty.$$

We then obtain $f \in L^2(\Omega, E_x)$ and $x \in D(\mathbb{I}(f))$.

(ii) Let $(M_n')_{n\in\mathbb{N}}$ be another limiting sequence for f. From items (i) and (iii) of Proposition 10.1, we have

$$\begin{aligned}\|\mathbb{I}(f\chi_{M_n})x - \mathbb{I}(f\chi_{M'_k})x\| &= \|f\chi_{M_n} - f\chi_{M'_k}\|_{L^2(\Omega,E_x)}\\ &\leq \|f\chi_{M_n} - f\|_{L^2(\Omega,E_x)} + \|f - f\chi_{M'_k}\|_{L^2(\Omega,E_x)} \to 0\end{aligned}$$

when $n, k \to \infty$. Therefore, the limit does not depend on the limiting sequence.

(iii) Let $x \in \mathcal{H}$. As $E(M_k) = \mathbb{I}(\chi_{M_k})$, it follows that

$$\mathbb{I}(f\chi_{M_k})x = \mathbb{I}(f\chi_{M_n}\chi_{M_k})x = \mathbb{I}(f\chi_{M_n})E(M_k)x = E(M_k)\mathbb{I}(f\chi_{M_n})x \tag{10.17}$$

for $n \geq k$. Since $\sup_n\{\|\mathbb{I}(f\chi_{M_n}E)E(M_k)x\|\} < \infty$, it follows that $E(M_k)x \in D(\mathbb{I}(f))$, where $\cup_k E(M_k)\mathcal{H} \subset D(\mathbb{I}(f))$.

From Eq. (10.17), for $n \geq k$ and arbitrary $x \in \mathcal{H}$, we can write $\mathbb{I}(f\chi_{M_k})x = \mathbb{I}(f\chi_{M_n})E(M_k)x$, where, by letting $n \to \infty$, it follows that $\mathbb{I}(f)E(M_k)x = \mathbb{I}(f\chi_{M_k})x$.

Also from (10.17), when we restrict to $x \in D(\mathbb{I}(f))$, by letting $n \to \infty$ in $\mathbb{I}(f\chi_{M_n})E(M_k)x = E(M_k)\mathbb{I}(f\chi_{M_n})x$, we obtain $E(M_k)\mathbb{I}(f)x = \mathbb{I}(f)E(M_k)x$. This proves that $E(M_k)\mathbb{I}(f) \subset \mathbb{I}(f)E(M_k) = \mathbb{I}(f\chi_{M_k})$, for $k \in \mathbb{N}$.

Finally, as $E(M_n)x \to x$ and $\mathbb{I}(f)E(M_n)x = E(M_n)\mathbb{I}(f)x \to \mathbb{I}(f)x$, for $x \in D(\mathbb{I}(f))$, it follows that $\cup_{n=1} E(M_n)\mathcal{H}$ is dense in $\mathbb{I}(f)$. □

Proposition 10.2 *Let $f, g \in \mathcal{S}$, $x \in D(\mathbb{I}(f))$, and $y \in D(\mathbb{I}(g))$. Then the following hold*

$$\langle \mathbb{I}(f)x, \mathbb{I}(g)y\rangle = \int_\Omega fg dE_{x,y}, \tag{10.18}$$

and

$$||\mathbb{I}(f)x||^2 = \int_\Omega |f|^2 dE_x. \tag{10.19}$$

Proof Notice that Eq. (10.19) follows from Eq. (10.18) when $f = g$ and $x = y$. Let us then prove the first equation. From Proposition 10.1 for $fg\chi_{M_n}$, we have

$$\int_\Omega f\overline{g}\chi_{M_n} dE_{x,y} = \langle \mathbb{I}(f\overline{g}\chi_{M_n})x, y\rangle = \langle \mathbb{I}(f\chi_{M_n})x, \mathbb{I}(g\chi_{M_n})y\rangle.$$

Since $x \in D(\mathbb{I}(f))$ and $y \in D(\mathbb{I}(g))$, it follows that $f \in L^2(\Omega, E_x)$ and $g \in L^2(\Omega, E_y)$. From Lemma 10.6, the integral $\int_\Omega f\overline{g}dE_{x,y}$ exists, whence

$$\begin{aligned}\left|\int_\Omega f\overline{g}\chi_{M_n} dE_{x,y} - \int_\Omega f\overline{g} dE_{x,y}\right| &= \left|\int_\Omega (f\chi_{M_n} - f)\overline{g} dE_{x,y}\right|\\ &\leq ||f\chi_{M_n} - f||_{L^2(\Omega,E_x)} ||g||_{L^2(\Omega,E_y)} \to 0\end{aligned}$$

when $n \to \infty$. □

Let us now state some "operational" properties of spectral integrals for functions in $\mathcal{S}(\Omega, \mathcal{U}, E)$.

Proposition 10.3 *For $f, g \in \mathcal{S}(\Omega, \mathcal{U}, E)$ and $\alpha, \beta \in \mathbb{C}$, we have:*

(i) $\mathbb{I}(\overline{f}) = \mathbb{I}(f)^*$.
(ii) $\mathbb{I}(\alpha f + \beta g) = \overline{\alpha \mathbb{I}(f) + \beta \mathbb{I}(g)}$.
(iii) $\mathbb{I}(fg) = \overline{\mathbb{I}(f)\mathbb{I}(g)}$.
(iv) $\mathbb{I}(f)$ *is a normal and closed operator in* $\mathcal{H}$ *and* $\mathbb{I}(f)^*\mathbb{I}(f) = \mathbb{I}(\overline{f}f) = \mathbb{I}(f)\mathbb{I}(f)^*$.
(v) $D(\mathbb{I}(f)\mathbb{I}(g)) = D(\mathbb{I}(g)) \cap D(\mathbb{I}(fg))$.

Proof To prove the statements, we will consider a limiting sequence $(M_n)_{n\in\mathbb{N}}$ for the functions f, g, and abbreviate $E_n = E(M_n)$, $h_n = h\chi_{M_n}$, $D_0 = \cup_n E_n\mathcal{H}$ and remember that $E_n x \to x$, for every $x \in \mathcal{H}$. Let us also note that the limiting sequence taken is also a limiting sequence for $f + g, fg, \overline{f}$.

(i) Let $x \in D(\mathbb{I}(f))$ and $y \in D(\mathbb{I}(\overline{f}))$. Using Eq. (10.15) twice, and also item (i) of Proposition 10.1 for $f_n = f\chi_{M_n} \in \mathcal{B}(\Omega, \mathcal{U})$, we obtain

$$\langle E_n\mathbb{I}(f)x, y\rangle = \langle \mathbb{I}(f_n)x, y\rangle = \langle x, \mathbb{I}(\overline{f_n})y\rangle = \langle x, E_n\mathbb{I}(\overline{f})y\rangle.$$

Hence, taking the limit $n \to \infty$, we obtain $\langle \mathbb{I}(f)x, y\rangle = \langle x, \mathbb{I}(\overline{f})y\rangle$, that is, $\mathbb{I}(\overline{f}) \subset \mathbb{I}(f)^*$. It remains to show that the domain $D(\mathbb{I}(f)^*) \subset D(\mathbb{I}(\overline{f}))$. Thus, taking $y \in D(\mathbb{I}(f)^*)$, and again from Eq. (10.15) and also item (i) of Proposition 10.1, we obtain

$$\langle x, E_n\mathbb{I}(f)^*y\rangle = \langle \mathbb{I}(f)E_n x, y\rangle = \langle \mathbb{I}(f_n)x, y\rangle = \langle x, \mathbb{I}(\overline{f_n})y\rangle, \quad \forall x \in \mathcal{H},$$

that is, $E_n\mathbb{I}(f)^*y = \mathbb{I}(\overline{f_n})y$, whence $\sup_n ||\mathbb{I}(\overline{f_n})y|| \leq ||\mathbb{I}(f)^*y||$. Hence, from Theorem 10.1, we obtain $y \in D(\mathbb{I}(\overline{f}))$, as desired.

It follows immediately from (i) that $\mathbb{I}(f) = \mathbb{I}(\overline{\overline{f}}) = \mathbb{I}(\overline{f})^*$, and, as the adjoint operator is closed, it follows that $\mathbb{I}(f)$ is always closed.

(ii) We will prove the case $\mathbb{I}(f + g) = \overline{(\mathbb{I}(f) + \mathbb{I}(g))}$. Initially, we need to show that it makes sense to write $\overline{(\mathbb{I}(f) + \mathbb{I}(g))}$, that is, that the operator $\mathbb{I}(f) + \mathbb{I}(g)$ is closable. Noting that, for two operators A, B we have $A^* + B^* \subset (A + B)^*$, then

$$D_0 \subset D(\mathbb{I}(\overline{f}) + \mathbb{I}(\overline{g})) = D(\mathbb{I}(f)^* + \mathbb{I}(g)^*) \subset D((\mathbb{I}(f) + \mathbb{I}(g))^*),$$

and thus, as D_0 is dense, and using the fact that an operator T is closable if and only if $D(T)$ is dense, it follows that $\mathbb{I}(f) + \mathbb{I}(g)$ is closable.

As $(M_n)_{n\in\mathbb{N}}$ is a limiting sequence of $f, g, f + g$, we can write, again using (10.15) and also item (i) of Proposition 10.1,

$$(\mathbb{I}(f) + \mathbb{I}(g))E_n = \mathbb{I}(f_n) + \mathbb{I}(g_n) = \mathbb{I}(f_n + g_n) = \mathbb{I}(f + g)E_n, \tag{10.20}$$

and also

$$E_n(\mathbb{I}(f)+\mathbb{I}(g)) = E_n\mathbb{I}(f)+E_n\mathbb{I}(g) \subset (\mathbb{I}(f)+\mathbb{I}(g))E_n = \mathbb{I}(f+g)E_n. \quad (10.21)$$

Using the fact that D_0 is a core for $\mathbb{I}(f+g)$, given $x \in D(\mathbb{I}(f+g))$, there exists a sequence $(x_n)_{n\in\mathbb{N}} \subset D_0$ such that $x_n \to x$ and $\mathbb{I}(f+g)x_n \to \mathbb{I}(f+g)x$. Therefore, taking $n \to \infty$ in (10.20), it follows from the fact that $\mathbb{I}(f+g)$ is closed that $\mathbb{I}(f+g) \subset \overline{\mathbb{I}(f)+\mathbb{I}(g)}$.

On other hand, as $\mathbb{I}(f+g)E_n$ is a bounded operator, it is also closed, whence, from (10.21), it follows that $E_n(\overline{\mathbb{I}(f)+\mathbb{I}(g)}) \subset \mathbb{I}(f+g)E_n$. Moreover, taking $x \in D(\overline{\mathbb{I}(f)+\mathbb{I}(g)})$, as $E_n z \to z$ for every $z \in \mathcal{H}$, the limit $n \to \infty$ of (10.21) implies $\overline{\mathbb{I}(f)+\mathbb{I}(g)} \subset \mathbb{I}(f+g)$. Combining the argument from the previous paragraph, it follows that $\mathbb{I}(f+g) = \overline{\mathbb{I}(f)+\mathbb{I}(g)}$, as desired.

(iii) As $(M_n)_{n\in\mathbb{N}}$ is a limiting sequence of $\overline{f}$, it follows that D_0 is a core for $\mathbb{I}(\overline{f})$, whence D_0 is contained in the domain of $\mathbb{I}(\overline{g})\mathbb{I}(\overline{f})$. Thus, from item (i), as

$$\mathbb{I}(\overline{g})\mathbb{I}(\overline{f}) = \mathbb{I}(g)^*\mathbb{I}(f)^* \subset (\mathbb{I}(f)\mathbb{I}(g))^*,$$

that is, $D_0 \subset D((\mathbb{I}(f)\mathbb{I}(g))^*)$, which implies that $D((\mathbb{I}(f)\mathbb{I}(g))^*)$ is dense in $\mathcal{H}$ and again using the fact that an operator T is closable if and only if $D(T)$ is dense, we conclude that $\mathbb{I}(f)\mathbb{I}(g)$ is closable. Note that

$$\mathbb{I}(fg)E_n = \mathbb{I}((fg)_n) = \mathbb{I}(f_n)\mathbb{I}(g_n) = \mathbb{I}(f)E_n\mathbb{I}(g_n) = \mathbb{I}(f)\mathbb{I}(g_n)E_n = \mathbb{I}(f)\mathbb{I}(g)E_n.$$

And so, taking the limit $n \to \infty$, and using the fact that D_0 is a core for $\mathbb{I}(fg)$, it follows that $\mathbb{I}(fg) \subset \overline{\mathbb{I}(f)\mathbb{I}(g)}$. Moreover, we can also write that

$$E_n\mathbb{I}(f)\mathbb{I}(g) \subset \mathbb{I}(f)\mathbb{I}(g)E_n = \mathbb{I}(fg)E_n,$$

whence, as $\mathbb{I}(fg)E_n$ is bounded, we obtain $E_n\overline{\mathbb{I}(f)\mathbb{I}(g)} \subset \mathbb{I}(fg)E_n$, and taking the limit $n \to \infty$, it finally follows that $\overline{\mathbb{I}(f)\mathbb{I}(g)} \subset \mathbb{I}(fg)$, which implies that $\mathbb{I}(fg) = \overline{\mathbb{I}(f)\mathbb{I}(g)}$.

(iv) From items (i) and (iii), we have $\mathbb{I}(f)^*\mathbb{I}(f) = \mathbb{I}(\overline{f})\mathbb{I}(f) \subset \mathbb{I}(\overline{f}f)$. As $\mathbb{I}(f)$ is closed, it follows from [17], Proposition 3.18, that $\mathbb{I}(f)^*\mathbb{I}(f)$ is self-adjoint. From (i), $\mathbb{I}(\overline{f}f)$ is self-adjoint. It follows that $\mathbb{I}(f)^*\mathbb{I}(f) = \mathbb{I}(\overline{f}f)$. Similarly, it is possible to obtain that $\mathbb{I}(f)\mathbb{I}(f)^* = \mathbb{I}(\overline{f}f)$. Therefore, we obtain $\mathbb{I}(f)\mathbb{I}(f)^* = \mathbb{I}(\overline{f}f) = \mathbb{I}(f)^*\mathbb{I}(f)$, which concludes the proof of this item.

(iv) From item (iii), we have $\mathbb{I}(f)\mathbb{I}(g) \subset \mathbb{I}(fg)$, and it follows that $D(\mathbb{I}(f)\mathbb{I}(g)) \subset D(\mathbb{I}(fg) \cap D(\mathbb{I}(g))$.

Conversely, let $x \in D(\mathbb{I}(fg) \cap D(\mathbb{I}(g))$, where, as $\mathbb{I}(fg)E_nx = \mathbb{I}(f)\mathbb{I}(g)E_nx$, when we take the limit $n \to \infty$, we obtain $\mathbb{I}(fg)E_nx \to \mathbb{I}(fg)x$, and as the operator $\mathbb{I}(f)$ is closed, it follows that $\mathbb{I}(g)x \in D(\mathbb{I}(f))$. This means that $x \in D(\mathbb{I}(f)\mathbb{I}(g))$, which concludes the proof. □

Proposition 10.4 *Let $f, g \in S(\Omega, \mathcal{U}, E)$. The following hold:*

(i) If $f = g$ E-a.e. in Ω, then $\mathbb{I}(f) = \mathbb{I}(g)$.
(ii) If f is a real function E-a.e. in Ω, then $\mathbb{I}(f)$ is self-adjoint.

(*iii*) *If f is nonnegative E-a.e. in Ω, then $\mathbb{I}(f)$ is positive and self-adjoint.*
(*iv*) *If g is nonnegative, E-a.e. in Ω, then $\mathbb{I}(\sqrt{g})$ is a positive, self-adjoint operator and $\mathbb{I}(\sqrt{g})^2 = \mathbb{I}(g)$.*

Proof For (i) and (iii), it suffices to apply Proposition 10.2 for $f - g$.

For (ii), from Proposition 10.3, item (i), it follows that $\mathbb{I}(f) = \mathbb{I}(\overline{f}) = \mathbb{I}(f)^*$, that is, $\mathbb{I}(f)$ is self-adjoint.

For (iv), as g is nonnegative, it follows from item (iii) that $\mathbb{I}(\sqrt{g})$ is self-adjoint and positively defined. Now, using the fact that the square of a self-adjoint operator is self-adjoint, we have that $\mathbb{I}(\sqrt{g})^2$ is also self-adjoint. Thus, $\mathbb{I}(\sqrt{g})^2 = (\mathbb{I}(\sqrt{g})^2)^* = \overline{\mathbb{I}(\sqrt{g})^2}$, and therefore, from Proposition 10.3, item (iii), it follows that

$$\mathbb{I}(g) = \overline{\mathbb{I}(\sqrt{g})\mathbb{I}(\sqrt{g})} = \overline{\mathbb{I}(\sqrt{g})^2} = \mathbb{I}(\sqrt{g})^2,$$

as desired. □

Proposition 10.5 *The operator $\mathbb{I}(f)$ is invertible if and only if $f(t) \neq 0$ E-almost everywhere in Ω. In this case, it also holds that $\mathbb{I}(f)^{-1} = \mathbb{I}(f^{-1})$, where f^{-1} denotes the function $f^{-1} = 1/f(t)$ in $\mathcal{S}$.*

Proof Consider the set $N(f) = \{t \in \Omega : f(t) = 0\}$. Then $\mathbb{I}(f)E(N(f)) = \mathbb{I}(f\chi_{N(f)}) = \mathbb{I}(0) = 0$, whence the kernel of $\mathbb{I}(f) \neq \{0\}$, that is, $\mathbb{I}(f)$ is not invertible. Conversely, suppose $E(N(f)) = 0$. Then, $f^{-1} \in \mathcal{S}$, whence, from the fact that $f^{-1}f = 1$ E-a.e. in Ω, we have $D(\mathbb{I}(f^{-1}f)) = D(\mathbb{I}(1)) = \mathcal{H}$, and therefore, $D(\mathbb{I}(f^{-1})\mathbb{I}(f)) = D(\mathbb{I}(f))$. Moreover, $\mathbb{I}(f^{-1})\mathbb{I}(f) \subset \mathbb{I}(f^{-1}f) = I$. We then conclude that $\mathbb{I}(f)$ is invertible and $\mathbb{I}(f)^{-1} \subset \mathbb{I}(f^{-1})$.

Repeating the same arguments by exchanging f with f^{-1} in the previous paragraph, we obtain $\mathbb{I}(f^{-1})^{-1} \subset \mathbb{I}(f)$, which implies that $(\mathbb{I}(f^{-1})^{-1})^{-1} = \mathbb{I}(f^{-1}) \subset \mathbb{I}(f)^{-1}$. Therefore, $\mathbb{I}(f)^{-1} = \mathbb{I}(f^{-1})$. □

Proposition 10.6 *The operator $\mathbb{I}(f)$ is bounded if and only if $f \in L^\infty(\Omega, E)$. Furthermore, it holds that $||\mathbb{I}(f)|| = ||f||_{L^\infty(\Omega,E)}$.*

Proof If $f \in L^\infty(\Omega, E)$, then from Eq. (10.19), for an arbitrary $x \in \mathcal{H}$, we can write

$$||\mathbb{I}(f)x||^2 = \int_\Omega |f|^2 dE_x \leq ||f||^2_{L^\infty(\Omega,E)}||x||^2,$$

from which we deduce that $\mathbb{I}$ is bounded and has norm less than $||f||_{L^\infty(\Omega,E)}$.

Conversely, let $\mathbb{I}(f)$ be bounded and $M_n = \{t \in \Omega : |f(t)| \geq ||\mathbb{I}(f)|| + 2^{-n}\}$, for $n \in \mathbb{N}$, and $M = \cup_n M_n$, from which it follows that $M = \{t \in \Omega : |f(t)| > ||\mathbb{I}(f)||\}$. Note then, from the boundedness of $\mathbb{I}(f)$, that $||\mathbb{I}(f)E(M_n)x|| \leq ||\mathbb{I}(f)||\,||E(M_n)x||$, and we can write

$$\begin{aligned} ||\mathbb{I}(f)||^2||E(M_n)x||^2 &\geq ||\mathbb{I}(f)E(M_n)x||^2 = ||\mathbb{I}(f\chi_{M_n})x||^2 \\ &= \int_\Omega |f\chi_{M_n}|^2 dE_x = \int_{M_n} |f|^2 dE_x \\ &\geq (||\mathbb{I}(f)|| + 2^{-n})^2||E(M_n)x||^2. \end{aligned}$$

However, this is only possible if $E(M_n)x = 0$, and from the arbitrariness of x, it follows that $E(M_n) = 0$, for every $n \in \mathbb{N}$. We thus conclude that $E(M) = 0$, i.e., $|f(t)| \leq ||\mathbb{I}(f)||$, E-almost everywhere, from which $||f||_{L^\infty(\Omega,E)} \leq ||\mathbb{I}(f)||$, and the result is proved. □

We can also obtain an interesting result that characterizes the spaces generated by eigenvectors of the operator $\mathbb{I}(f)$ through spectral measures acting on sets related to the behavior of f.

Proposition 10.7 *Let $f \in S(\Omega, \mathcal{U}, E)$. $\lambda \in \mathbb{C}$ is an eigenvalue of $\mathbb{I}(f)$ if and only if $E(\{t \in \Omega : f(t) = \lambda\}) \neq 0$. Moreover, $E(\{t \in \Omega : f(t) = \lambda\})$ is the projection onto the space generated by the eigenvectors associated with the eigenvalue λ. Finally, we have*

$$\sigma(\mathbb{I}(f)) = \{\lambda \in \mathbb{C} : E(\{t \in \Omega : |f(t) - \lambda| < \epsilon\}) \neq 0, \forall \epsilon > 0\}. \tag{10.22}$$

Proof For simplicity, by changing f to $f - \lambda$, we can assume $\lambda = 0$. Given $x \in \mathcal{H}$, let $\mathcal{N}(f) = \{t \in \Omega : f(t) = 0\}$. From Proposition 10.2, we know that $\mathbb{I}(f) = 0$ if and only if f is null E_x-almost everywhere in Ω, which is equivalent to $\langle E(\Omega \setminus \mathcal{N}(f))x, x\rangle = 0$, which is also equivalent to $E(\Omega \setminus \mathcal{N}(f))x = x - E(\mathcal{N}(f))x = 0$. Therefore, $x \in \mathcal{N}(\mathbb{I}(f))$ if and only if $x = E(\mathcal{N}(f))x$. Thus, $E(\mathcal{N}(f))$ is the projection onto $\mathcal{N}(\mathbb{I}(f))$. In particular, $\mathcal{N}(\mathbb{I}(f)) \neq \{0\}$ if and only if $E(\mathcal{N}(f)) \neq 0$.

Now, to prove (10.22), note that, by definition, $0 \in \rho(\mathbb{I}(f))$ if and only if $\mathbb{I}(f)$ has a bounded inverse defined over $\mathcal{H}$. However, from Propositions 10.5 and 10.6, $0 \in \rho(\mathbb{I}(f))$ if and only if $f \neq 0$ E-almost everywhere and $f^{-1} \in L^\infty(\Omega, E)$, or, equivalently, there exists $c > 0$ such that $E(\{t \in \Omega : |f(t)| < c\}) = 0$. Therefore, $0 \in \sigma(\mathbb{I}(f))$ if and only if for every $\epsilon > 0$, $E(\{t \in \Omega : |f(t)| < \epsilon\}) \neq 0$. □

Proposition 10.8 *A bounded operator $T : \mathcal{H} \to \mathcal{H}$ commutes with a spectral measure E (that is, $TE(M) = E(M)T$ for every $M \in \mathcal{U}$) if and only if $T\mathbb{I}(f) \subseteq \mathbb{I}(f)T$ for every $f \in S(\Omega, \mathcal{U}, E)$.*

Proof Let T be a bounded operator that satisfies $T\mathbb{I}(f) \subseteq \mathbb{I}(f)T$ for all $f \in S(\Omega, \mathcal{U}, E)$. In particular, for $M \in \mathcal{U}$, we know that $\mathbb{I}(\chi_M) = E(M)$, where $TE(M) \subseteq E(M)T$, with $E(M)$ also being bounded. Since $TE(M)$ and $E(M)T$ are bounded operators, equality holds, i.e., T commutes with all projections $E(M)$.

Conversely, let T be a bounded operator that commutes with E (i.e., T commutes with $E(M)$ for every $M \in \mathcal{U}$). Given $x \in \mathcal{H}$, it follows that

$$\begin{aligned} E_{Tx}(M) &= \|E(M)Tx\|^2 = \|TE(M)x\|^2 \\ &\leq \|T\|E_x(M). \end{aligned}$$

Therefore, according to (10.13), $Tx \in D(\mathbb{I}(f))$ if $x \in D(\mathbb{I}(f))$. Since T commutes with E, then T commutes with $\mathbb{I}(g)$ for every simple function g. Moreover, for a function $h \in \mathcal{B}$, we can take a sequence of simple functions converging to h, where it is also possible to obtain the convergence of T with $\mathbb{I}(h)$. In particular, T commutes with all operators $\mathbb{I}(f\chi_M)$. Finally, as $T(D(\mathbb{I}(f))) \subseteq D(\mathbb{I}(f))$, it follows from item (ii) of Theorem 10.1 that $T\mathbb{I}(f) \subseteq \mathbb{I}(f)T$. □

To conclude this section, note that, given $(\Omega, \mathcal{U}, E)$ and an application $\varphi : \Omega \to \Omega_0$, we can construct a σ-algebra $\mathcal{U}_0 = \{M \subset \Omega_0 : \varphi^{-1}(M) \in \mathcal{U}\}$ and a measure $F(M) = E(\varphi^{-1}(M))$.

Proposition 10.9 *If $h \in \mathcal{S}(\Omega_0, \mathcal{U}_0, F)$, then $h \circ \varphi \in \mathcal{S}(\Omega, \mathcal{U}, E)$ and*

$$\int_{\Omega_0} h dF = \int_{\Omega} h \circ \varphi dE. \tag{10.23}$$

Proof The first statement is immediate. From the change of variables formula for scalar measures, it follows that

$$\int_{\Omega_0} |h|^2 dF_x = \int_{\Omega} |h \circ \varphi|^2 dE_x, \tag{10.24}$$

$$\int_{\Omega_0} h dF_x = \int_{\Omega} (h \circ \varphi) dE_x. \tag{10.25}$$

Note that the first equation shows us that $D(\mathbb{I}_F(h)) = D(\mathbb{I}_E(h \circ \varphi))$, where $\mathbb{I}_F$ and $\mathbb{I}_E$ denote the spectral integrals with respect to the measures F and E, respectively. From Proposition 10.2, it follows that $\langle \mathbb{I}_F(h)y, y\rangle = \langle \mathbb{I}_E(h \circ \varphi)y, y\rangle$, for all $y \in D(\mathbb{I}_F(h)) = D(\mathbb{I}_E(h \circ \varphi))$. Therefore, from the fact that, if $\langle z, w\rangle = 0$ for all w in a dense set implies that $z = 0$, it follows that $\mathbb{I}_F(h) = \mathbb{I}_E(h \circ \varphi)$. □

10.3 Spectral Theorem

In order to prove the Spectral Theorem for self-adjoint operators, we initially focus on a version for bounded operators. Thus, let us consider a bounded and self-adjoint operator A that gives us $\sigma(A) \subset J$, with J being a compact interval on the real line (see [10], Theorem 9.2.1). In what follows, for a function $f \in C(J)$, i.e., a continuous function on interval J, we consider the norm $||f||_J = \sup\{|f(t)|; t \in J\}$, and $\mathbf{C}[t]$ will be used to denote the set of polynomials of one variable over the field of complex numbers.

Lemma 10.8 *$\sigma(p(A)) \subset p(\sigma(A))$ for $p \in \mathbb{C}[t]$.*

Proof Let $\gamma \in \sigma(p(A))$ and $n = \deg(p) > 0$. From the Fundamental Theorem of Algebra, there exist complex numbers $\alpha_1, \ldots, \alpha_n$ such that $p(t) - \gamma = a_n(t - \alpha_1)\ldots(t - \alpha_n)$. Hence,

$$p(A) - \gamma I = a_n(A - \alpha_1 I)\ldots(A - \alpha_n I).$$

Suppose for contradiction that all $\alpha_i \in \rho(A)$. Then each term on the right-hand side of the above equation would have a bounded inverse, implying that $p(A) - \gamma I$ would also have a bounded inverse, leading to $\gamma \in \rho(p(A))$, a contradiction. Thus, we conclude that there exist some α_j in $\sigma(A)$. It follows that $p(\alpha_j) = \gamma$, hence $\gamma \in p(\sigma(A))$. □

Lemma 10.9 *For $p \in \mathbb{C}[t]$, we have $||p(A)|| \leq ||p||_J$.*

Proof

$$\begin{aligned}||p(A)||^2 &= ||p(A)^* p(A)|| = ||(\overline{p}p)(A)|| = \sup\{|\gamma|; \gamma \in \sigma((\overline{p}p)(A))\}\\ &\leq \sup\{|\overline{p}p(\lambda)|; \lambda \in \sigma(A)\} \leq ||\overline{p}p||_J = ||p||_J^2.\end{aligned}$$

□

Lemma 10.10 *Given a continuous linear functional F defined on $(\mathbb{C}[t], ||\cdot||_J)$, there exists a unique regular Borel complex measure μ on J such that*

$$F(p) = \int_J p\, d\mu, \quad p \in \mathbb{C}[t].$$

Furthermore, $|\mu(M)| \leq ||F||$ for every $M \in \mathcal{B}(J)$.

Proof From the density of polynomials in $(C(J), ||\cdot||_J)$, every continuous linear functional in $(\mathbb{C}[t], ||\cdot||_J)$ has a continuous extension to $C(J)$. The existence and uniqueness of the measure μ follows from Riesz Representation Theorem for measures (see, for instance, [15], Theorem 6.19). □

Theorem 10.2 (Spectral Theorem for Bounded Self-adjoint Operators) *Let A be a bounded self-adjoint operator on $\mathcal{H}$ and J a compact interval containing $\sigma(A)$. There exists a unique spectral measure E over the Borel algebra $\mathcal{B}(J)$ such that*

$$A = \int_J \lambda\, dE. \tag{10.26}$$

Proof Given $x, y \in \mathcal{H}$, define the linear functional $F_{x,y} : \mathbb{C}[t] \to \mathbb{C}$, such that $p \mapsto F_{x,y}(p) = \langle p(A)x, y\rangle$. Since

$$|F_{x,y}(p)| \leq ||p(A)||\, ||x||\, ||y|| \leq ||p||_I\, ||x||\, ||y||,$$

$F_{x,y}$ is continuous in $(\mathbb{C}[t], ||\cdot||_J)$ and $||F_{x,y}|| \leq ||x||\, ||y||$. From Lemma 10.10, there exists a unique regular Borel complex measure $\mu_{x,y}$ on $\mathcal{B}(J)$ such that

$$F_{x,y}(p) = \langle p(A)x, y\rangle = \int_J p\, d\mu_{x,y}, \quad p \in \mathbb{C}[t]. \tag{10.27}$$

Our goal is to show the existence of a spectral measure E on the Borel σ-algebra $\mathcal{B}(J)$ such that $\mu_{x,y}(M) = \langle E(M)x, y\rangle$, for $x, y \in \mathcal{H}$ and $M \in \mathcal{B}(J)$.

For $\alpha_1, \alpha_2 \in \mathbb{C}$ and $x_1, x_2 \in \mathcal{H}$, we can write

$$\int_J p\, d\mu_{\alpha_1 x_1+\alpha_2 x_2, y} = \alpha_1 \int_J p\, d\mu_{x_1,y} + \alpha_2 \int_J p\, d\mu_{x_2,y}$$
$$= \int_J p\, d(\alpha_1 \mu_{x_1,y} + \alpha_2 \mu_{x_2,y}), \ \ p \in \mathbb{C}[t].$$

From the uniqueness of the measure associated with the functional $F_{\alpha_1 x_1+\alpha_2 x_2, y}$, it follows that

$$\mu_{\alpha_1 x_1+\alpha_2 x_2, y}(M) = \alpha_1 \mu_{x_1,y}(M) + \alpha_2 \mu_{x_2,y}(M)$$

for $M \in \mathcal{B}(J)$. Similarly, for $y_1, y_2 \in \mathcal{H}$, we obtain

$$\mu_{x,\alpha_1 y_1+\alpha_2 y_2}(M) = \overline{\alpha_1}\mu_{x,y_1}(M) + \overline{\alpha_2}\mu_{x,y_2}(M).$$

Now, from the bound given by Lemma 10.10, we have $|\mu_{x,y}(M)| \le ||F_{x,y}|| \le ||x||\,||y||$. Thus, given $M \in \mathcal{B}(J)$, the application $(x, y) \mapsto \mu_{x,y}(M)$ is a continuous sesquilinear form in $\mathcal{H}$. Hence, there exists a bounded linear operator $E(M)$ defined on $\mathcal{H}$ such that

$$\mu_{x,y}(M) = \langle E(M)x, y\rangle. \tag{10.28}$$

Setting $p = 1$ in Eq. (10.27), we obtain $\langle x, y\rangle = \mu_{x,y}(J) = \langle E(J)x, y\rangle$, for all $x, y \in \mathcal{H}$, i.e., $E(J) = I$. From Lemma 10.4, it follows that we need only to show that $E(M)$ is a projection. For this, given $x, y \in \mathcal{H}$ and $p \in \mathbb{C}[t]$, we have

$$\int_J p\, d\mu_{x,y} = \langle p(A)x, y\rangle = \overline{\langle \overline{p}(A)y, x\rangle} = \overline{\int_J \overline{p}\, d\mu_{y,x}} = \int_J p\, d\overline{\mu}_{x,y},$$

hence we conclude that $\mu_{x,y}(M) = \overline{\mu_{y,x}(M)}$, for $M \in \mathcal{B}(J)$, and, from this, it follows that $E(M) = E(M)^*$. To prove that $E(M)^2 = E(M)$, initially observe that $(pq)(A) = p(A)q(A)$, for $p, q \in \mathbb{C}[t]$, and hence

$$\int_J p\, d\mu_{q(A)x,y} = \langle p(A)q(A)x, y\rangle = \langle (pq)(A)x, y\rangle = \int_J pq\, d\mu_{x,y}.$$

From the uniqueness given by Lemma 10.10, it follows that $d\mu_{q(A)x,y} = d\mu_{x,y}$, hence $\langle E(M)q(A)x, y\rangle = \int_M d\mu_{x,y}$, for $M \in \mathcal{B}(J)$. As $E(M)^* = E(M)$, it follows that

$$\int_J q\, d\mu_{x,E(M)y} = \langle q(A)x, E(M)y\rangle = \langle E(M)q(A)x, y\rangle = \int_J q\chi_M\, d\mu_{x,y},$$

and again from the uniqueness given by Lemma 10.10, we have $d\mu_{x,E(M)y} = \chi_M\, d\mu_{x,y}$, i.e., $\mu_{x,E(M)y}(N) = \int_N \chi_M\, d\mu_{x,y} = \mu_{x,y}(M \cap N)$, for $N \in \mathcal{B}(J)$. Therefore, $\langle E(N)x, E(M)y\rangle = \langle E(M \cap N)x, y\rangle$. Thus, it follows that $E(M \cap N) = E(M)E(N)$. Setting $N = M$, it follows that $E(M)^2 = E(M)$, as desired. Finally, given $x, y \in \mathcal{H}$ and $p \in \mathbb{C}[t]$, we have

$$\langle p(A)x, y\rangle = \int_J p\, dE_{x,y}, \tag{10.29}$$

and in particular, for $p(\lambda) = \lambda$, it follows that $\langle Ax, y\rangle = \int_J \lambda\, dE_{x,y}$, and hence the equation (10.26) is derived. □

Corollary 10.1 *Let A be a positive self-adjoint and bounded operator on $\mathcal{H}$. Then there exists a positive and self-adjoint operator B defined on $\mathcal{H}$, denoted by $A^{1/2}$, such that $B^2 = A$. Moreover, if $(p_n)_{n\in\mathbb{N}}$ is a sequence of polynomials such that $p_n \to \lambda^{1/2}$ uniformly on $[0, b]$, with $\sigma(A) \subset [0, b]$, then* $\lim_{n\to\infty} ||p_n(A) - A^{1/2}|| = 0$.

Proof Since A is positive and bounded, it follows that $\sigma(A) \subset [0, b]$ for some $b \in \mathbb{R}$. From the Spectral Theorem 10.2, there exists a spectral measure E such that $A = \int_{[0,b]} \lambda\, dE$. Setting $B = \int_{[0,b]} \lambda^{1/2}\, dE$, it follows that B possesses the properties stated in the present corollary. Moreover, from Proposition 10.1, item (iv), it immediately follows that $||p_n(A) - B|| \le ||p_n - \lambda^{1/2}||_{[0,b]} \to 0$ as $n \to \infty$. □

We now aim to extend the result of the Spectral Theorem to unbounded self-adjoint operators. Consider T a densely defined and closed operator on $\mathcal{H}$. From [17], Proposition 3.18, the operator $C_T = (I + T^*T)^{-1}$ is positive and bounded, from which, by Corollary 10.1, there exists an operator $C_T^{1/2}$. The operator $Z_T = TC_T^{1/2}$ is called the bounded transformation of T.

Lemma 10.11 *If T is a closed and densely defined operator on $\mathcal{H}$, then:*

(i) *Z_T is a bounded operator, defined on $\mathcal{H}$, satisfying*

$$||Z_T|| \le 1 \text{ and } C_T = (I + T^*T)^{-1} = I - Z_T^* Z_T. \tag{10.30}$$

(ii) *$(Z_T)^* = Z_{T^*}$. In particular, Z_T is self-adjoint if T is self-adjoint.*
(iii) *If T is normal, then Z_T is also normal.*

Proof We abbreviate $C = C_T$, $Z = Z_T$, $C_* = C_{T^*}$, and $Z_* = Z_{T^*}$:

(i) We have $C\mathcal{H} = D(I + T^*T) = D(T^*T)$. For any $x \in \mathcal{H}$,

$$\begin{aligned} ||TC^{1/2}C^{1/2}x||^2 = ||TCx||^2 = \langle TCx, TCx\rangle = \langle T^*TCx, Cx\rangle \\ \le \langle T^*TCx, Cx\rangle + \langle Cx, Cx\rangle \\ = \langle (I + T^*T)Cx, Cx\rangle = \langle x, Cx\rangle = ||C^{1/2}x||^2, \end{aligned}$$

implying $||Zy|| = ||TC^{1/2}y|| \leq ||y||$, for $y \in C^{1/2}\mathcal{H}$. Since $C^{1/2}\mathcal{H}$ is dense in $\mathcal{H}$, Z is defined on $\mathcal{H}$ with $||Z|| \leq 1$.

(ii) Recall that $C = (I + T^*T)^{-1}$ and $C_* = (I + TT^*)^{-1}$. For $x \in D(T^*)$, setting $y = C_*x$ gives $x = (I + T^*T)y$ and $T^*x = T^*(I + TT^*)y = (I + T^*T)T^*y$, and hence $C_*x \in D(T^*)$ and $CT^*x = T^*y = T^*C_*$. By induction, one can show that $p(C_*)x \in D(T^*)$ and $p(C)T^*x = T^*p(C_*)x$ for any polynomial p.

From the Weierstrass Theorem (see, for instance, [14], Theorem 7.26), there exists a sequence of polynomials $(p_n)_{n\in\mathbb{N}}$ such that $p_n(t) \to t^{1/2}$ uniformly on [0, 1]. Then

$$\lim_{n\to\infty} ||p_n(C) - C^{1/2}|| = \lim_{n\to\infty} ||p_n(C_*) - (C_*)^{1/2}|| = 0.$$

Taking the limit yields $p(C)T^*x = T^*p_n(C_*)x$, and as T^* is closed, $C^{1/2}T^*x = T^*(C_*)^{1/2}x$ for $x \in D(T^*)$. Since $C^{1/2}T^* \subset (TC^{1/2})^* = Z^*$, we have $Z^*x = C^{1/2}T^*x = T^*(C_*)^{1/2}x = Z_*x$ for $x \in D(T^*)$. As $D(T^*)$ is dense in $\mathcal{H}$, $Z^* = Z_*$ in $\mathcal{H}$.

(iii) Assuming T is normal, i.e., $T^*T = TT^*$, it follows that

$$I - Z^*Z = (I + T^*T)^{-1} = (I + TT^*)^{-1} = I - ZZ^*,$$

implying Z is normal.

□

Theorem 10.3 (Spectral Theorem for Unbounded Self-adjoint Operators) *Let A be a self-adjoint operator defined on the Hilbert space $\mathcal{H}$. Then there exists a unique spectral measure $E_A = E$ on the Borel σ-algebra $\mathcal{B}(\mathbb{R})$ such that*

$$A = \int_{\mathbb{R}} \lambda \, dE_A. \tag{10.31}$$

Proof Let us abbreviate $C = C_A$ and $Z = Z_A$. From Lemma 10.11, it follows that $Z = AC^{1/2}$ is bounded and self-adjoint with $||Z|| \leq 1$, so that, by [10], Theorem 9.2.1 and Theorem 9.2.2, we have $\sigma(Z) \subset [-1, 1]$. Then $I - Z^2 = C = (I + A^2)^{-1}$, and from the Spectral Theorem for bounded operators, there exists a unique spectral measure F on $\mathcal{B}([-1, 1])$ such that

$$Z = \int_{[-1,1]} z \, dF.$$

Since $F(\{1\}) = \mathcal{N}(I - Z)$ and $F(\{-1\}) = \mathcal{N}(I + Z)$, we have $F(\{1\})\mathcal{H} + F(\{-1\})\mathcal{H} = \mathcal{N}((I + A^2)^{-1}) = \{0\}$, and it follows that $F(\{1\}) = F(\{-1\}) = 0$. Therefore, $\varphi(z) = z/(1 - z^2)^{-1/2}$ is finite F-almost everywhere in $[-1, 1]$.

Let us prove that $A = \mathbb{I}(\varphi)$. Since $D(\mathbb{I}(\varphi)) = D(\mathbb{I}((1 - z^2)^{-1/2})$, we have $\mathbb{I}(\varphi) = \mathbb{I}(z)\mathbb{I}((1 - z^2)^{-1/2})$, from items (iii) and (v) of Theorem 10.1. From Proposition 10.5, it follows that $\mathbb{I}((1 - z^2)^{-1/2}) = \mathbb{I}((1 - z^2)^{1/2})^{-1}$. From the uniqueness of the square root, we also obtain $C^{1/2} = (I - Z^2)^{1/2} = \mathbb{I}((1 - z^2)^{1/2})$. Thus, we have proved so far that $\mathbb{I}(\varphi) = Z(C^{1/2})^{-1}$.

Since $Z = AC^{1/2}$ is defined on the whole $\mathcal{H}$, $C^{1/2}\mathcal{H} \subset D(A)$, and therefore, $\mathbb{I}(\varphi) = Z(C^{1/2})^{-1} \subset A$. As $\mathbb{I}(\varphi)$ and A are self-adjoint, $\mathbb{I}(\varphi) = A$. Applying Proposition 10.9 using the mapping φ, there exists a measure $E(M) = F(\varphi^{-1}(M))$, $M \in \mathcal{B}(\mathbb{R})$, and we obtain

$$\int_{\mathbb{R}} \lambda \, dE = \int_{[-1,1]} \varphi \, dF = \mathbb{I}(\varphi) = A. \tag{10.32}$$

□

Given a self-adjoint linear operator $A\colon D(A) \subset \mathcal{H} \to \mathcal{H}$, from the Spectral Theorem 10.3, there exists a spectral measure E_A defined on the Borel σ-algebra $\mathcal{B}(\mathbb{R})$, such that

$$A = \int_{\mathbb{R}} \lambda d E_A.$$

From this spectral measure E_A, and given a function $f \in S(\mathbb{R}, \mathcal{B}(\mathbb{R}), E_A)$, we can construct the operator $f(A) := \mathbb{I}(f)$ (in the measure E_A), as in Theorem 10.1, where

$$D(f(A)) = \left\{ x \in \mathcal{H} \colon \int_{\mathbb{R}} |f(t)|^2 dE_x < \infty \right\}.$$

Thus, the mapping

$$\begin{aligned} \varphi : S(\mathbb{R}, \mathcal{B}(\mathbb{R}), E_A) &\longrightarrow \mathcal{L}(\mathcal{H}) \\ f &\longmapsto f(A) \end{aligned} \tag{10.33}$$

with $\mathcal{L}(\mathcal{H})$ being the set of all linear operators $T\colon D(T) \subset \mathcal{H} \to \mathcal{H}$ is called the functional calculus associated with A. From the construction, it follows that the operators $f(A)$ possess the properties already obtained for operators in the form $\mathbb{I}(f)$.

Proposition 10.10 *Given $f, g \in S(\mathbb{R}, \mathcal{B}(\mathbb{R}), E_A)$, $\alpha, \beta \in \mathbb{C}$, and $x, y \in D(f(A))$, we have:*

(i) $\langle f(A)x, y\rangle = \int_{\mathbb{R}} f(\lambda) d\langle E_A(\lambda)x, y\rangle$.
(ii) $\|f(A)x\|^2 = \int_{\mathbb{R}} |f(\lambda)|^2 d\langle E_A(\lambda)x, x\rangle$. *In particular, if $f(t) = g(t)$, E_A-almost everywhere in $\mathbb{R}$, then $f(A) = g(A)$.*
(iii) *If $f \in L^\infty(\mathbb{R}^n, E_A)$, then $\|f\|_{L^\infty} = \|f(A)\|$.*
(iv) $\overline{f}(A) = f(A)^*$. *In particular, if $f(x) \in \mathbb{R}$ E_A-almost everywhere in $\mathbb{R}^n$, then $f(A)$ is self-adjoint.*

(v) $(\alpha f + \beta g)(A) = \overline{\alpha f(A) + \beta g(A)}$.
(vi) $\chi_M(A) = E_A(M)$ *for* $M \in \mathcal{B}(\mathbb{R})$.

Theorem 10.4 *Let A be a self-adjoint operator and* E_A *its spectral measure, and let* $M \in \mathcal{B}(\mathbb{R}^n)$ *be such that* $M \cap \sigma(A) = \emptyset$. *Then* $E_A(M) = 0$.

Proof Initially, observe that, from Proposition 10.7, specifically from Eq. (10.22)

$$\rho(A) = \{\lambda \in \mathbb{C} : \exists c > 0 \text{ such that } E(\{t \in \mathbb{R}, |t - \lambda| < c\}) = 0\}. \tag{10.34}$$

Thus, as $M \cap \sigma(A) = \emptyset$, it follows that $M \subset \rho(A)$. Hence, if $\lambda \in M$, then there exists $c_\lambda > 0$ such that $E_A(B(\lambda, c_\lambda)) = 0$. Since $M \subset \bigcup_{\lambda \in M} B(\lambda, c_\lambda)$, by the Lindelöf's theorem, we can extract an enumerable sequence $(\lambda_i)_{i \in \mathbb{N}}$, such that $M \subset \bigcup_{i \in \mathbb{N}} B(\lambda_i, c_{\lambda_i})$ and, therefore, $E_A(M) \subset E_A(\bigcup_{i \in \mathbb{N}} B(\lambda_i, c_{\lambda_i})) = 0$. □

10.4 Decomposition of $H^1(\mathbb{R}^n)$

In this section, following some ideas presented in [5], we aim to discuss a decomposition of $H^1(\mathbb{R}^n)$ that arises naturally from the spectral analysis of the so-called Schrödinger operators.

Let $V : \mathbb{R}^n \to \mathbb{R}$, $V \in L^\infty(\mathbb{R}^n) \cap C(\mathbb{R}^n, \mathbb{R})$, satisfying

$$\lim_{n \to \infty} V(x) = v_\infty > 0. \tag{10.35}$$

We define

$$\begin{aligned} S : H^2(\mathbb{R}^n) &\longrightarrow L^2(\mathbb{R}^n) \\ u &\longmapsto -\Delta u + Vu, \end{aligned} \tag{10.36}$$

as the Schrödinger operator. As the Laplacian operator in $H^2(\mathbb{R}^n)$ is self-adjoint (see for instance [3], Example 3.23, page 49), and $M_V : L^2(\mathbb{R}^n) \to L^2(\mathbb{R}^n)$, acting as $M_V u = V(x)u(x)$ is bounded and self-adjoint operator, it follows (from [3], Lemma 3.27, page 51) that the Schrödinger operator is self-adjoint.

It is well known that (see, for instance, [19], Theorem 3.15, page 44 and [8], Theorem 30, page 150), for V satisfying (10.35), we can characterize the spectrum of S by

$$\begin{aligned} &\sigma_{ess}(S) \subset [v_\infty, +\infty), \;\; \sigma(S) \cap (-\infty, v_\infty) = \sigma_d(S) \cap (-\infty, v_\infty) \;\; \text{and} \\ &\sigma(S) \subset [-M, +\infty), \end{aligned} \tag{10.37}$$

where $-M = \inf_{x \in \mathbb{R}} V(x)$. Let us also consider that S satisfies

$$\sup[\sigma(S) \cap (-\infty, 0)] = \sigma^- < 0 < \sigma^+ = \inf[\sigma(S) \cap (0, \infty)]. \tag{10.38}$$

By the Spectral Theorem 10.3 applied to the Schrödinger operator S, there exists a spectral measure E such that

$$S = \int_{\mathbb{R}} \lambda \, dE.$$

From functional calculus, given the function $f : [0, \infty) \to \mathbb{R}$, $f(x) = |x|^{1/2}$, we can construct the operator

$$A =: f(S) = |S|^{1/2} = \int_{\mathbb{R}} |\lambda|^{1/2} \, dE, \tag{10.39}$$

and, from Theorem 10.1, it has the domain set

$$D(A) = \left\{ u \in L^2(\mathbb{R}^n); \int_{\mathbb{R}} |\lambda| \, d\langle E(\lambda)u, u\rangle < \infty \right\}. \tag{10.40}$$

Furthermore, as S is self-adjoint, and f is real, it follows from Proposition 10.10, item (iv), that $f(S) = A$ is self-adjoint, and consequently, it is closed. From the fact that the operator A is closed, it is possible to show that the set $E := D(A)$ is a Hilbert space when seen as a vector space equipped with the inner product

$$\langle u, v\rangle_E = \langle u, v\rangle_{L^2} + \langle Au, Av\rangle_{L^2} \tag{10.41}$$

and norm

$$||u||_E = (||u||_{L^2}^2 + ||Au||_{L^2}^2)^{1/2} = \left(\int_{\mathbb{R}} (1 + |\lambda|) \, d\langle E(\lambda)u, u\rangle \right)^{1/2}. \tag{10.42}$$

We can decompose $\mathbb{R} = \mathbb{R}_- \cup \{0\} \cup \mathbb{R}_+$, where, denoting by $P_+ = \chi_{\mathbb{R}_+}(S)$, $P_- = \chi_{\mathbb{R}_-}(S)$, and $P_0 = \chi_0(S)$ and using Proposition 10.10, item (vi) on functional calculus, that $\chi_M = E(M)$ (recalling that $E(M)$ is an orthogonal projection on the set M), it follows that the projections P_+, P_-, and P_0 are orthogonal, and furthermore, as $\mathbb{R}_+$, $\mathbb{R}_-$, $\{0\}$ are pairwise disjoint sets, it follows from Lemma 10.9 that P_+, P_-, P_0 are mutually orthogonal. Also, notice that from the additivity of spectral measures, it follows that $I = E(\mathbb{R}_+ \cup \{0\} \cup \mathbb{R}_-) = P_+ + P_- + P_0$. Thus, we also have the natural decomposition of E given by

$$E = E_+ \oplus E_- \oplus E_0, \tag{10.43}$$

where $E_+ = P_+(E)$, $E_- = P_-(E)$, and $E_0 = P_0(E)$.

Lemma 10.12 *The spaces E_+, E_- and E_0 are invariant under A, i.e., $A(E_i) \subset E_i$ for $i \in \{+, 0, -\}$.*

Proof Initially note that, from Proposition 10.8, given $M \in \mathcal{B}(\mathbb{R})$, it follows that, as $E(M)$ commutes with E (in the sense of commuting with all projections $E(N)$, $N \in \mathcal{U}$), we have

$$P_i A = E(M)A \subset AE(M) = AP_i, \quad i \in \{+, -, 0\}.$$

Thus, given $u \in E_i$, then $u = P_i u$, and it follows that

$$P_i Au = P_i AP_i u = AP_i^2 u = AP_i u = Au, \quad \forall i \in \{+, 0, -\},$$

i.e., $P_i Au = Au$, for $i \in \{+, 0, -\}$, and the assertion follows. □

Let us show that $dim(E_0), dim(E_-) < \infty$. Note that, by (10.37), we have $\mathbb{R}_- \cap \sigma(S) = (\lambda_n)_{n=1}^N$, a finite sequence of eigenvalues, for some $N \in \mathbb{N}$, and corresponding eigenvectors $(\varphi_n)_{n=1}^{N'}$. Furthermore, according to (10.37), $\sigma_{ess}(S) \subset [v_\infty, \infty) \subset (0, \infty)$. Moreover, according to Theorem 10.4, along with Proposition 10.7

$$\begin{aligned} E(\mathbb{R}_-)(H^1(\mathbb{R}^n)) &= E(\sigma(S) \cap \mathbb{R}_-)(H^1(\mathbb{R}^n)) \\ &= E\left(\bigcup_{n=1}^N \{\lambda_n\}\right)(H^1(\mathbb{R}^n)) \\ &= \sum_{n=1}^N E(\{\lambda_n\})(H^1(\mathbb{R}^n)) \\ &= \text{span}\{\varphi_n\}_{n=1}^{N'}. \end{aligned}$$

From which we conclude that $E_- = \text{span}\{\varphi_n\}_{n=1}^{N'}$.

Also note that, from hypothesis (10.38), we have $0 \in (\sigma^-, \sigma^+)$, from which we can conclude that 0 is not an accumulation point of $\sigma(S)$. In fact, taking $\epsilon = \min\{|\sigma^+|, |\sigma^-|\}$, it follows that $\sigma(S) \cap (-\epsilon, 0) = \emptyset$ and $\sigma(S) \cap (0, \epsilon) = \emptyset$. Indeed, for the first case, if there existed $u \in \sigma(S) \cap (0, \epsilon)$, then $\sup[\sigma(S) \cap (-\infty, 0)] = \sigma^- < u$, which is a contradiction. Thus, we conclude that 0 is not an accumulation point of $\sigma(S)$.

Therefore, 0 is either an isolated point of $\sigma(S)$ or 0 does not belong to $\sigma(S)$. In other words,

$$0 \notin \sigma(S) \text{ or } 0 \in \sigma_d(S). \tag{10.44}$$

Note that, if $0 \notin \sigma(S)$, it follows that $E(\sigma(S) \cap \{0\}) = E(\emptyset) = 0$, and clearly $E_0 = E(\emptyset)(H^1(\mathbb{R}^n)) = \{0\}$. Now, from Proposition 10.7 if $0 \in \sigma_d(S)$, it follows that $E_0 = E(\{0\})(H^1(\mathbb{R}^n))$ is the space generated by the eigenvectors associated with 0, and as $0 \in \sigma_d(S)$, 0 has finite multiplicity, and we obtain $\dim(E_0) < \infty$.

Consider the norm $||\cdot||_X$ defined by

$$||u||_X = (||AP_+u||_{L^2}^2 + ||AP_-u||_{L^2}^2 + ||P_0u||_{L^2}^2)^{1/2}. \tag{10.45}$$

We can verify that $||\cdot||_X$ and $||\cdot||_E$ are equivalent.

Proposition 10.11 *Let A be the operator defined in* (10.39) *and $E = D(A)$ as defined in* (10.40). *Then the norms $||\cdot||_E$ and $||\cdot||_X$, as defined in* (10.42) *and* (10.45), *respectively, are equivalent.*

Proof First, we show that for any $u \in E$, $||u||_X \leq ||u||_E$. From the decomposition (10.43) and Lemma 10.12, it follows that

$$||u||_{L^2}^2 = ||P_+u||_{L^2}^2 + ||P_-u||_{L^2}^2 + ||P_0u||_{L^2}^2,$$

and

$$||Au||_{L^2}^2 = ||AP_+u||_{L^2}^2 + ||AP_-u||_{L^2}^2 + ||AP_0u||_{L^2}^2,$$

and hence, it follows that

$$\begin{aligned}||u||_X^2 &= ||AP_+u||_{L^2}^2 + ||AP_-u||_{L^2}^2 + ||P_0u||_{L^2}^2\\ &\leq ||P_+u||_{L^2}^2+||P_-u||_{L^2}^2+||P_0u||_{L^2}^2+||AP_+u||_{L^2}^2+||AP_-u||_{L^2}^2+||AP_0u||_{L^2}^2\\ &= ||u||_E^2.\end{aligned}$$

Conversely, let us observe that we need to analyze two cases:

(i) $0 \notin \sigma(S)$. Then we have $0 \notin \sigma(A)$. Therefore, $0 \in \rho(A)$, from which A^{-1} exists and is bounded. Thus, given $u = u_- + u_+ + u_0 \in E$, it follows that

$$\begin{aligned}||u||_{L^2} &= ||A^{-1}Au||_{L^2} = ||A^{-1}||||Au||_{L^2}\\ &\leq ||A^{-1}||(||Au_+||_{L^2} + ||Au_-||_{L^2})\\ &\leq ||A^{-1}||(||Au_+||_{L^2} + ||Au_-||_{L^2} + ||u_0||_{L^2})\\ &= ||A^{-1}||(||AP_+u||_{L^2} + ||AP_-u||_{L^2} + ||P_0u||_{L^2})\\ &\leq ||A^{-1}||||u||_X,\end{aligned}$$

and combining with

$$\begin{aligned}||Au||_{L^2} &\leq ||AP_+u||_{L^2} + ||AP_-u||_{L^2}\\ &\leq ||AP_+u||_{L^2} + ||AP_-u||_{L^2} + ||P_0u||_{L^2}\\ &\leq ||u||_X,\end{aligned}$$

it immediately follows that $||u||_E^2 = ||u||_{L^2}^2 + ||Au||_{L^2}^2 \leq C||u||_X^2$.

(ii) If $0 \in \sigma_d(S)$, then $\dim E_0 < \infty$. Thus, $A|_{E_0}$ is a continuous operator, from which $||A|_{E_0}|| \leq a_0$ for some real and positive a_0. Therefore, if $u \in E_0$, it follows that $u = P_0 u$, where

$$\begin{aligned} ||u||_E^2 &= ||u||_{L^2} + ||Au||_{L^2} \\ &\leq (1 + a_0)||u||_{L^2}^2 \\ &\leq (1 + a_0)||P_0 u||_{L^2}^2. \end{aligned}$$

Now consider the operator S restricted to the set $E_- \oplus E_+$, and as the spaces E_+ and E_- are invariant by S (by an analogous argument made in Proposition 10.12), we can write $S|_{E_- \oplus E_+} : E_- \oplus E_+ \to E_- \oplus E_+$.

Claim: $0 \notin \sigma(S|_{E_- \oplus E_+})$, and furthermore, $0 \notin \sigma(A|_{E_- \oplus E_+})$. To prove this claim, let us initially observe that $S|_{E_- \oplus E_+} : E_- \oplus E_+ \to L^2(\mathbb{R}^n)$ is injective. Indeed, given $v \in \mathcal{N}(S|_{E_- \oplus E_+})$, then $v = v_+ + v_-$, where $v_- \in E_-$ and $v_+ \in E_+$. Thus

$$\begin{aligned} &0 = S|_{E_- \oplus E_+} v = S|_{E_- \oplus E_+} v_+ + S|_{E_- \oplus E_+} v_- \\ &\Longrightarrow E_+ \ni S|_{E_- \oplus E_+} v_+ = -S|_{E_- \oplus E_+} v_- \in E_-, \end{aligned}$$

but as $E_+ \cap E_- = \{0\}$, it follows that $S|_{E_- \oplus E_+} v_+ = S|_{E_- \oplus E_+} v_- = 0$. Furthermore, as $S|_{E_- \oplus E_+} u = Su$, for $u \in E_- \oplus E_+$, it follows that $Sv_- = Sv_+ = 0$, whence $v_-, v_+ \in \mathcal{N}(S) = E_0$. Again, as $v_- \in E_- \cap E_0 = \{0\}$ and $v_+ \in E_+ \cap E_0 = \{0\}$, it follows that $v = v_- + v_+ = 0$. Therefore, $S|_{E_- \oplus E_+}$ is injective.

Let us show that $S|_{E_- \oplus E_+}$ is self-adjoint. Noting that, from S being symmetric, it follows that $S|_{E_- \oplus E_+}$ is also symmetric, and it suffices to show that $S|_{E_- \oplus E_+} - iI$ and $S|_{E_- \oplus E_+} + iI$ are surjective. Given $v \in E_- \oplus E_+$, it follows from the fact that S is self-adjoint that $S - iI$ is surjective, whence there exists $u \in E$ such that $(S - iI)u = v$. As $P_i S \subset S P_i$ for $i \in \{+, 0, -\}$, it follows that $(S|_{E_- \oplus E_+} - iI)(P_- + P_+)u = (S - iI)(P_- + P_+)u = (P_- + P_+)v = v$, and it follows that $(S|_{E_- \oplus E_+} - iI)$ is surjective. The same argument applies for $(S|_{E_- \oplus E_+} + iI)$. Thus, $S|_{E_- \oplus E_+}$ is self-adjoint and consequently closed. Therefore, we can define (where $S|_{E_- \oplus E_+}$ is surjective) the inverse $(S|_{E_- \oplus E_+})^{-1} : R(S|_{E_- \oplus E_+}) \to E_+ \oplus E_-$. From $S|_{E_- \oplus E_+}$ being closed, it follows that $(S|_{E_- \oplus E_+})^{-1}$ is closed, and then, from the Closed Graph Theorem, $(S|_{E_- \oplus E_+})^{-1}$ is bounded, whence $0 \notin \sigma(S|_{E_- \oplus E_+})$, as desired.

From this assertion, let us note that $0 \in \rho(A|_{E_- \oplus E_+})$, from where there exists $(A|_{E_- \oplus E_+})^{-1}$ and such operator is bounded $||A|_{E_- \oplus E_+}|| \leq a_0$. Thus, for $u = u_- + u_+ \in E_- \oplus E_+$ it follows that

$$\begin{aligned} ||u||_{L^2} &= ||A^{-1}|_{E_- \oplus E_+} A|_{E_- \oplus E_+} u||_{L^2} \\ &\leq ||A^{-1}|_{E_- \oplus E_+}||(||A|_{E_- \oplus E_+} u_-||_{L^2} + ||A|_{E_- \oplus E_+} u_+||_{L^2}) \\ &\leq ||A^{-1}|_{E_- \oplus E_+}||(||A|_{E_- \oplus E_+} P_- u||_{L^2} + ||A|_{E_- \oplus E_+} P_+ u||_{L^2}) \\ &\leq a_0(||AP_+ u||_{L^2} + ||AP_- u||_{L^2}), \end{aligned}$$

and in a manner entirely analogous to case (i), it follows that $||Au||_{L^2} \leq ||AP_+u||_{L^2} + ||AP_-u||_{L^2}$. Joining, for $u \in E_+ \oplus E_-$, it follows that

$$\begin{aligned} ||u||_E &= ||u||_{L^2} + ||Au||_{L^2} \\ &\leq (a_0 + 1)(||AP_+u||_{L^2} + ||AP_-u||_{L^2}). \end{aligned}$$

Finally, given $u \in E$, $u = u_0 + u'$, where $u_0 \in E_0$, and $u' \in E_- \oplus E_+$, and then

$$\begin{aligned} ||u||_E &= ||u_0||_E + ||u'||_E \\ &\leq (1 + a_0)(||AP_+u||_{L^2} + ||AP_-u||_{L^2} + ||P_0u||_{L^2}), \end{aligned}$$

as desired. □

In this context, we can state the most important point of this section:

Proposition 10.12 *The set $E = D(A)$ coincides with the Sobolev Space $H^1(\mathbb{R}^n)$, and also, $|| \cdot ||_X$ is equivalent to the norm $|| \cdot ||_{H^1}$. Moreover, it holds that*

$$||P_+u||_X^2 - ||P_-u||_X^2 = \int_{\mathbb{R}^n} (|\nabla u(x)|^2 + V(x)u(x)^2)\, dx, \quad \forall u \in H^1(\mathbb{R}^n). \tag{10.46}$$

Proof The proof will be constructed as follows: We will build an auxiliary norm $|| \cdot ||_M$, which we will show to be equivalent to $|| \cdot ||_E$ and $|| \cdot ||_{H^1}$, from which it will be possible to conclude that $|| \cdot ||_E$ and $|| \cdot ||_{H^1}$ are equivalent. Finally, from the equivalence between $|| \cdot ||_X$ and $|| \cdot ||_E$, the result will be proved.

Before we delve into the technical aspects of the proof, let us mention that, from Theorem 10.4, it follows that, when applying Spectral Theorem 10.3 for an operator A, we can change the integrand from $\mathbb{R}$ to $\sigma(A)$ without causing any effect on the integral. We will employ this change on integrand without explicit mentions.

Let us start by constructing the norm $|| \cdot ||_M$. For this purpose, consider $-M := \inf_{x \in \mathbb{R}} V(x)$, which, according to (10.37), gives us that $\sigma(S) \subset [-M, \infty)$. Let us define the inner product (and consequently the norm)

$$\langle u, v \rangle_M = \langle u, v \rangle_{L^2} + \int_{\sigma(S)} (\lambda + 2M)\, d\langle E(\lambda)u, v \rangle, \quad \forall u, v \in E, \tag{10.47}$$

from where we can define the norm

$$||u||_M^2 = ||u||_{L^2}^2 + \int_{\sigma(S)} (\lambda + 2M)\, d\langle E(\lambda)u, u \rangle. \tag{10.48}$$

We can then write

$$\begin{aligned}||u||_M^2 = ||u||_{L^2}^2 + &\int_{\lambda\in\sigma(S),\lambda>0} (\lambda + 2M)\, d\langle E(\lambda)u, u\rangle + \\ &+ \int_{\lambda\in\sigma(S),\lambda<0} (\lambda + 2M)\, d\langle E(\lambda)u, u\rangle + \\ &+ \left(\int_{\lambda\in\sigma(S),\lambda<0} -\lambda\, d\langle E(\lambda)u, u\rangle - \int_{\lambda\in\sigma(S),\lambda<0} -\lambda\, d\langle E(\lambda)u, u\rangle\right),\end{aligned}$$

where we can regroup the terms to obtain

$$\begin{aligned}||u||_M^2 = ||u||_{L^2}^2 + &\int_{\lambda\in\sigma(S)} |\lambda|\, d\langle E(\lambda)u, u\rangle + \\ &+ \int_{\lambda\in\sigma(S),\lambda>0} 2M\, d\langle E(\lambda)u, u\rangle + \int_{\lambda\in\sigma(S),\lambda<0} (2\lambda + 2M)\, d\langle E(\lambda)u, u\rangle.\end{aligned}$$

It is immediate that $||u||_E^2 \le ||u||_M^2$. Now, since $-M \le \lambda$ for $\lambda < 0$, $\lambda \in \sigma(S)$, it follows that $2M(1-\lambda) > 2M > 2M + 2\lambda$ when $\lambda < 0$, $\lambda \in \sigma(S)$. Therefore, we can write

$$\begin{aligned}\int_{\lambda\in\sigma(S),\lambda<0} (2M + 2M|\lambda|)\, d\langle E(\lambda)u, u\rangle &= \int_{\lambda\in\sigma(S),\lambda<0} (2M+2M(-\lambda))\, d\langle E(\lambda)u, u\rangle \\ &= \int_{\lambda\in\sigma(S),\lambda<0} (2M(1-\lambda))\, d\langle E(\lambda)u, u\rangle \\ &\ge \int_{\lambda\in\sigma(S),\lambda<0} (2M + 2\lambda)\, d\langle E(\lambda)u, u\rangle.\end{aligned} \tag{10.49}$$

Thus, if we add $||u||_{L^2}^2 + \int_{\sigma(S)} |\lambda|\, d\langle E(\lambda)u, u\rangle + \int_{\lambda\in\sigma(S_s),\lambda>0} 2M\, d\langle E(\lambda)u, u\rangle$ on both sides of (10.49), we obtain $||u||_M^2 \le (2M+1)||u||_E^2$. Linking the ideas, we have then that $||u||_E^2 \le ||u||_M^2 \le (2M+1)||u||_E^2$. In other words, $||.||_E$ is equivalent to $||.||_M$.

Now, taking $\phi \in C_0^\infty(\mathbb{R}^n)$, we have

$$\begin{aligned}||\phi||_M^2 &= \int_{\mathbb{R}} \lambda\, d\langle E(\lambda)\phi, \phi\rangle + (2M+1)||\phi||_{L^2}^2 \\ &= \langle S\phi, \phi\rangle + (2M+1)||\phi||_{L^2}^2 \\ &= \int_{\mathbb{R}^n} (|\nabla\phi(x)|^2 + V(x)\phi(x)^2 + (2M+1)\phi(x)^2)\, dx,\end{aligned}$$

from where there exists a constant $C > 0$ such that

$$||\phi||_{H^1}^2 \leq ||\phi||_M^2 \leq C||\phi||_{H^1}^2, \quad \forall \phi \in C_0^\infty(\mathbb{R}^n),$$

i.e., $||.||_M$ defines a norm equivalent to $||.||_{H^1}$ on $C_0^\infty(\mathbb{R}^n)$. Let H be the closure of $C_0^\infty(\mathbb{R}^n)$ with respect to the norm $||.||_M$, and we can conclude that $H = H^1(\mathbb{R}^n)$. However, we still need to conclude that $E = H^1(\mathbb{R}^n)$. Initially observing that, as $||.||_E \leq ||.||_M \leq C||.||_E$, it follows that $H \subset E$. Assuming by contradiction that $H \neq E$, there would exist $u_0 \in E$, $u_0 \neq 0$, such that

$$\langle \phi, u_0 \rangle_M = 0, \quad \forall \phi \in C_0^\infty(\mathbb{R}^n),$$

and therefore

$$0 = \langle S\phi, u_0 \rangle_{L^2} + (2M+1)\langle \phi, u_0 \rangle_{L^2} = \langle (S+(2M+1)I)\phi, u_0 \rangle_{L^2}, \quad \forall \phi \in C_0^\infty(\mathbb{R}^n).$$

However, from the density of $C_0^\infty(\mathbb{R}^n)$ in $H^1(\mathbb{R}^n)$, it follows that $u_0 \in D(S^*) = D(S)$, and moreover, $(S+(2M+1)I)u_0 = 0$. But this would mean that $-(2M+1) \in \sigma(S)$, which contradicts the fact that $\sigma(S) \subset [-M, \infty)$, from where it follows that $E = H^1(\mathbb{R}^n)$.

Finally, given $\phi \in C_0^\infty(\mathbb{R}^n)$, we can write

$$\begin{aligned} ||P_+\phi||_X^2 - ||P_-\phi||_X^2 &= \int_{\lambda>0} |\lambda| \, d\langle E(\lambda)\phi, \phi \rangle - \int_{\lambda<0} |\lambda| \, d\langle E(\lambda)\phi, \phi \rangle \\ &= \int_{\sigma(S)} \lambda \, d\langle E(\lambda)\phi, \phi \rangle = \langle S\phi, \phi \rangle \\ &= \int_{\mathbb{R}^n} (|\nabla \phi(x)|^2 + V(x)\phi(x)^2) \, dx. \end{aligned}$$

Thus, from the density of $C_0^\infty(\mathbb{R}^n)$ in $H^1(\mathbb{R}^n)$, it follows the desired result. □

Acknowledgments E. Toon was supported by CNPq grant 312833/2022-9.

References

1. R.G. Bartle, *Elements of Integration and Lebesgue Measure* (Wiley, 1966)
2. V. Benci, D. Fortunato, *Variational Methods in Nonlinear Field Equations. Solitary Waves, Hylomorphic Solitons and Vortices*, Springer Monographs in Mathematics (Springer, Cham, 2014)
3. D. Borthwick, *Spectral Theory - Basic Concepts and Applications*, Graduate Texts in Mathematics, vol. 284 (Springer, Cham, 2020)

4. J.A. Cardoso, J.M.B. do Ó, E. Medeiros, Hamiltonian elliptic systems involving nonlinear Schröodinger equations with critical growth. Z. Angew. Math. Phys. ZAMP **66**, 2237–2254 (2015)
5. D.G. Costa, H. Tehrani, Existence and multiplicity results for a class of Schrödinger equations with indefinite nonlinearities. Adv. Differ. Equ. **8**, 1319–1340 (2003)
6. D.G. Costa, H. Tehrani, M. Ramos, Non-zero solutions for a Schrödinger equation with indefinite linear and nonlinear terms. Proc. R. Soc. Edinb. Sect. A Math. **134**(2), 249–258 (2004)
7. G.S. do Amaral Costa, G.M. Figueiredo, J.C.D.O. Junior, A non-periodic indefinite variational problem in $\mathbb{R}^N$ with critical exponent. Proc. Edinb. Math. Soc. **66**(2), 579–612 (2023)
8. Y. Egorov, V. Kondratriev, *On Spectral Theory of Elliptic Operators*, vol. 89 (Birkhäuser, 1996)
9. M.F. Furtado, R. de Marchi, Asymptotically periodic superquadratic Hamiltonian systems. J. Math. Anal. Appl. **433**, 712–731 (2016)
10. E. Kreyszig (ed.), *Introductory Functional Analysis with Applications* (Wiley, New York, 1978)
11. L.A. Maia, M. Soares, An abstract linking theorem applied to indefinite problems via spectral properties. Adv. Nonlinear Stud. **19**(3), 545–567 (2019)
12. L.A. Maia, M. Soares, Spectral theory approach for a class of radial indefinite variational problems. J. Differ. Equ. **266**(11), 6905–6923 (2019)
13. L.A. Maia, M. Soares, An indefinite elliptic problem on RN autonomous at infinity: the crossing effect of the spectrum and the nonlinearity. Calc. Var. Partial Differ. Equ. **59**(1), 41 (2020)
14. W. Rudin, *Principles of Mathematical Analysis* (McGraw-Hill, 1976)
15. W. Rudin, *Real and Complex Analysis* (McGraw-Hill, 1987)
16. R. Schilling, *Measures, Integrals and Martingales* (Cambridge University Press, 2005)
17. K. Schmüdgen, *Unbounded Self-adjoint Operators on Hilbert Space*, Graduate Texts in Mathematics (2014)
18. B.S. Sirakov, On the existence of solutions of Hamiltonian elliptic systems in $\mathbb{R}^N$. Adv. Differ. Equ. **5**(10–12), 1445–1464 (2000)
19. C. Stuart, *An Introduction to Elliptic Equations on RN*, Trieste Notes (1999)
20. X.H. Tang, New conditions on nonlinearity for a periodic Schrödinger equation having zero as spectrum. J. Math. Anal. Appl. **413**(1), 392–410 (2014)
21. E. Toon, P. Ubilla, Existence of positive solutions of Schrödinger equations with vanishing potentials. Discrete Contin. Dyn. Syst. **40**(10), 5831–5843 (2020)
22. E. Toon, P. Ubilla, Hamiltonian systems of Schrödinger equations with vanishing potentials. Commun. Contemp. Math. **24**(01), 2050074 (2022)
23. L. Xia, J. Zhang, F. Zhao, Ground state solutions for superlinear elliptic systems on $\mathbb{R}^N$. J. Math. Anal. Appl. **401**, 518–525 (2013)
24. X. Xu, Q. Kuang, Y. Gong, Infinitely many solutions for superlinear periodic Hamiltonian elliptic systems. Bound. Value Probl. **14** (2015). https://doi.org/10.1186/s13661-014-0274-4
25. Y. Yang, *Solitons in Field Theory and Nonlinear Analysis*, Springer Monographs in Mathematics (Springer, New York, 2001)
26. W. Zhang, J. Zhang, F. Zao, Multiple solutions for asymptoticaly quadratic and superquadratic elliptic system of Hamiltonian type. J. Appl. Math. Comput. **263**, 36–46 (2015)
27. W. Zhang, X. Xie, H. Mi, Ground state solutions for Hamiltonian elliptic system with sign-changing potential. Electron. J. Differ. Equ. **2017**(64), 1–13 (2017)
28. L. Zhao, F. Zhao, On ground state solutions for superlinear Hamiltonian elliptic systems. Z. Angew. Math. Phys. ZAMP **64**, 403–418 (2013)

Chapter 11
Periodic Solutions of a Class of Perturbed Implicit Equations with Memory

Marco Spadini

Abstract We study the harmonic solutions of periodically perturbed implicit integrodifferential equations of a particular form obtaining, under appropriate topological conditions, a "branching-type" result for the nontrivial periodic solutions. The proof employs a combination of techniques from the theory of Topological Degree, Functional Differential Equations, and Differential-Algebraic Equations.

Keywords Periodic solutions · Memory effects · Differential-algebraic equations · Functional differential equations · Topological degree

2020 MSC 34K13, 34A09, 34K27

11.1 Introduction

In this chapter, we consider implicit integrodifferential equations in the following particular form:

$$\dot{x}(t) = f\left(x(t), \dot{x}(t), \int_{-\infty}^{t} \gamma_a^b(t-s)\varphi\big(x(s), \dot{x}(s)\big)\,\mathrm{d}s\right), \tag{11.1}$$

where $f\colon \mathbb{R}^n \times \mathbb{R}^n \times \mathbb{R}^k \to \mathbb{R}^n$ is C^∞ and $\varphi\colon \mathbb{R}^n \times \mathbb{R}^n \to \mathbb{R}^k$ is continuous. The integral kernel γ_a^b, for $a > 0$ and $b \in \mathbb{N} \setminus \{0\}$, is the gamma probability distribution with mean b/a and variance b/a^2, that is,

$$\gamma_a^b(s) = \frac{a^b s^{b-1} \mathrm{e}^{-as}}{(b-1)!} \quad \text{for } s \geq 0, \qquad \gamma_a^b(s) = 0 \ \text{ for } s < 0. \tag{11.2}$$

M. Spadini (✉)
Dipartimento di Matematica e Informatica "Ulisse Dini", Università degli Studi di Firenze, Firenze, Italy
e-mail: marco.spadini@unifi.it

P. Benevieri, J. G. Mesquita (eds.), *Functional Differential Equations and Dynamic Equations on Time Scales*, Advances in Continuum Mechanics 52,
https://doi.org/10.1007/978-3-031-83327-4_11

The product inside the integral must be understood component-wise.

We will focus on the structure of periodic solutions of Eq. (11.1) when subject to a periodic perturbation in the "memory part" as follows:

$$\dot{x}(t) = f\left(x(t), \dot{x}(t), \int_{-\infty}^{t} \gamma_a^b(t-s)\varphi\big(x(s), \dot{x}(s)\big)\,\mathrm{d}s \; + \lambda h(t)\right), \qquad \lambda \geq 0, \tag{11.3}$$

where $h\colon \mathbb{R} \to \mathbb{R}^k$ is a C^1 function T-periodic in t.

The problem that we take on in this chapter is indeed an extension, albeit limited to the case when the right-hand side has smooth regularity and the perturbing term is less general, of those studied in [20] for first order equations and of [5, 14] for the second order case, see also [4].

Furthermore, throughout this chapter, we make the following crucial assumptions on the right-hand side function f:

Assumption 1 *For all* $(u, v, w) \in \mathbb{R}^n \times \mathbb{R}^n \times \mathbb{R}^k$ *the Fréchet partial derivative* $\partial_2 f(u, v, w)$ *with respect to the second variable is well defined and has no eigenvalue equal to* 1. *That is, the map* $v \mapsto \partial_2\big(v - f(u, v, w)\big) = \mathrm{Id} - \partial_2 f(u, v, w)$ *is invertible for each* $(u, v, w) \in \mathbb{R}^n \times \mathbb{R}^n \times \mathbb{R}^k$*; here* "Id" *denotes the identity on* $\mathbb{R}^n$.

We point out that Assumption 1 does *not* imply that (11.1) can be solved globally with respect to $\dot{x}(t)$ even in the special case when f does not depend explicitly on the second variable; as an example of this fact, for arbitrary $u = (u_1, u_2)$, $v = (v_1, v_2) \in \mathbb{R}^2$, $w \in \mathbb{R}$, consider $f(u, v, w) = \big(v_1 - \mathrm{e}^{v_1}\cos(v_2) + u_1\,,\, v_2 - \mathrm{e}^{v_1}\sin(v_2) + u_1\big)$, and observe that for any (u, w) there are infinitely many values of $v = (v_1, v_2)$ such that $v = f(u, v, w)$ (one for each plane $v_2 = \frac{\pi}{4} + \nu\pi$, $\nu \in \mathbb{Z}$). Notice also that even when (11.1) *can* be solved with respect to $\dot{x}(t)$, the resulting expression may be very difficult to use as, for instance, it happens for scalar equations of the form

$$\dot{x}(t) = -\dot{x}^7(t) + g\left(x(t), \int_{-\infty}^{t} \gamma_a^b(t-s)\psi\big(x(s)\big)\,\mathrm{d}s\right),$$

where $g\colon \mathbb{R} \times \mathbb{R} \to \mathbb{R}$ and $\psi\colon \mathbb{R} \to \mathbb{R}$ are some given continuous functions. Here, clearly f is taken as $f(u, v, w) = -v^7 + g(u, w)$ and $\varphi(u, v) = \psi(u)$.

It is also worth noticing that Assumption 1 is automatically satisfied when f does not depend explicitly on $\dot{x}$.

Equation (11.3) may look unusual. To understand its relevance observe that it can be used to describe physical phenomena that depend on the complete history of the phase space of the system as illustrated by the following example:

Example 11.1 Consider the following equation:

$$\ddot{x}(t) = -x(t) - \int_{-\infty}^{t} \mathrm{e}^{-(t-s)}\ddot{x}(s)\,\mathrm{d}s + \lambda h(t), \tag{11.4}$$

which describes an oscillator whose restoring force has a fading memory of past acceleration. To put (11.4) in the form (11.3), observe that for $a = b = 1$ one has $\gamma_1^1(t-s) = \mathrm{e}^{-(t-s)}$; we then rewrite (11.4) as a system

$$\begin{cases} \dot{x}_1(t) = x_2(t), \\ \dot{x}_2(t) = -x_1(t) - \int_{-\infty}^{t} \gamma_1^1(t-s)\dot{x}_1(s)\,\mathrm{d}s + \lambda h(t), \end{cases} \tag{11.5}$$

set $n = 2$, $k = 1$, and introduce the functions $f\colon \mathbb{R}^2 \times \mathbb{R}^2 \times \mathbb{R} \to \mathbb{R}^2$ and $\varphi\colon \mathbb{R}^2 \times \mathbb{R}^2 \to \mathbb{R}$ given by $f(u_1, u_2; v_1, v_2; w) = (u_2, -u_1 - w)$ and $\varphi(u_1, u_2; v_1, v_2) = v_1$. Thus, introducing $\xi = (x_1, x_2)$, (11.5) becomes

$$\dot{\xi}(t) = f\left(\xi(t), \dot{\xi}(t), \int_{-\infty}^{t} \gamma_1^1(t-s)\varphi\big(\xi(s), \dot{\xi}(s)\,\mathrm{d}s - \lambda h(t)\big)\right).$$

Roughly speaking, our main result, Theorem 11.2, yields the existence of a connected set with noncompact closure, consisting of T-periodic solutions of (11.3), which, in a sense that will become clear in the following, "emanates" from the set

$$\big\{p \in \mathbb{R}^n : f\big(p, 0_n, \varphi(p, 0_n)\big) = 0_n\big\},$$

regarded as a set of constant functions, whenever the degree of a certain vector field in $\mathbb{R}^{2n+kb}$ is nonzero. Here and in what follows, by 0_r, $r = 2, 3, \ldots$, we mean the origin of $\mathbb{R}^r$.

To achieve this goal we use a slightly modified version of the so-called *linear chain trick* (see, e.g., [3, 8, 9, 18, 20]) to reduce equation (11.3) to a Differential-Algebraic Equation (DAE) of the form considered, e.g., in [1, 19] whose set of periodic solutions can be studied with topological degree methods. See, e.g., [12] for the general theory of DAEs.

11.2 Preliminary Facts

11.2.1 Topological Degree

Topological degree theory plays a fundamental role in our result. Despite the fact that many references are easily accessible on this classical topic (see, e.g., [7] and also [13, 15], and [6]), we provide below a short essential introduction for the sake of completeness.

Let p be a point in $\mathbb{R}^s$ and f be an $\mathbb{R}^s$-valued function defined in a neighborhood of the closure $\overline{U}$ of an open subset U of $\mathbb{R}^s$, when $f^{-1}(p) \cap U$ is compact, the triple (f, U, p) is called *admissible* in U with respect to p or, and for simplicity, we say that f is *admissible*. In this case, it is defined an integer $\deg(f, U, p)$, called the *degree of f in U with respect to p*. In its essence, $\deg(f, U, p)$ is an algebraic count of the elements of $f^{-1}(p) \cap U$ to each of which a sign $+1$ or -1 is assigned according to whether the linearization of f at that point preserves or inverts the orientation of $\mathbb{R}^s$. This assertion requires some explanation: When f is C^1 and p is a regular value of f (i.e., either $f^{-1}(p) = \emptyset$ or the Jacobian matrix $J_q f$ of f at q is nonsingular for each $q \in f^{-1}(p)$), then $f^{-1}(p)$ is a finite set. Then, we define

$$\deg(f, U, p) = \sum_{q \in f^{-1}(p)} \operatorname{sign} \det J_q f. \tag{11.6}$$

It is seen that replacing f with a sufficiently close approximating C^1 function and replacing p by a near enough regular value, the value of the degree as defined in (11.6) remains constant. Also, by Sard's theorem (see, e.g., [15]) the set of regular values is dense in $\mathbb{R}^s$. Thus, formula (11.6) provides a basis for an effective definition of degree: Set for an arbitrary admissible triple (f, U, p)

$$\deg(f, U, p) := \deg(g, U, q), \tag{11.7}$$

where g is a C^1 function and q is a regular value sufficiently close to f and p, respectively. It is worth mentioning, even though we will not use this fact, that there are other computation-oriented approaches to topological degree. We only bring up one based on surface integrals (see, e.g., [2] and [17]) and one on volume integrals (see, e.g., [11] and [16]).

Below, the "target" point p will always be the origin 0_s of $\mathbb{R}^s$, so, for the sake of brevity, we simply write $\deg(f, U)$ instead of $\deg(f, U, 0_s)$. Indeed, $\deg(f, U)$ can also be seen as the degree (sometimes called "characteristic" or "rotation") of the map f regarded as a tangent vector field on $\mathbb{R}^s$ (see, e.g., [10] for an axiomatic approach).

11.2.2 Harmonic Solutions of Periodically Perturbed DAEs

We now recall a result from [19] that plays a central role in this chapter.

Consider in $\mathbb{R}^r \times \mathbb{R}^m$ the following perturbed DAE system in semi-explicit form:

$$\begin{cases} \dot{\xi} = F(\xi, \eta) + \lambda H(t, \xi, \eta), & \lambda \geq 0, \\ G(\xi, \eta) = 0_m, \end{cases} \tag{11.8}$$

where $G\colon \mathbb{R}^r \times \mathbb{R}^m \to \mathbb{R}^m$ and $F\colon \mathbb{R}^r \times \mathbb{R}^m \to \mathbb{R}^k$ are continuous maps defined on an open connected set $U \subseteq \mathbb{R}^r \times \mathbb{R}^m$, with $G \in C^\infty$ and the partial derivative of G with respect to the second variable, $\partial_2 G(p, q)$, invertible for each $(p, q) \in \mathbb{R}^r \times \mathbb{R}^m$. The perturbing map $H\colon \mathbb{R} \times \mathbb{R}^r \times \mathbb{R}^m \to \mathbb{R}^m$ is assumed to be continuous and T-periodic in t. Let also $U \subseteq \mathbb{R}^r \times \mathbb{R}^m$ be open.

We wish to study the set of T-periodic solutions of (11.8). In order to do so in a systematic way, we introduce some definition. Let $C_T(U)$ be the metric subspace of the Banach space $C_T(\mathbb{R}^r \times \mathbb{R}^m)$ consisting of all the continuous T-periodic functions that take values in U. We say that $(\mu; x, y) \in [0, \infty) \times C_T(U)$ is a *T-pair* of (11.8) if (x, y) satisfies (11.8) for $\lambda = \mu$; here (x, y) is thought of as a single element of $C_T(U)$.

Furthermore, given any point $\mathfrak{q}$ in some arbitrary $\mathbb{R}^d$, $d \in \mathbb{N}$, with the symbol $\overline{\mathfrak{q}} \colon \mathbb{R} \to \mathbb{R}^d$, we indicate the constant map $\overline{\mathfrak{q}}(t) \equiv \mathfrak{q}$. By abuse of notation, given any $(p, q) \in \mathbb{R}^r \times \mathbb{R}^m$, we write $(\overline{p}, \overline{q})$ for the map in $C_T(\mathbb{R}^r \times \mathbb{R}^m)$ that is constantly equal to (p, q), instead of $\overline{(p, q)}$. A T-pair of the form $(0; \overline{p}, \overline{q})$ is called *trivial.*

We point out that the set of trivial T-pairs of (11.8) corresponds bijectively with

$$\big\{(p, q) \in U : F(p, q) = 0_r,\ G(p, q) = 0_m\big\}.$$

Finally, given $\Omega \subseteq [0, \infty) \times C_T(U)$, it is convenient to introduce a notation for the set of those points of U that correspond to a constant function of Ω; that is,

$$\Omega_U = \big\{(p, q) \in U : (0, \overline{p}, \overline{q}) \in \Omega\big\}.$$

Theorem 11.1 ([19, Thm. 5.1]) *Let $U \subseteq \mathbb{R}^r \times \mathbb{R}^m$ be open and connected, and let $G \colon \mathbb{R}^r \times \mathbb{R}^m \to \mathbb{R}^m$, $F \colon \mathbb{R}^r \times \mathbb{R}^m \to \mathbb{R}^r$, $H \colon \mathbb{R} \times \mathbb{R}^r \times \mathbb{R}^m \to \mathbb{R}^r$, and $T > 0$ be such that F and H are continuous, and G is C^∞ with $\partial_2 G(p, q)$ invertible for all $(p, q) \in U$. Define $\mathfrak{F} \colon U \to \mathbb{R}^r \times \mathbb{R}^m$ by*

$$\mathfrak{F}(p, q) = \big(F(p, q), G(p, q)\big). \tag{11.9}$$

Given an open set $\Omega \subseteq [0, \infty) \times C_T(U)$, suppose that $\deg(\mathfrak{F}, \Omega_U)$ *is well defined and nonzero. Then, there exists a connected set Γ of nontrivial solution pairs of* (11.8) *whose closure in* $[0, \infty) \times C_T(U)$ *intersects the set*

$$\big\{(0, \overline{p}, \overline{q}) \in \Omega : \mathfrak{F}(p, q) = (0_r, 0_m)\big\}$$

and is not contained in any compact subset of Ω.

11.2.3 Linear Chain Trick

Equation (11.3) is transformed in the following system by the introduction of two new appropriate functions y and u:

$$\dot{x}(t) = y(t), \tag{11.10a}$$

$$u(t) = \int_{-\infty}^{t} \gamma_a^b(t - s)\varphi\big(x(s), y(s)\big)\, \mathrm{d}s + \lambda h(t), \tag{11.10b}$$

$$y(t) - f\big(x(t), y(t), u(t)\big) - 0_n. \tag{11.10c}$$

Roughly speaking we want to replace Eq. (11.10b) with a system of differential equations so that we may apply Theorem 11.1 to the resulting enlarged system. The so-called *linear chain trick* allows us to do precisely that. The following lemma contains a slightly modified version of the linear chain trick, and its proof is indebted to [14, 18, 20].

Lemma 11.1 *Given a continuous function $\zeta : \mathbb{R} \to \mathbb{R}^k$ and a C^1 function $\sigma : \mathbb{R} \to \mathbb{R}^k$, let, for $i = 1, \dots, b$,*

$$v_i(t) = \int_{-\infty}^{t} \gamma_a^i(t-s)\zeta(s)\,\mathrm{d}s + \sigma(t). \tag{11.11}$$

Then,

(i) The map $t \mapsto \big(v_1(t), \dots, v_b(t)\big)$ satisfies the system

$$\begin{cases} \dot v_1(t) = a\big(\zeta(t) - v_1(t)\big) - a\sigma(t) + \sigma'(t), \\ \dot v_i(t) = a\big(u_{i-1}(t) - u_i(t)\big) + \sigma'(t), \qquad i = 2, \dots, b. \end{cases} \tag{11.12}$$

(ii) When ζ and σ are assumed to be T-periodic, system (11.12) *admits a unique T-periodic solution given by* (11.10).

Proof Let us prove (i). For $i = 1$, we have

$$\begin{aligned} \frac{\mathrm{d}}{\mathrm{d}t} v_1(t) &= \frac{\mathrm{d}}{\mathrm{d}t} \int_{-\infty}^{t} a\mathrm{e}^{-a(t-s)}\zeta(s)\,\mathrm{d}s + \sigma'(t) \\ &= a\zeta(t) - a\int_{-\infty}^{t} a\mathrm{e}^{-a(t-s)}\zeta(s)\,\mathrm{d}s + \sigma'(t) \\ &= a\big(\zeta(t) - v_1(t)\big) - a\sigma(t) + \sigma'(t). \end{aligned}$$

For $i = 2, \dots, b$, we find instead

$$\begin{aligned} \frac{\mathrm{d}}{\mathrm{d}t} v_i(t) &= \gamma_a^i(0)\zeta(t) + \int_{-\infty}^{t} \frac{\mathrm{d}}{\mathrm{d}t}\gamma_a^b(t-s)\zeta(s)\,\mathrm{d}s + \sigma'(t) \\ &= a\int_{-\infty}^{t} \gamma_a^{i-1}(t-s)\zeta(s) - \gamma_a^i(t-s)\zeta(s)\,\mathrm{d}s + \sigma'(t) \\ &= a\big(v_{i-1}(t) - \sigma(t) - v_i(t) + \sigma(t)\big) + \sigma'(t) \\ &= a\big(v_{i-1}(t) - v_i(t)\big) + \sigma'(t). \end{aligned}$$

Thus Eq. (11.12) holds.

Let us now prove (ii). Observe first that if ζ and σ are T-periodic, then they are bounded and σ' is bounded (and T-periodic) as well. The matrix $A \in \mathbb{R}^{kb\times kb}$ of the homogeneous part of (11.12), written in block-matrix form, is

$$A = \begin{pmatrix} -I_k & 0 & \dots & & 0 \\ I_k & -I_k & 0 & . & 0 \\ 0 & I_k & -I_k & \ddots & \vdots \\ \vdots & \ddots & \ddots & \ddots & 0 \\ 0 & \dots & 0 & I_k & -I_k \end{pmatrix},$$

where I_k denotes the $k \times k$ identity matrix. Clearly, the only eigenvalue of A is -1, with multiplicity kb. Thus the unique bounded solution of the homogeneous part is the map which is constantly equal to 0_{kb}. As a consequence, we have that (11.12) has a unique bounded solution. This follows by observing that the difference of two bounded solution is indeed a bounded solution of the homogeneous part, so it must be constantly zero.

The fact that $t \mapsto \big(v_1(t), \dots, v_b(t)\big)$, with the v_i's given by (11.11), is a solution of (11.12) can be checked directly, see the computation for part (i). To verify that this solution is T-periodic, observe that for $i = 1, \dots, b$, using the change of variable $\tau = s - T$, we have

$$\begin{aligned} \int_{-\infty}^{t+T} \gamma_a^i(T + t - s)\zeta(s)\,\mathrm{d}s &= \int_{-\infty}^{t} \gamma_a^i(t - \tau)\zeta(\tau + T)\,\mathrm{d}\tau \\ &= \int_{-\infty}^{t} \gamma_a^i(t - \tau)\zeta(\tau)\,\mathrm{d}\tau. \end{aligned} \tag{11.13}$$

Since we are assuming σ to be T-periodic, by (11.13) we have

$$\begin{aligned} v_i(t + T) &= \int_{-\infty}^{t+T} \gamma_a^i(t - s)\zeta(s)\,\mathrm{d}s + \sigma(t + T) \\ &= \int_{-\infty}^{t} \gamma_a^i(t - s)\zeta(s)\,\mathrm{d}s + \sigma(t) = v_i(t). \end{aligned}$$

The proof is complete. □

Applying Lemma 11.1 to Eq. (11.10b) with $\zeta(t) = \varphi\big(x(t), y(t)\big)$ and $\sigma(t) = \lambda h(t)$, we see that $u(t) = u_b(t)$, where u_b is determined by the system

$$\begin{aligned} \dot{u}_1(t) &= a\big(\varphi\big(x(t), y(t)\big) - u_1(t)\big) + \lambda\big(h'(t) - ah(t)\big), \\ \dot{u}_i(t) &= a\big(u_{i-1}(t) - u_i(t)\big) + \lambda h'(t), \qquad i = 2, \dots, b. \end{aligned}$$

Thus, system (11.10) leads us to

$$\begin{cases} \dot{x}(t) = y(t), \\ \dot{u}_1(t) = a\big(\varphi\big(x(t), y(t)\big) - u_1(t)\big) + \lambda\big(h'(t) - ah(t)\big), \\ \dot{u}_i(t) = a\big(u_{i-1}(t) - u_i(t)\big) + \lambda h'(t), \qquad i = 2, \dots, b. \\ y(t) - f\big(x(t), y(t), u_b(t)\big) = 0_n \end{cases} \tag{11.14}$$

11.3 The Main Result

We obtain our main result applying Theorem 11.1 to (11.14) and taking into account Remark 11.1. This is made possible by the following fact:

Remark 11.1 Observe that, by our construction, if x is a T-periodic solution of (11.3), then setting $y(t) = \dot{x}(t)$ and

$$u_i(t) = \int_{-\infty}^{t} \gamma_a^i(t-s)\varphi\big(x(t), y(t)\big)\,\mathrm{d}s + \lambda h(t), \tag{11.15}$$

for $i = 1, \ldots, b$, we have that $t \mapsto \big(x(t), y(t), u_1(t), \ldots, u_b(t)\big)$ is a T-periodic solution of (11.14). Conversely, again by construction, if $t \mapsto \big(x(t), y(t), u_1(t), \ldots, u_b(t)\big)$ is a T-periodic solution of (11.14), then $x(t)$ is a T-periodic solution of (11.3).

Let us introduce vectors $\eta \in \mathbb{R}^n$ and $\xi \in \mathbb{R}^{n+kb}$ with $\xi = (\xi_0, \xi_1, \ldots, \xi_b)$, where

$$\xi_0 \in \mathbb{R}^n, \qquad \text{and } \xi_1, \ldots, \xi_b \in \mathbb{R}^k.$$

Define the vector fields $F\colon \mathbb{R}^{n+kb} \times \mathbb{R}^n \to \mathbb{R}^{n+kb}$ and $H\colon \mathbb{R} \times \mathbb{R}^{n+kb} \times \mathbb{R}^n \to \mathbb{R}^{n+kb}$ given by

$$F(\xi, \eta) = \Big(\eta, a\big(\varphi(\xi_0, \eta) - \xi_1\big), a(\xi_2 - \xi_3), \ldots, a(\xi_{b-1} - \xi_b)\Big), \tag{11.16}$$

$$H(t, \xi, \eta) = \Big(0_n,\ h'(t) - ah(t),\ \underbrace{h'(t), \ldots, h'(t)}_{b-1 \text{ times}}\Big). \tag{11.17}$$

Observe that although H does not actually depend on ξ and η, we keep these two variables in the arguments of H in order to maintain compatibility with Theorem 11.1. This will come in handy when later we consider an extension of Eq. (11.3).

Let us also define $G\colon \mathbb{R}^{n+kb} \times \mathbb{R}^n \to \mathbb{R}^n$ by

$$G(\xi, \eta) = \big(\eta - f(\xi_0, \eta, \xi_b)\big), \tag{11.18}$$

With these definitions, (11.14) can be written in the form (11.8) with $r = n + kb$ and $m = n$, that is,

$$\begin{cases} \dot{\xi} = F(\xi, \eta) + \lambda H(t, \xi, \eta), & \lambda \geq 0, \\ G(\xi, \eta) = 0_n. \end{cases} \tag{11.19}$$

Definition 11.1 We say that $(\lambda, x) \in [0, \infty) \times C_T(\mathbb{R}^n)$ is a *T-pair for* (11.3) if x is a T-periodic solution of (11.3) corresponding to λ. A T-pair of the form $(0, \overline{p})$ is called *trivial*.

Here, as in the previous section, $\overline{p}$ denotes the constant function $\overline{p} = p \in \mathbb{R}^n$.

Remark 11.2 Observe that $(0, \overline{p})$ is a trivial T-pair for (11.3) if and only if

$$f\big(p, 0_n, \varphi(p, 0_n)\big) = 0_n. \tag{11.20}$$

In fact, if $(0, \overline{p})$ is a trivial T-pair, substituting in (11.3)

$$\begin{aligned} 0_n &= f\left(p, 0_n, \int_{-\infty}^{t} \gamma_a^b(t-s)\varphi(p, 0_n)\,\mathrm{d}s\right) \\ &= f\left(p, 0_n, \varphi(p, 0_n)\int_0^{+\infty} \gamma_a^b(t-s)\,\mathrm{d}s\right) = f\big(p, 0_n, \varphi(p, 0_n)\big). \end{aligned}$$

Conversely, when $p \in \mathbb{R}^n$ satisfies (11.20), then $\overline{p}$ is a constant solution of (11.3) for $\lambda = 0$, and thus $(0, \overline{p})$ is a trivial T-pair.

Remark 11.3 In view of Remark 11.1, we can establish a bijective correspondence between the T-pairs of (11.3) and those of (11.19) that respects the notion of triviality, in the sense that makes the set of trivial T-pairs correspond to each other. Such correspondence is obtained by a map $\mathfrak{P}$ that associates with any T-pair $(\lambda; x, y, u_1, \ldots, u_b)$ of (11.19) the T-pair (λ, x) of (11.3). Conversely, given any T-pair (λ, x) of (11.3), we construct the map $\mathfrak{P}^{-1}$ that associates with it the T-pair $(\lambda; x, y, u_1, \ldots, u_b)$ of (11.19), where $y = \dot{x}$ and u_i is given by (11.15). One immediately checks that both $\mathfrak{P}$ and $\mathfrak{P}^{-1}$ are continuous, so that $\mathfrak{P}$ is a homeomorphism, and both $\mathfrak{P}$ and $\mathfrak{P}^{-1}$ send trivial T-pairs into trivial T-pairs.

In order to state concisely our main result, we need some further notation. Given $U \subseteq \mathbb{R}^n$ open, we define $\widehat{U} = U \times \mathbb{R}^{n+kb}$. Also, given an open set $\Omega \subseteq [0, \infty) \times C_T(U)$, we let $\widehat{\Omega} = \Omega \times C_T(\mathbb{R}^{n+kb})$ which, by abuse of notation, identify as an open subset of $[0, \infty) \times C_T(U \times \mathbb{R}^{n+kb})$, and, similarly to the previous section, we put $\Omega_U = \{p \in U : (0, \overline{p}) \in \Omega\}$ and $\widehat{\Omega}_{\widehat{U}} = \{(p, q) \in \widehat{U} : (0, \overline{p}, \overline{q}) \in \widehat{\Omega}\}$.

We are finally in a position to state and prove our main result.

Theorem 11.2 *Let $U \subseteq \mathbb{R}^n$ be open and connected, and let f, φ, and h be as in* (11.3). *Let G and F be as in* (11.18) *and* (11.16), *respectively. Define $\mathcal{F}\colon \mathbb{R}^{n+kb} \times \mathbb{R}^n \to \mathbb{R}^{n+kb} \times \mathbb{R}^n$ by*

$$\mathcal{F}(p, q) = \big(F(p, q), G(p, q)\big).$$

Given $\Omega \subseteq [0, \infty) \times C_T(U)$ open, suppose that $\deg(\mathcal{F}, \widehat{\Omega}_{\widehat{U}})$ *is well defined and nonzero. Then, there exists a connected set Γ of nontrivial solution pairs of* (11.3) *whose closure in $[0, \infty) \times C_T(U)$ intersects the set*

$$\big\{(0, \overline{p}) \in \Omega : f\big(p, 0_n, \varphi(p, 0_n)\big) = 0_n\big\}$$

of trivial T-pairs and is not contained in any compact subset of Ω.

Proof We apply Theorem 11.1, with $r = n + kb$ and $m = n$, to the map $\mathcal{F}$ (i.e., the map $\mathfrak{F}$, introduced in (11.9), but constructed with the particular choice of F and G given by (11.18) and (11.16), respectively).

Let H be as in (11.17). By the hypotheses on f, in particular Assumption 1, and on the degree of $\mathcal{F}$, we have that Theorem 11.1 yields a connected set Υ of nontrivial solution pairs of (11.19) whose closure in $[0, \infty) \times C_T(U \times \mathbb{R}^{kb})$ intersects the set

$$\left\{(0, \overline{p}, \overline{q}) \in \widehat{\Omega} : \mathcal{F}(p, q) = (0_{n+kb}, 0_n)\right\}$$

of trivial T-pairs of (11.19) and is not contained in any compact subset of $\widehat{\Omega}$.

Take $\Gamma = \mathfrak{P}(\Upsilon)$. By Remark 11.3, Γ is a connected set of nontrivial solution pairs of (11.3) whose closure in $[0, \infty) \times C_T(U)$ intersects the set of trivial solution pairs of (11.3) that, by Remark 11.2, is given by

$$\left\{(0, \overline{p}) \in \Omega : f\big(p, 0_n, \varphi(p, 0_n)\big) = 0_n\right\}.$$

The proof is complete. □

The vector field $\mathcal{F}$ in Theorem 11.2 may be difficult to figure out. To help the reader picture this field in the mind, we write it down explicitly in coordinates. Below, we indicate with $f_1, \ldots, f_n$ the components of f and $\varphi_1, \ldots, \varphi_k$ those of φ. We write

$$p = (p_{0,1}, \ldots, p_{0,n}, p_{1,1}, \ldots, p_{1,k}, \ldots, p_{b,1}, \ldots, p_{b,k}) \in \mathbb{R}^{n+kb},$$

and

$$q = (q_1, \ldots, q_n) \in \mathbb{R}^n.$$

Then,

$$\begin{aligned}\mathcal{F}(p, q) = \Big(q_1, \ldots, q_n, a\big(\varphi_1(p_{0,1}, \ldots, p_{0,n}) - p_{1,1}\big), \ldots,&\\ a\big(\varphi_k(p_{0,1}, \ldots, p_{0,n}) - p_{1,k}\big),&\\ a(p_{2,1} - p_{3,1}), \ldots, a(p_{2,k} - p_{3,k}), \ldots, (p_{b-1,1} - p_{b,1}), \ldots, a(p_{b-1,k} - p_{b,k}),&\\ q_1 - f_1(p_{0,1}, \ldots, p_{0,n}, q_1, \ldots, q_n, p_{b,1}, \ldots, p_{b,k}), \ldots&\\ q_n - f_n(p_{0,1}, \ldots, p_{0,n}, q_1, \ldots, q_n, p_{b,1}, \ldots, p_{b,k})\Big) \in \mathbb{R}^{2n+kb}.&\end{aligned}$$

An explicit example is in order.

Example 11.2 Let a, b, f, and φ be as in Example 11.1. Then the vector field $\mathcal{F}$ relative to Eq. (11.5) becomes

$$\mathcal{F}(p_{0,1}, p_{0,2}, p_{1,1}, q_1, q_2) = (q_1, q_2, q_1 - p_{0,1}, q_1 - p_{0,2}, q_2 + p_{0,1} + p_{1,1}).$$

Fix $T > 0$. Let $U = \mathbb{R}$ and $\Omega = [0, \infty) \times C_T(\mathbb{R})$, and then $\widehat{U} = \mathbb{R}^5$, $\widehat{\Omega} = [0, \infty) \times C_T(\mathbb{R}^5)$, and $\widehat{\Omega}_{\widehat{U}} = \mathbb{R}^5$. It is not difficult to check that $\deg(\mathcal{F}, \widehat{\Omega}_{\widehat{U}}) = 1$. Thus, by Theorem 11.2, there exists a connected set of nontrivial T-pairs for Eq. (11.4) whose closure in $[0, \infty) \times C_T(U)$ intersects the set

$$\{(0, \overline{p}) \in \Omega : f(p, 0_n, \varphi(p, 0_n)) = 0_n\}$$

of trivial T-pairs and is not contained in any compact subset of Ω.

Remark 11.4 Observe that given an $\mathbb{R}^k$-valued, T-periodic function ψ, define $h(t) = \int_{-\infty}^{t} \gamma_a^b(t-s)\psi(s)\,ds$. As in (11.13) one has that $h(t+T) = h(t)$. Thus, for f and φ as in (11.1), we have

$$f\left(x(t), \dot{x}(t), \int_{-\infty}^{t} \gamma_a^b(t-s)\Big(\varphi\big(x(s), \dot{x}(s)\big) + \lambda\psi(s)\Big)\,ds\right) =$$
$$= f\left(x(t), \dot{x}(t), \int_{-\infty}^{t} \gamma_a^b(t-s)\varphi\big(x(s), \dot{x}(s)\big)\,ds + \lambda h(t)\right).$$

This shows that Theorem 11.2 can also be applied to equations of the form

$$\dot{x}(t) = f\left(x(t), \dot{x}(t), \int_{-\infty}^{t} \gamma_a^b(t-s)\Big(\varphi\big(x(s), \dot{x}(s)\big) + \lambda\psi(s)\Big)\,ds\right),$$

where the perturbation is placed inside the integral.

11.4 An Extension

The techniques developed so far can be applied also in the case when the perturbing function is allowed to depend also on $x(t)$, provided that h is C^1 and the norm of its partial derivative with respect to the second variable $\|\partial_2 h(t, p)\|$ is a bounded function of (t, p). That is, (11.3) can be replaced by

$$\dot{x}(t) = f\left(x(t), \dot{x}(t), \int_{-\infty}^{t} \gamma_a^b(t-s)\varphi\big(x(s), \dot{x}(s)\big)\,ds + \lambda h\big(t, x(t)\big)\right), \quad \lambda \geq 0. \tag{11.21}$$

To understand why, observe that in this case (11.15) is replaced by

$$u_i(t) = \int_{-\infty}^{t} \gamma_a^i(t-s)\varphi\big(x(t), y(t)\big)\,ds + \lambda h\big(t, x(t)\big), \tag{11.22}$$

where as in Sect. 11.2.3 we have put $y(t) = \dot{x}(t)$. By lemma 11.1, with $\sigma(t) = h\big(t, x(t)\big)$ proceeding as for Eq. (11.14), we are led to the following system:

$$\begin{cases} \dot{x}(t) = y(t), \\ \dot{u}_1(t) = a\big(\varphi\big(x(t), y(t)\big) - u_1(t)\big) + \\ \qquad\qquad +\lambda\Big(\partial_1 h\big(t, x(t)\big) + \partial_2 h\big(t, x(t)\big)y(t) - ah\big(t, x(t)\big)\Big), \\ \dot{u}_i(t) = a\big(u_{i-1}(t) - u_i(t)\big) + \\ \qquad\qquad +\lambda\Big(\partial_1 h\big(t, x(t)\big) + \partial_2 h\big(t, x(t)\big)y(t)\big)\Big), \qquad i = 2, \dots, b. \\ y(t) - f\big(x(t), y(t), u_b(t)\big) = 0_n. \end{cases} \tag{11.23}$$

Proceeding as in Sect. 11.3, we rewrite (11.23) as (11.19) with the only difference being the form taken by H that in this case is

$$H(t, \xi, \eta) = \Big(0_n,\ \partial_1 h(t, \xi) + \partial_2 h(t, \xi)\eta - ah(t, \xi), \\ \underbrace{\partial_1 h(t, \xi) + \partial_2 h(t, \xi)\eta, \dots, \partial_1 h(t, \xi) + \partial_2 h(t, \xi)\eta}_{b-1 \text{ times}}\Big).$$

Since the particular form of H is irrelevant for Theorem (11.1), the same argument used in Sect. 11.3, in particular Remarks 11.1, 11.3, and 11.2, shows that the statement of Theorem 11.2 holds also for (11.21).

References

1. L. Bisconti, M. Spadini, On a class of differential-algebraic equations with infinite delay. Electron. J. Qual. Theory Differ. Equ. (81), 21 (2011)
2. F.E. Browder, Fixed point theory and nonlinear problems. Bull. Amer. Math. Soc. (N.S.) **9**, 1–39 (1983)
3. T.A. Burton, *Stability and Periodic Solutions of Ordinary and Functional Differential Equations* (Dover Publications, Inc., Mineola, 2005). Corrected version of the 1985 original
4. A. Calamai, M. Spadini, Periodic perturbations of constrained motion problems on a class of implicitly defined manifolds. Commun. Contemp. Math. **17**(2), 1450027 (2015), 19pp.
5. A. Calamai, M.P. Pera, M. Spadini, Periodic perturbations of reducible scalar second order functional differential equations. Electron. J. Qual. Theory Differ. Equ. Paper No. 18, 23 (2023)
6. K. Deimling, *Nonlinear Functional Analysis* (Springer, Berlin, 1985)
7. G. Dinca, J. Mawhin, *Brouwer Degree - The Core of Nonlinear Analysis*. Progress in Nonlinear Differential Equations and their Applications, vol. 95 (Birkhäuser/Springer, Cham, 2021)
8. D. Fargue, Réductibilité des systèmes héréditaires à des systèmes dynamiques (régis par des équations différentielles ou aux dérivées partielles. C. R. Acad. Sci. Paris Sér. A-B **277**, B471–B473 (1973)
9. D. Fargue, Réductibilité des systèmes héréditaires. Int. J. Non-Linear Mech. **9**, 331–338 (1973)
10. M. Furi, M.P. Pera, M. Spadini, A set of axioms for the degree of a tangent vector field on differentiable manifolds. Fixed Point Theory Appl. **11**, Art. n. 845631 (2010)

11. O. Kavian, *Introduction à la Théorie des Points Critiques et Applications aux Problèmes elliptiques*. Mathématiques & Applications (Berlin) [Mathematics & Applications], vol. 13 (Springer, Paris, 1993)
12. P. Kunkel, V. Mehrmann, *Differential Algebraic Equations, analysis and numerical solutions*. EMS Textbooks in Mathematics (European Mathematical Society, Zürich, 2006)
13. N.G. Lloyd, *Degree Theory*. Cambridge Tracts in Mathematics, vol. 73 (Cambridge University Press, Cambridge, 1978)
14. P.M. Mariano, M. Spadini, Periodic solutions to perturbed nonlinear oscillators with memory. Phys. D **445**, Paper No. 133635, 8pp. (2023)
15. J. Milnor, *Topology from the Differentiable Viewpoint* (The University Press of Virginia, Charlottesville, 1965)
16. L. Nirenberg, *Topics in Nonlinear Functional Analysis*. Courant Lecture Notes in Mathematics, vol. 6 (American Mathematical Society, Providence, 2001). Chapter 6 by E. Zehnder, Notes by R. A. Artino. Revised reprint of the 1974 original
17. E. Rosset, Topological degree in $\mathbb{R}^n$. Rendiconti dell'Istituto di Matematica dell'Università di Trieste. Int. J. Math. **20**, 319–329 (1988). Available at http://hdl.handle.net/10077/4865
18. H. Smith, *An Introduction to Delay Differential Equations with Applications to the Life Sciences* (Springer, New York, 2011)
19. M. Spadini, A note on topological methods for a class of differential-algebraic equations. Nonlinear Anal. **73**(4), 1065–1076 (2010)
20. M. Spadini, Periodic perturbations of a class of functional differential equations. J. Dyn. Diff. Eq. **34**, 535–553 (2022)

Index

P. Benevieri, J. G. Mesquita (eds.), *Functional Differential Equations and Dynamic Equations on Time Scales*, Advances in Continuum Mechanics 52,
https://doi.org/10.1007/978-3-031-83327-4

The manufacturer's authorised representative in the EU is Springer Nature Customer Service Centre GmbH, Europaplatz 3, 69115 Heidelberg, Germany. If you have any concerns regarding our products, please contact ProductSafety@springernature.com

Printed and bound by CPI Group (UK) Ltd, Croydon, CR0 4YY
30/09/2026
02225584-0007